Integrated Circuits and Systems for Smart Sensory Applications

Integrated Circuits and Systems for Smart Sensory Applications

Editors

Francesc Serra-Graells
Michele Dei
Kyoungrok Cho

MDPI • Basel • Beijing • Wuhan • Barcelona • Belgrade • Manchester • Tokyo • Cluj • Tianjin

Editors
Francesc Serra-Graells
Department of Microelectronics and Electronic Systems, Universitat Autònoma de Barcelona
Spain

Michele Dei
Department of Information Engineering, University of Pisa
Italy

Kyoungrok Cho
Department of Information and communication Engineering, Chungbuk National University
Korea

Editorial Office
MDPI
St. Alban-Anlage 66
4052 Basel, Switzerland

This is a reprint of articles from the Special Issue published online in the open access journal *Sensors* (ISSN 1424-8220) (available at: https://www.mdpi.com/journal/sensors/special_issues/IC_Smart_Sensory).

For citation purposes, cite each article independently as indicated on the article page online and as indicated below:

LastName, A.A.; LastName, B.B.; LastName, C.C. Article Title. *Journal Name* **Year**, *Volume Number*, Page Range.

ISBN 978-3-0365-3264-6 (Hbk)
ISBN 978-3-0365-3265-3 (PDF)

Contents

About the Editors

Francesc Serra-Graells received his PhD degree in Electronics Engineering at the Universitat Politècnica de Catalunya, Spain, in 2001. He is currently Associate Professor at the Department of Microelectronics and Electronic Systems of the Universitat Autònoma de Barcelona, Spain. Since 1994, he is also with the Integrated Circuits and Systems group of the Institut de Microelectrònica de Barcelona, Centro Nacional de Microelectrónica, IMB-CNM(CSIC), Spain. His research interests are very low-power, both low-voltage and low-current, design techniques for analog and mixed-signal CMOS circuits, like programmable amplifiers with automatic gain control, phase-locked loops, analog-to-digital and digital-to-analog data converters, oscillators and power managment units, among others. The results of his research activities are application-specific integrated circuits (ASICs) and intellectual-property (IP) cores targeting smart sensing systems, like hearing aids, IR and X-ray imagers, NEMS interfaces, chemical sensors and biosensors, space instrumentation and neural recording systems. He is co-author of more than 60 publications and has participated in more than 30 research and industrial ASIC projects.

Michele Dei received the Laurea degree cum Laude in Electronic Engineering and the Ph.D degree in Information Engineering at the University of Pisa (Italy) in 2007 and 2010, respectively, working on the design of smart sensors for detection of physical quantities. He is currently working at the Information Engineering Department of the University of Pisa (Italy), through of a Marie Skłodowska-Curie Individual Fellowship contract (H2020-MSCA-IF-2019/893544) as Principal Investigator of the SWeaT project regarding wearable electro-chemical instrumentation. Previously he worked within the IMB-CNM (Barcelona, Spain) and with the School of Information and Communication Technology at KTH (Stockholm, Sweden). His engineering background involves low-power analog-mixed signal integrated circuit for sensor interfacing, data converters in the context of a number of high-impact research and industrial projects.

Kyoungrok Cho received a B.S. degree in Electronic Engineering from Kyoungpook National University, Taegu, Korea, in 1977, and an M.S. and Ph.D. degree in Electrical Engineering from the University of Tokyo, Japan, in 1989 and 1992, respectively. From 1979 to 1986, he was with TV research center of LG Electronics in Korea. He is currently a Distinguished Professor at the College of Electrical and Computer Eng. of Chungbuk National University, Korea. His research interests are in the field of high-speed and low-power circuit design, SoC and platform design for the communication system, and prospective IOT sensor networks. He has been a director of a regional center of IDEC(IC Design Education Center) in Korea covering Chungcheong provinces and Taejon city from January 2004. In 1999 and again in 2006, he spent a year at Oregon State University, USA, as a visiting scholar.

Preface to "Integrated Circuits and Systems for Smart Sensory Applications"

Connected intelligent sensing reshapes our society by empowering people with increasing new forms of mutual interactions. As integration technologies keep their scaling roadmap, the horizon of sensory applications is rapidly widening, thanks to myriad light-weight low-power or, in some cases, even self-powered smart devices with high-connectivity capabilities. CMOS integrated circuits technology is the best candidate to supply the required intelligence and to pioneer these emerging sensory systems. As a result, new challenges are arising around the design of these integrated circuits and systems for sensory applications in terms of low-power edge computing, power management strategies, low-range wireless communications, and integration with sensing devices. In this Special Issue, recent advances in application-specific integrated circuits (ASIC) and systems for smart sensory applications are presented via the following five emerging topics:

I. Dedicated short-range communications transceivers for radio communications in intelligent transportation systems [1,6];

II. Digital smart sensors featuring digital compensations [2], re-programmable fuse trimming [5] and all-digital implementation for ultra-low power operation [10];

III. Implantable neural interfaces, including a review on closed-loop implantable electronics for epilepsy control [3] and an AC-coupled neural amplifier design with improved linearity [7];

IV. Power Management Strategies in wireless sensor nodes, both from the hardware [4] and the software [8] perspectives;

V. Neuromorphic hardware systems for image recognition [9].

Dedicated short-range communications (DSRC) services designate one-way or two-way wireless communication for automotive use involving vehicle-to-vehicle and vehicle-to-infrastructure communications. In 2020, The United States Federal Communications Commission (FCC) allocated the 5.895-5.925 GHz band for Intelligent Transportation System (ITS) services. Equipment in the DSRC Service comprises On-Board Units (OBUs) and Roadside Units (RSUs). An OBU is a transceiver (which is normally battery operated) fitted inside the vehicle. In such scenarios, a high-speed reliable radio link need to be established under power-constrained requirements. Ali et al. [1] presents a new wake-up receiver (WuRx) to support he main RF receiver of the OBU by providing wake-up detection to exit the OBU hibernation state. The proposed ASIC features an intelligent digital controller (IDC) for improving WuRx reliability and for replacing complex and power-hungry analog blocks such as band-pass filters and frequency detectors. The IDC implements a number of smart power-management techniques such as: (i) self-hibernation of the IDC and range communication (RC) oscillator; (ii) digital hysteresis for accommodating wake-up signal frequency variation and enhancing WuRx accuracy; (iii) a watch-dog timer for IDC self-recovery to avoid uncertain conditions during poor and false wake-up; (iv) configurable wake-up signal cycles before enabling a power-hungry RF transceiver. The IDC prototype in 130 nm CMOS technology occupies a modest silicon area (94×82 μm^2). The resulting WuRx shows comparable sensitivity (-46 dBm) with the state-of-the-art, while outperforming it in terms of power consumption (2.48 μW), thus demonstrating the effectiveness of the proposed IDC. DSRC transceivers also comprise analog-to-digital converters

(ADC) to allow the transceiver to communicate with the digital base-band electronics, with speed and resolution requirements of few tens of MS/s and 10 bits, respectively. To meet these requirements, Shehzad at al. [6] proposed a 12 bit 20 MS/s successive approximation register (SAR) ADC fabricated in 65 nm CMOS with an active area of 0.14 mm^2, performing a SNDR of 65.44 dB while consuming only 472.2 μW with 1 V power supply. Low-power operation has been achieved employing various circuital techniques such as specific capacitor-switching strategy, asynchronous control logic, and custom modification of the dynamic-latch comparator.

Digital smart sensors are at the heart of IoT development, ranging from consumer gadgets, sensor networks, and image sensors to biomedical instruments, thanks to their digital-ready output and their added functionalities over classical analog-domain transducers, with almost zero impact on the device cost. Such added functionalities comprise sensor compensation, trimming and/or full-digital CMOS implementation for rapid IP embedding on complex system on chips (SoC). Ali et al. [2] propose polynomial-based adaptive digital temperature compensation for piezoresistive-type (PRT) pressure sensors. Such sensors have gained attention in a variety of applications due to their simplicity, low cost, miniature size, and ruggedness; however, their electrical behavior is temperature dependent and highly nonlinear. To avoid severe impairment of measurement accuracy, the authors propose an ASIC fabricated in 180 nm CMOS technology, delivering compensation accuracy within ±0.068% of full scale when temperature varies from −40 to +150 °C. On the other hand, low-cost and low-power temperature sensors, often targeting challenging requirements in terms of accuracy, precision, and linearity, are ubiquitously demanded. When cost per unit is considered, higher accuracy for a thermal sensor is based on a trade-off between the production costs for calibration and the required precision. Vasile et al. [5] show a trimmed digital sensor with a +1.5/−1.0 °C inaccuracy in the temperature range of −20 to +125 °C using a 180-nm CMOS EEPROM process by one-point calibration. Finally, Cicalini et al [10] show the implementation of an innovative, all-digital 180 nm CMOS capacitive-to-digital converter adequate for medium-low resolution body sweat-flow rate in wearable applications. They demonstrate the operation of 10.3 effective-number-of-bits resolution readout interface at 0.9 V-supply for a 0–250 pF input capacitance featuring $\geq$ 1.884 nJ/conversion, excellent linearity and robustness against process/temperature corners, while using only 0.0192 mm^2 of silicon area.

Implantable neural interfaces concern multi-electrode systems routinely designed as Application Specific Integrated Circuits (ASIC), which comprise hundreds or thousands of recording amplifiers on a single chip. To this aim, the signal recording chain must be very carefully designed so as to operate in low power and low latency, while enhancing the probability of correct event detection. High-fidelity recording of neuronal signals, comprising the action potentials (AP) range (300 Hz–10 kHz) and the local field potential (LFP) (1–300 Hz) range, requires strict noise (>10 μVRMS) and a total harmonic distortion (THD) (>2%) specifications, when dealing with signals up to 10 mVpp. Minimization of the dissipated power and silicon area is also critical for the design of neuronal interfaces with a very large number of recording channels. The large DC voltage at the input of the amplifier results from electrochemical interactions between the electrode and the tissue. The recording circuit must cut off this DC electrode voltage with high-pass filter (i.e., AC coupling) with a lower cutoff frequency, typically in the order of 1 Hz, and amplify the remaining AC signals with a gain in the order of 40 dB. Ranjandish and Schmid [3] report the latest advances in closed-loop implantable electronics for epilepsy control, focusing on both implantable and external commercial systems and pointing out the following research challenges: (i) size and weight; (ii) power consumption and temperature elevation; (iii) battery powering and rechargeability; (iv) the biocompatibility of the package and enclosure; (v) data compression and storage. Trzpil-Jurgielewicz [7] propose a linearity-enhancing circuital technique for AC-coupled neural amplifiers to remove the electrode DC voltage. A prototype preamplifier fabricated in 180 nm CMOS technology shows THD values are below 1.17% for signal frequencies 1 Hz–10 kHz and signal amplitudes up to 10 mV peak to peak. While using only 0.0046 mm^2 of silicon area, the prototype has an input-referred noise of 8.3 μVRMS in the 1 Hz–10 kHz range, while consuming 7.2 μW per channel.

Power Management Strategies in wireless sensor nodes (WSN) involves both hardware and software techniques. Smart sensor nodes perform a set of tasks, often corresponding to different load states. The predictable transition schedule allows the software to proactively reconfigure the voltage converter to supply required amount of current for various load conditions. Under varying load conditions, however, a large Switched Capacitor (SC) converter often provides poor efficiency at smaller loads. An output voltage ripple is often alleviated by having a large output capacitor and/or high switching frequency, which represents another challenge faced by modern SC converters. Arslan et al. [4] propose a voltage converter whose switching frequency and output voltage are proactively adjusted to maintain high conversion efficiency based on the schedules of load current demanded by the target load. Multiphase operation is also implemented to provide low-output ripple. The ASIC prototype, fabricated in 130 nm CMOS technology, supports a load current range between 10 μA and 10 mA for switching frequencies ranging from 100 kHz to 200 MHz, while providing an efficiency of above 80%. The area of the converter is 0.59 mm^2, operating a 1.5-V supply; it delivers a tunable output voltage between 0.4 and 1.1 V with maximum ripple of 56 mV. Apart from transistor-level power optimization, software control over the operating states of WSN is important for overall power saving of the battery-constrained system. You et al. [8] proposes a novel power management method (PMM) that leads to less energy consumption in an idle state than conventional PMMs. While conventional PMMs rely on operation between Sleep, Idle, and Run modes, the proposed approach splits the Sleep mode into three different modes: Deep-Sleep and Semi-Idle Sensor and Semi-Idle WuRx. The proposed PMM strategy has been tested on a gas-sensing WSN using the commercial Mica2 platform and compared to conventional PMM. The proposed approach offers power savings between 2 and 74.2% depending on event rate, thus demonstrating its effectiveness in low-event-probability WSNs.

Neuromorphic hardware systems differ from classical von Neumann processor architectures since they are naturally configured for parallel information processing. Mimicking the vastly and densely connected neurons of the human brain, neuromorphic architectures encode information in sequences of action potentials called spike trains, theoretically promising a breakthrough in energy efficiency for signal processing. In this sense, specific and power-optimized hardware implementation of such neural networks is a hot topic in this research area. Asghar et al. [9] present an CMOS implementation of a Spiking Neural Network (SNN) for real-time 9×9 pixel input image for pattern recognition. Fabricated using 180 nm CMOS process, the proposed chip achieves a classification accuracy of 94.66% for the MNIST dataset while occupying 3.6 mm^2 chip core area and presenting an average power consumption of 1.06 mW.

Francesc Serra-Graells, Michele Dei, Kyoungrok Cho
Editors

sensors

MDPI

Article

A Highly Reliable, 5.8 GHz DSRC Wake-Up Receiver with an Intelligent Digital Controller for an ETC System

Imran Ali, Muhammad Asif, Muhammad Riaz Ur Rehman, Danial Khan, Huo Yingge, Sung Jin Kim, YoungGun Pu, Sang-Sun Yoo and Kang-Yoon Lee *

College of Information and Communication Engineering, Sungkyunkwan University (SKKU), Suwon 16419, Korea; imran.ali@skku.edu (I.A.); m.asif@skku.edu (M.A.); riaz@skku.edu (M.R.U.R.); danialkhan@skku.edu (D.K.); yingge@skku.edu (H.Y.); sun107ksj@skku.edu (S.J.K.); hara1015@skku.edu (Y.P.); rapter@kaist.ac.kr (S.-S.Y.)

* Correspondence: klee@skku.edu; Tel.: +82-31-299-4954

Received: 18 June 2020; Accepted: 17 July 2020; Published: 19 July 2020

Abstract: In this article, a highly reliable radio frequency (RF) wake-up receiver (WuRx) is presented for electronic toll collection (ETC) applications. An intelligent digital controller (IDC) is proposed as the final stage for improving WuRx reliability and replacing complex analog blocks. With IDC, high reliability and accuracy are achieved by sensing and ensuring the successive, configurable number of wake-up signal cycles before enabling power-hungry RF transceiver. The IDC and range communication (RC) oscillator current consumption is reduced by a presented self-hibernation technique during the non-wake-up period. For accommodating wake-up signal frequency variation and enhancing WuRx accuracy, a digital hysteresis is incorporated. To avoid uncertain conditions during poor and false wake-up, a watch-dog timer for IDC self-recovery is integrated. During wake-up, the digital controller consumes 34.62 nW power and draws 38.47 nA current from a 0.9 V supply. In self-hibernation mode, its current reduces to 9.7 nA. It is fully synthesizable and needs 809 gates for its implementation in a 130 nm CMOS process with a 94 × 82 μm^2 area. The WuRx measured power consumption is 2.48 μW, has −46 dBm sensitivity, and a 0.484 mm^2 chip area.

Keywords: wake-up receiver; digital controller; reliability; electronic toll collection (ETC) system; dedicated short range communication (DSRC)

1. Introduction

Recently, the radio frequency (RF) wake-up receiver has become an attractive research area for battery-operated transceivers in a variety of applications such as electronic toll collection (ETC) systems, wireless sensor networks (WSNs), wireless body area networks (WBANs), internet-of-things (IoTs), and wearable devices [1–4]. Nowadays, the ETC system (ETCS) is rapidly being adopted as an intelligent transportation solution in automotive vehicles. It uses a 5.8-GHz dedicated short range communication (DSRC) for a high speed radio link between a road side equipment (RSE) fixed at the toll gate and on-board unit (OBU) fitted inside the vehicle [1,5–7], as shown in Figure 1. Without stopping the vehicles, the toll is paid automatically and it saves time and eliminates traffic congestion on the roads. In ETCS, a wake-up receiver (WuRx) is an auxiliary RF receiver, additionally to the main RF transceiver, as shown in Figure 2, and is mandated due to the battery powered OBU. The WuRx is a pure asynchronous communication scheme and it maximizes data transceiver sleep time. This not only reduces OBU energy dissipation but also diminishes network latency. Figure 3 shows the asynchronous communication between RSE and OBU with WuRx. The reliability, false wake-up, power dissipation, and sensitivity are key considerations in WuRx design. In WuRx, designed with low power dissipation and good sensitivity, false and poor wake-up turns on main transmitter and receiver modules, which reduces battery life and degrades the overall WuRx performance. The numerous WuRx circuits have been investigated in literature to optimize power consumption, maximize sensitivity, and improve reliability [8–11]. The various WuRx architectures are summarized in Figure 4. Based on the type of the power source, the WuRx is categorized in active and passive wake-up circuits. The active WuRx circuits are powered from a battery fitted on an OBU to monitor a possible wake-up signal. The energy for the wake-up circuit is harvested from the incident RF signal in passive receivers. Mostly, the active circuits adopt either RF envelope detector (RFED) structures [1,12,13], shown in Figure 4a,b, or frequency conversion architectures [14,15], depicted in Figure 4c. The frequency conversion based wake-up receivers offer higher sensitivity due to RF amplification before RF envelope detection [16], or local oscillator (LO) generation for down-conversion before amplification and envelope detection at intermediate frequency (IF) [14]. These circuits dissipate more power and occupy a larger area due to a power hungry phase-locked loop (PLL) and automatic channel scanning circuits for two channel receptions [15]. The majority of wake-up circuits are implemented with RFED because of its low power consumption using Schottky diodes [12,13] or metal–oxide–semiconductor field-effect transistor (MOSFET) operating in the weak-inversion region. The WuRx structure shown in Figure 4a uses an analog to digital converter (ADC) after a programmable gain amplifier (PGA) which occupies more area and increases power consumption. The WuRx shown in Figure 4b uses an analog band pass filter (BPF) at the output interface which requires more chip area. Figure 4c shows an envelope detector based WuRx structure with a front end amplifier and bulk acoustic wave (BAV) input network [17,18]. The WuRx in [19] also incorporates ADC at the output; however, in order to achieve better sensitivity and to reduce receiver noise, it incorporates a low noise amplifier (LNA) before an envelope detector and uses a double sampling technique. It offers better sensitivity at the cost of increased power consumption and larger chip area for WuRx. The passive WuRx structure [20] incorporates radio frequency to direct current (RF-DC) converter for harvesting energy from incident RF signals as shown in Figure 4d. For this purpose, an RF-DC converter [21,22] is employed to produce the envelope of the on–off keying (OOK) wake-up message signal, and at the same time, it efficiently converts the RF carrier to a DC voltage in order to supply the comparator and the other WuRx circuits. Although this architecture is power efficient, it exhibits low sensitivity. Furthermore, the deficiency of false and poor wake-up filtering is vulnerable.

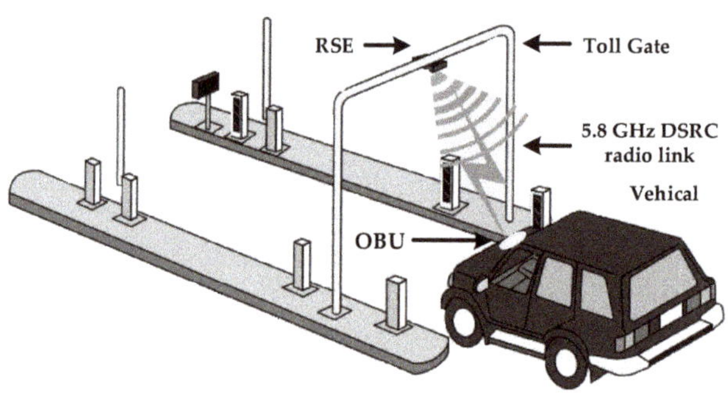

Figure 1. The dedicated short range communication (DSRC) system overview with road side equipment (RSE) fitted at toll gate, battery operated on-board unit (OBU) fixed inside the vehicle, and a 5.8 GHz DSRC radio link as communication channel.

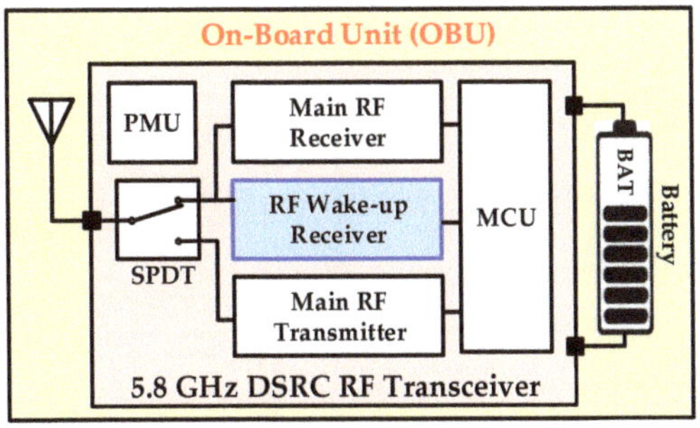

Figure 2. The DSRC OBU system level block diagram.

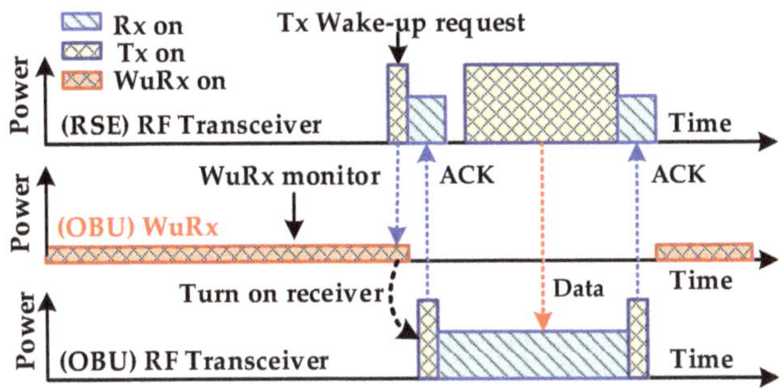

Figure 3. The road side equipment (RSE) and OBU pure asynchronous communication with the RF wake-up receiver (WuRx).

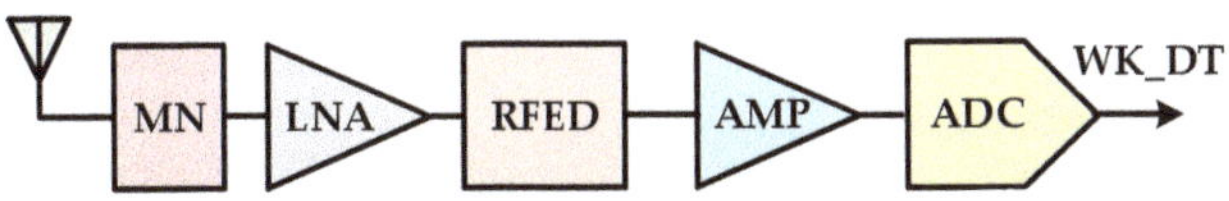

(a)

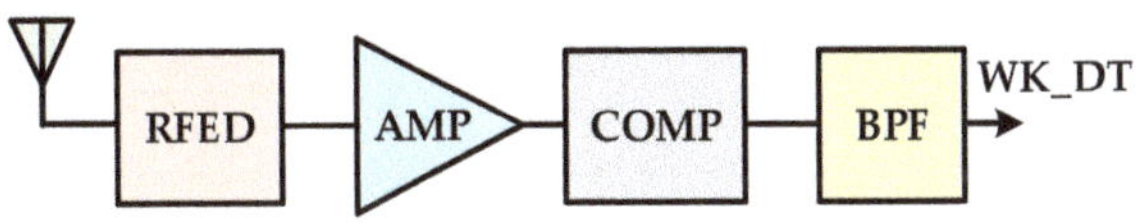

(b)

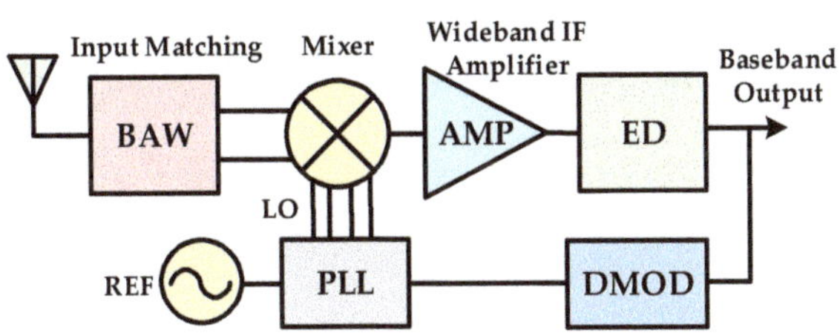

(c)

Figure 4. *Cont.*

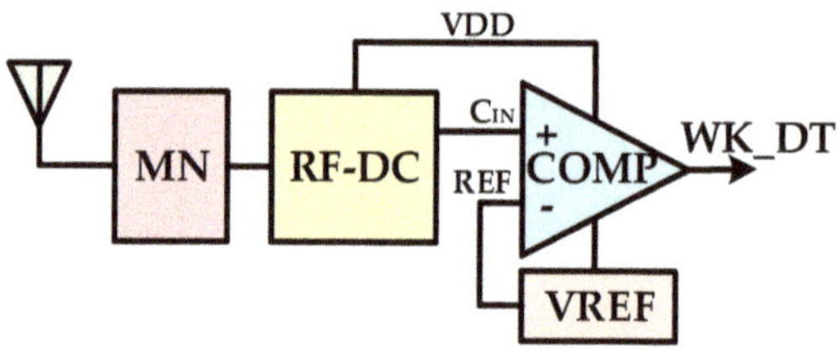

(d)

Figure 4. The previous wake-up receiver architecture: (**a**) RF envelope detector (RFED) based with analog to digital converter (ADC); (**b**) RFED based with band pass filter (BPF); (**c**) frequency conversion with LO; (**d**) passive circuit with RF-DC.

The wake-up signal is categorized as a single wake-up tone of a bit sequence [23]. The bit sequence signal, also called identity-based wake-up, is widely used in WSN and WBAN for addressing a particular destination sensor node for unicasting. For broadcasting, single wake-up tone is adopted [23]. The DSRC wake-up signal is a 14 kHz OOK single tone signal of 15–17 cycles which is modulated with a 5.8 GHz carrier frequency [9].

The precise sensitivity control is essential for ETCS wake-up circuits [1]. Neither a very low nor very high sensitivity is intended. The available communication time will be insufficient when WuRx sensitivity is lower than a minimum level. Similarly, if WuRx has very high sensitivity, it will turn on OBU and start communication even if vehicle is far away from a RSE and toll area. Moreover, very high sensitivity will result in communication failure and interrupts other OBU devices on the road. Therefore, WuRx sensitivity must be in a range, robust, insensitive to process, voltage, and temperature (PVT) variations [11].

The key characteristic of WuRx is to detect and ensure the presence of wake-up request in the received RF signal and decide whether to turn on the power hungry transceiver on OBU. In most previous studies, the WuRx circuits only identify the signal amplitude which is inadequate in ETC systems. If false and poor wake-up signals are not identified and filtered out in the WuRx circuit, these signals turn on power hungry main RF transceiver modules and the battery performance is degraded. In the past, numerous WuRx architectures have been investigated for improving the sensitivity and reducing the power consumption. However, false and poor wake-up problems have been left unaddressed. This paper presents a RFED based highly reliable WuRx. An intelligent digital controller (IDC) is proposed to ensure the WuRx reliability and accuracy by identifying and rejecting unwanted, false, and poor wake-up signals. It also replaces complex blocks such as ADC and BPF after amplifier and comparator [8–10,24] and reduces current consumption and area. Due to its digital nature, it is fully synthesizable, immune to noise and PVT variations [25], offers system flexibility, a wide dynamic range for wake-up and oscillator frequencies, and is adaptive to technology scaling.

The rest of the paper is organized as follows: Section 2 presents a proposed WuRx architecture overview. The detailed design of the proposed digital controller is described in Section 3. The tunable range communication (RC) oscillator is presented in Section 4. The RF front end and baseband analog processing is included in Section 5. Section 6 describes the experimental results. Lastly, the paper is concluded in Section 7.

2. Proposed Wake-Up Receiver Architecture

Figure 5 shows the proposed 5.8-GHz RF WuRx in which IDC is integrated to ensure its reliability and accuracy. An antenna receives the incoming RF signal and sends it to an off-chip pi-matching network. The matching network is an essential passive circuit to transfer the maximum RF signal power to the receiver circuit. The pi-matching network matches the antenna equivalent impedance with the input impedance of the proposed WuRx and ensures the maximum power transfer from the antenna to the WuRx circuit. Unlike [8,24], the chip internal matching network in addition to off-chip pi-matching receives a RF wake-up signal, boosts voltage, and improves sensitivity and $|S_{11}|$. The high gain RF envelope detector recovers the baseband wake-up signal and improves signal-to-noise ratio (SNR) without additional current consumption. The RFED is a critical circuit and it interfaces WuRx with the antenna, down-converts the amplitude of the modulated RF 5.8 GHz signal, and generates a 14 kHz baseband wake-up signal. The PGA provides flexibility to improve gain and amplify the baseband signal significantly. The main control unit (MCU) enables/disables and configures different programmable parameters of RFED, baseband analog (BBA), comparator (COMP), and IDC. It can either be an on/off chip modem, externally controllable registers, or an external microcontroller. The comparator (COMP) generates digital output for IDC processing. An ultra-low power range communication (RC) oscillator (OSC) with a dynamic tuning range generates a configurable clock for the IDC block. The digital controller is proposed for ensuring WuRx reliability and accuracy by identifying and filtering non-wake-up signals. It is a fully synthesizable block, consumes very low power, and needs a very small chip area. The digital controller also replaces complex power consuming and large area interface blocks, such as ADC, BPF.

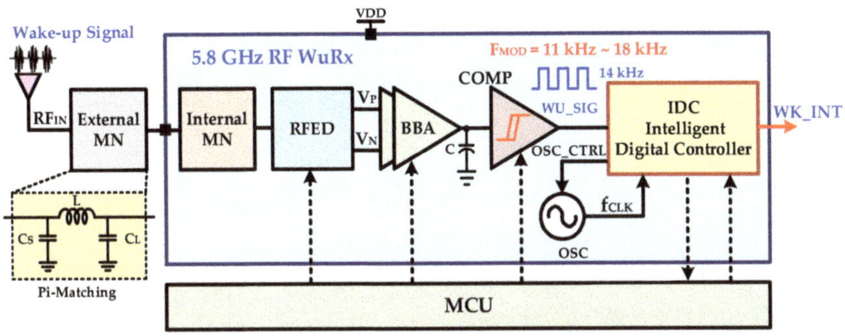

Figure 5. Proposed RF wake-up receiver architecture.

3. Multi-Mode, Configurable Intelligent Digital Controller (IDC)

In the DSRC WuRx design, other than high sensitivity and low power consumption, the false and poor wake-up signals identification is very crucial in order to extend OBU battery lifetime. This is achieved by filtering non-wake-up and noise signals and prohibiting turning on main power hungry RF transceiver. For this purpose, a novel, multi-mode, and configurable intelligent digital controller is proposed for ensuring WuRx reliability and accuracy. This controller is also a low power, small area digital replacement of complex, high power analog blocks such as ADC and a band pass filter. Different parameters of wake-up, self-hibernation, digital hysteresis, wake-on, watch-dog timer, and self-test are fully configurable, which make the controller architecture very flexible and adaptive. The simplified architecture of a WuRx digital controller is illustrated in Figure 6 and the timing diagram is elaborated in Figure 7 for DSRC applications. The signal selection multiplexer (SSM) selects either a baseband wake-up signal, *WU_SIG*, from the comparator output in normal operation or self-test signal *st_sig*, generated from self-test pattern generator (STPG) during test mode. The signal positive edge generator (SPEG) detects rising transitions in the final selected signal *wusig* and generates a pulse signal *wu_pe*. The finite state machine controller (FSMC) is the key building block of IDC which is designed as control unit and data path. It mainly senses, ensures, and generates the wake-up interrupt and filters unwanted signals. It controls other blocks such as the adaptive frequency measurement unit (AFMU), configurable watch-dog timer (CWDT), and wake-on generator (WOG). When enabled by FSMC (*fm_en* = 1), the AFMU measures the frequency of the wake-up signal and determines by generating signal *fm_det* if the input signal value is either within the configured range or not. It also ensures the valid successive number of configured (*WU_N*) wake-up signal cycles. The CWDT, when enabled by signal *wdt_en* from FSMC, starts a timer. The timer duration is configurable from the

WDTN parameter. It enhances IDC reliability and helps to avoid any halt situation during frequency measuring and signal ensuring states. If there is any abnormal situation, CWDT resets FSMC to its initial state when the configured timer expires. The WOG implements a pseudo-synchronous interrupt generation. When it is enabled, the wake-on interrupt WO_INT is generated instead of the wake-up interrupt. The STPG enhances WuRx reliability by verifying IDC operation in self-test mode. When enabled (*st_en* = 1), it is capable of generating a variety of valid and invalid signals with different frequencies and number of cycles. The output multiplexer (OM) outputs *WU_INT* by selecting either internally generated interrupt signal *int_r* or external manually control interrupt WU_EXT. The control decoder (CDEC) decodes interrupt *int_ctrl*, mode *mode_ctrl*, and monitor *monitor_ctrl* control signals from the external CTRL input.

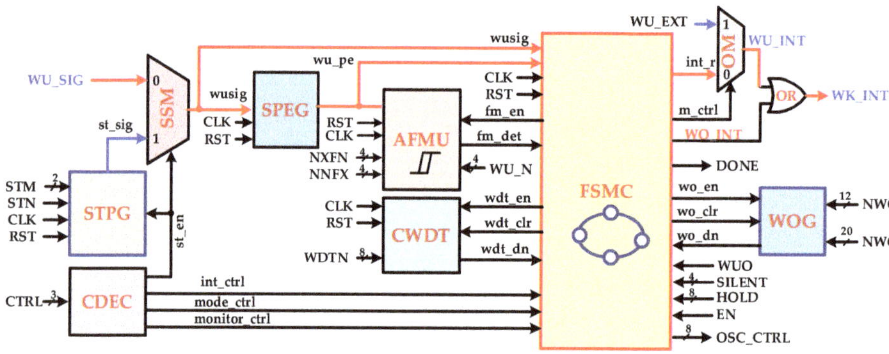

Figure 6. RF WuRx intelligent digital controller (IDC) architecture.

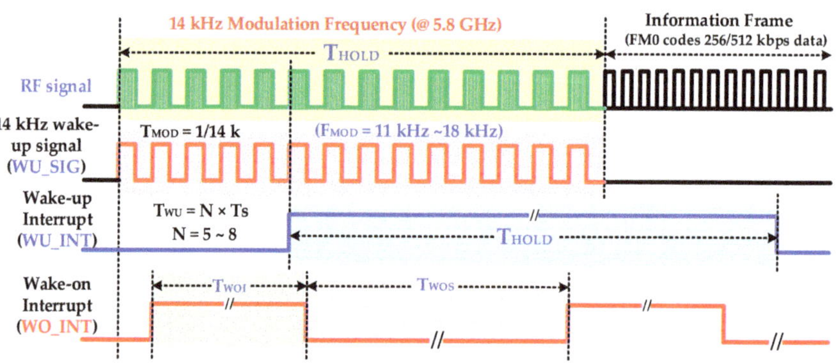

Figure 7. RF wake-up and wake-on interrupt timing diagram.

3.1. Self-Hibernation for Low Power Consumption

A self-hibernation methodology with dynamic frequency and voltage scaling is introduced for reducing the IDC and oscillator power consumption during the non-wake-up interval. The average power consumption P_{AV} in the CMOS circuit is the sum of dynamic P_{DYN}, short circuit P_{SHORT}, leakage $P_{LEAKAGE}$, and static P_{STATIC} power consumptions [26] as explained in (1) as follows:

$$P_{AV} = P_{DYN} + P_{SHORT} + P_{LEAKAGE} + P_{STATIC}. \tag{1}$$

In CMOS circuits, P_{DYN} is the dominant power consumption component. It is the linear function of operating frequency f and the quadratic function of the supply voltage V_{DD} of the circuit as given in (2) as follows:

$$P_{DYN} = KCfV_{DD}^2, \quad (2)$$

where K is switching activity factor, C is loading capacitance, and V_{DD} is supply voltage [26].

From Equation (2) it becomes obvious that if the switching frequency reduces, the power consumption of the CMOS circuit is reduced significantly. Furthermore, if the supply voltage is minimized, the dynamic power is reduced. The OSC frequency is configurable from IDC. The IDC programs the OSC at a relatively higher frequency, f_{WU}, and performs WuRx signal identification during the active interval. After the wake-up interval, it operates at a relatively very low frequency, f_{SH}, during the sleep period after configuring the OSC for the slowest frequency. Since the communication between RSE and OBU is only for a very short duration and the OBU is in sleep mode for most of the time, the self-hibernation proves its significant impact for reducing WuRx power consumption and extending battery lifetime. The supply voltage is also reduced from 1.2 V to 0.9 V to save battery power.

3.2. Built-In Self-Test for IDC Reliability

For ensuring IDC reliability and accuracy, a built-in self-test technique is integrated. It verifies the IDC full operation and functional accuracy without the presence of an external RF wake-up signal. In the presented self-test scheme, a configurable self-test pattern generator module generates a variety of configurable wide range valid wake-up signals with a frequency described as follows in Equation (3):

$$f_{ST} = \frac{N_{ST}}{f_{WU}}, \quad (3)$$

where f_{ST} is test wake-up signal *st_sig* frequency, which is programmable from the N_{ST} parameter. It is also capable of generating non-wake-up, false and poor wake-up, and noise signals during test mode and guarantees IDC functional accuracy and enhances overall WuRx reliability.

3.3. Configurable Modes

In the proposed IDC structure, two fully configurable modes are explored for WuRx. The wake-up mode (WUM) is for purely asynchronous wake-up signal detection from RSE with reduced latency. It processes the baseband recovered signal from the envelope detector after the comparator for identifying the wake-up signal. On the other hand, the wake-on mode (WOM) is an auxiliary pseudo-synchronous mode. When it is enabled, it turns on the main receiver for a very small configurable time interval, listens to any possible request from RSE, and keeps off for a relatively long duration. The mode control signal wake-up/-on (WUO) chooses the current selected mode. When WUO is low, a wake-up mode is enabled, which is the default mode. For enabling WOM, the WUO is configured as high.

3.3.1. Wake-Up Mode

The FSMC is key building block of IDC, and its flow chart is elaborated in Figure 8. The wake-up mode with self-hibernation, digital hysteresis, and wake-up interrupt (WU-I) is the default flow in the wake-up period for ensuring WuRx reliability and accuracy. On power up, FSMC is in a wake-up self-hibernation (WUSH) state. The proposed self-hibernation technique reduces the power consumption of IDC and OSC significantly in the non-wake-up interval by configuring OSC to its lowest self-hibernation frequency, f_{SH}. The dynamic power of a circuit is directly proportional to its operating frequency, as shown in (2). If the frequency reduces, the power consumption also reduces. When IDC detects high assertion on the *wusig* signal during self-hibernation, it configures OSC to its normal wake-up frequency, f_{WU}, and waits for OSC settling in the WUSH state. The f_{WU} is a much higher frequency than f_{SH} for achieving higher wake-up signal measuring accuracy. The controller starts sensing a wake-up signal in the wake-up signal sense (WUSS) state. The SPEG detects *wusig* every rising edge and generates a *wu_pe* pulse signal which is sensed in the WUSS state. The controller enables CWDT by asserting a high *wdt_en* signal. The CWDT provides a self-recovery mechanism for FSMC and it is enabled to avoid uncertain situations and improve IDC reliability. If *WU_SIG* is a noise pulse or glitch, it is identified and filtered out at this stage and FSMC moves back to WUSH for self-hibernation. After the sensing stage, the controller clears CWDT by asserting a high *wdt_clr* signal for one clock cycle and moves to next state. The IDC verifies the *WU_SIG* signal in the wake-up signal assurance (WUSA) state. The AFMU evaluates *WU_SIG* each cycle and confirms if its frequency f_{SIG} is in a configured range. The integrated configurable digital hysteresis technique accomplishes this task and accommodates wake-up frequency variations to improve reliability. The *WU_SIG* is a valid signal if its frequency fulfills the following condition described in (4) as follows:

$$f_{MOD.MIN} \geq f_{SIG} \leq f_{MOD.MAX}, \tag{4}$$

where $f_{MOD.MIN}$ and $f_{MOD.MAX}$ are the lower and upper limits of the valid WuRx signal modulation frequencies. These limits are configurable by parameters N_{NFX} and N_{XFN}, respectively, and are described as follows in (5):

$$f_{MOD.MIN} = \frac{f_{WU}}{N_{XFM}}, \tag{5a}$$

$$f_{MOD.MAX} = \frac{f_{WU}}{N_{NFX}}. \tag{5b}$$

The configurable successive number of wake-up signal cycles, WU_N, in the allowed frequency bandwidth are ensured in the WUSA state. If the *WU_SIG* WuRx signal frequency is not in the allowed range as described in (4) or the signal cycles are less than WU_N, then it means the signal is not a valid wake-up signal and a main transceiver must remain off to save power. The false and poor wake-up signals are identified and filtered out in the WUSA state by AFMU and CWDT successfully. If a non-wake-up signal is identified in this state, the controller moves back to a self-hibernation state. After sensing and assurance, the wake-up interrupt *WU_INT* is initiated for configurable T_{HOLD} duration in the wake-up interrupt generation (WUIG) state. The interrupt hold duration is defined as follows in (6):

$$T_{HOLD} = \frac{N_{HOLD}}{f_{WU}}, \tag{6}$$

where N_{HOLD} is the configurable parameter for defining wake-up interrupt hold duration. After the interrupt generation, FSMC remains silent for the T_{SILENT} interval in the wake-up silent (WUS) state and moves back to the WUSH state. The WUS state prohibits the WuRx to detect the current wake-up signal again if WU_N and T_{HOLD} parameters are configured to smaller values. The silent interval is programmable from parameter N_{SILENT} according to (7) as follows:

$$T_{SILENT} = \frac{N_{SILENT}}{f_{WU}}. \tag{7}$$

If wake-up monitoring (WU-M) is enabled by signal *monitor_ctrl*, then the FSMC moves to the wake-up interrupt enable (WUIE) state after confirming WU_N wake-up signal cycles in the WUSA state. It asserts *WU_INT* high and moves to the wake-up monitoring (WUSM) state. The controller continuously evaluates *WU_SIG* for the presence of a valid wake-up signal and it additionally provides *WU_INT* for the entire duration of the wake-up signal instead of T_{HOLD}. It gives more space to MCU to detect interrupt and trigger an acknowledge signal to RSE at the end of the wake-up signal. The proposed WUM guarantees to pass only a valid wake-up signal and it definitely identifies and filters out all non-wake-up signals. It ensures the accuracy and reliability of WuRx. The IDC turns on a main heavy powered transceiver at OBU only with a valid wake-up request from RSE.

Figure 8. WuRx intelligent digital controller (IDC) finite state machine controller (FSMC) flow diagram.

3.3.2. Wake-On Mode

The pseudo-synchronous wake-on mode (WOM) enhances overall reliability of transceiver in case of an issue in the main WuRx path. When WOM is enabled, the IDC configures OSC to desire frequency f_{WO} based on configured parameters in the wake-on oscillator configuration (WOOC) state, as shown in Figure 8. The wake-on interrupt T_{WOI} and wake-on sleep T_{WOS} intervals are computed according to (8) and (9), respectively, as follows:

$$T_{WOI} = \frac{N_{WOI}}{f_{WO}}, \quad (8)$$

$$T_{WOS} = \frac{N_{WOS}}{f_{WO}}. \quad (9)$$

After frequency configuration and OSC settling, the controller moves to a wake-on interrupt (WOI) state. In this state, *WO_INT* is asserted high for the T_{WOI} duration and MCU turns on the main receiver for intercepting any communication request from RSE. After the T_{WOI} interval, FSMC jumps to a wake-on sleep (WOS) state. The *WO_INT* is asserted low and MCU turns off the main receiver for the programmed T_{WOS} duration. After a sleep interval, the controller moves back to the WOI state and periodically generates configurable wake-on interrupt *WO_INT* for receiver.

4. Ultra-Low Power Configurable RC Oscillator

In the proposed WuRx, an ultra-low power configurable RC oscillator is integrated, which is the clock source for IDC. The RC oscillator structure is adopted rather than the crystal oscillator due to its low cost, less power consumption, fast start-up interval, and easy on-chip integration [27]. The fabricated oscillator has a wide controllable tuning range, $f_{CLK.MIN}$~$f_{CLK.MAX}$, and IDC configures it for different frequencies in wake-up, self-hibernation, and wake-on mode by controlling its capacitance values. For ultra-low power applications, the circuits are preferred to be operated in a weak inversion region, also known as the sub-threshold region [28,29]. Therefore, the oscillator is designed to operate in a sub-threshold region. Figure 9 shows the ultra-low power RC oscillator [27]. The configurability for various frequencies is achieved by altering resistance R and capacitance C values from IDC. This oscillator is composed of current reference, start-up, capacitor charge/discharge sensing circuits, and a frequency generation part. The current source or sink circuit is preferable to function in a weak inversion region for low power operation. The MOS transistor is operating in a sub-threshold region when the drain current I_D flows and the gate to source voltage V_{GS} is less than the threshold voltage V_{TH}. The diffusion current between source and drain mainly contributes to this current. The start-up circuit prohibits self-biased circuits to work at a zero biasing point. To enhance current sink or source output resistance, the current mirrors in a cascade structure are used. This generated current is mirrored by the current mirror and fed to capacitor, hysteresis controller M_1, and current-starved invertors for clock generation. The drain current I charges the capacitor C, and when V_C becomes equal to hysteresis controlling transistor M_1 V_{TH} value, the M_1 turns on. For clock frequency generator circuit, the voltage V_C across capacitor C linearly increases with the increase in current I when constant current flows into the capacitor, as described in (10) as follows:

$$\Delta V_C = \frac{I \Delta t}{C}. \tag{10}$$

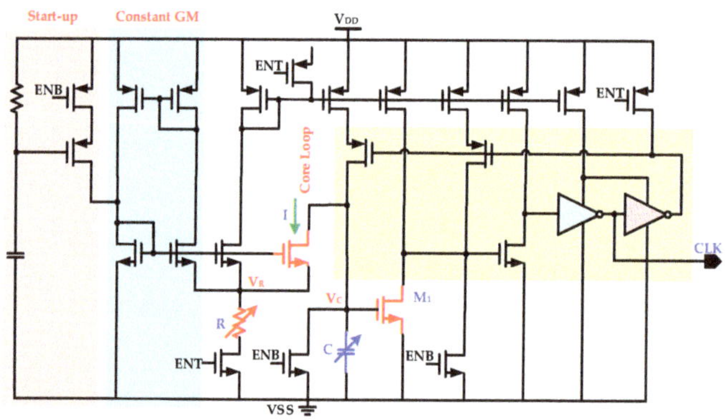

Figure 9. Ultra-low power configurable range communication (RC) oscillator.

The transistor M_1 logical V_{TH} also controls capacitor voltage. Hence, the capacitor C charging and discharging duration is controlled by current I and transistor M_1, and a triangular voltage waveform is generated for the capacitor. From (10), single charging or discharging cycle time Δt is given in (11):

$$\Delta t = \frac{C \Delta V}{I}. \tag{11}$$

The current mirror builds a constant current source as a current generator part. The voltage V_R is always stabilized by the feedback path. Hence, according to Ohm's law, the resistance R decides the amount of current. Thus, from (11), the generated clock period T_{CLK} is described in (12) as follows:

$$T_{CLK} = 2RC\frac{\Delta V_C}{V_R}. \quad (12)$$

The RC oscillator output frequency f_{CLK} is given in (13) as follows:

$$f_{CLK} = \frac{1}{T_{CLK}} = \frac{1}{2RC}\cdot\frac{V_R}{\Delta V_C}. \quad (13)$$

The capacitance C is designed as unit weighted capacitor bank, and IDC configures this capacitor back at different values for generating frequencies f_{WU}, f_{SH}, and f_{WO} for the wake-up mode, self-hibernation, and wake-on mode, respectively.

5. RF Front End and Baseband Processing

The RF front end in the proposed WuRx is composed of a high sensitivity RF envelope detector with an embedded internal matching network. The RFED-based approach is most common for designing a WuRx circuit for its low current consumption. With this scheme, the requirement of a LO generation for frequency down-conversion and RF amplification is also eliminated at the cost of reduced sensitivity. The internal matching improves sensitivity of the circuit. The RFED is the key building block in the WuRx circuit for interfacing with an antenna through a matching network and generating a baseband output signal by down-converting the input amplitude-modulated RF signal. Figure 10 illustrates the proposed RF envelope detector circuit with internal matching network [18,30]. The circuit is mainly composed of nonlinear transistor element M_1, input signal DC blocking capacitor C_{ac}, self-biasing feedback resistor R_f, impedance matching network with gate inductor L_g, and excess capacitor C_{ex}. The impedance matching network provides passive voltage amplification. With large self-biasing resistance, when RF input signal is not present, the M_1 gate to source voltage V_{gs} is closed to transistor threshold voltage V_{TH} with a very negligible biasing current. In this way, the limited sensitivity issue due to V_{TH} loss is compensated. The R_f is designed with a pseudo-resistor for large resistance with minimized parasitic capacitance and a small area. It perfectly isolates output voltage V_0 from RF input signal V_{RF} and prevents envelope detector loading. When V_{RF} is applied, the drain current I_1 exponentially increases while the biasing current supplied from M_2 is almost constant.

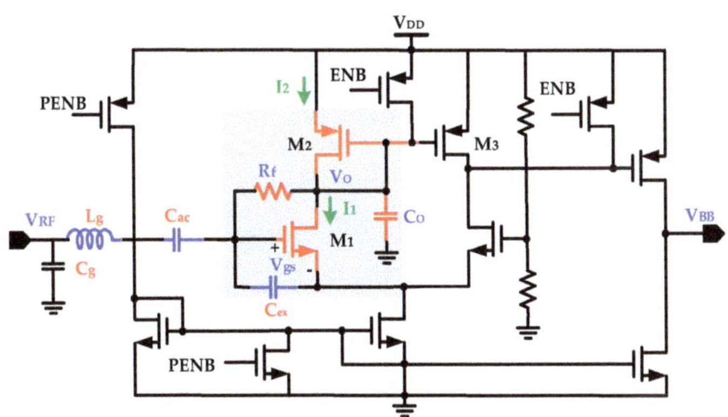

Figure 10. RF envelope detector with internal matching and self-biasing feedback resistor.

The capacitor C_O discharges and this discharge current decreases V_O until I_1 and I_2 become almost equal. This generates a 180° phase shift between V_{RF} and V_O. The convergence gain G_C, defined as the ratio of output baseband signal voltage amplitude to the RF input signal voltage amplitude, is evaluated as follows in (14):

$$G_C = \frac{1}{4}\frac{Q^2 I_1}{(nU_T)^2} r_0, \tag{14}$$

where Q is quality factor of the matching network, n is sub-threshold slope factor, U_T is thermal voltage, and r_0 is intrinsic output impedance.

The baseband signal V_{BB} produced by the RF envelope detector is subsequently processed for enhancing amplitude by a programmable gain baseband amplifier and a comparator before it is fed to IDC for digital processing for generating an interrupt signal. Figure 11 shows the PGA circuit that is a baseband amplification stage. It has high input impedance and its gain is configurable from external MCU. It provides flexibility for improving gain and amplifies the baseband signal V_{BB} significantly.

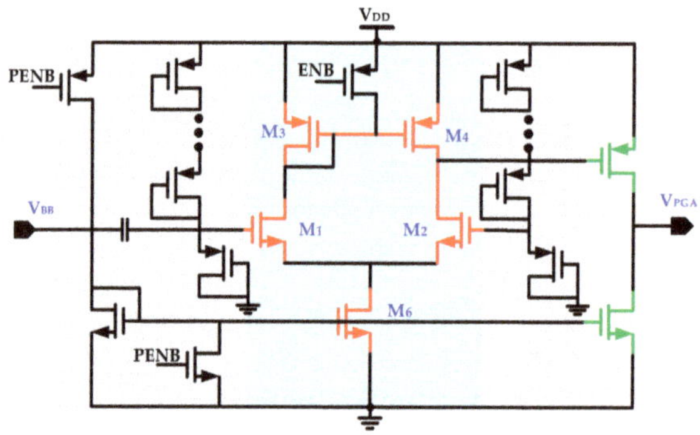

Figure 11. Programmable gain amplifier.

The hysteresis comparator, shown in Figure 12, is the final stage of baseband processing to generate a digital signal for IDC processing. It is composed of a positive feedback circuit with an amplifier, AMP. A two stage amplifier with output inverter [31] is used as a low power CMOS amplifier. With feedback resistor R_2, hysteresis upper and lower threshold values are configured for eliminating multiple transitions caused by noise. The three stage amplifier is composed of a differential amplifier, common source amplifier, and an output inverter. The analog differential input signals *IN−* and *IN+* are applied at differential pair M_1 and M_2. The differential pair transistor width is increased to reduce input offset voltage and increase gain. To minimize the propagation delay and reduce common source transistor M_7 gate parasitic capacitance, the M_7 is designed with a small size. The final inverter stage also enhances gain and improves the comparator slew rate. The final digital wake-up comparator output signal *WU_SIG* is fed to IDC for digital processing. The RF front end and analog baseband processing processes all signals and it does not filter or remove any non-wake-up signal. The IDC differentiates between actual wake-up signals and unwanted signals.

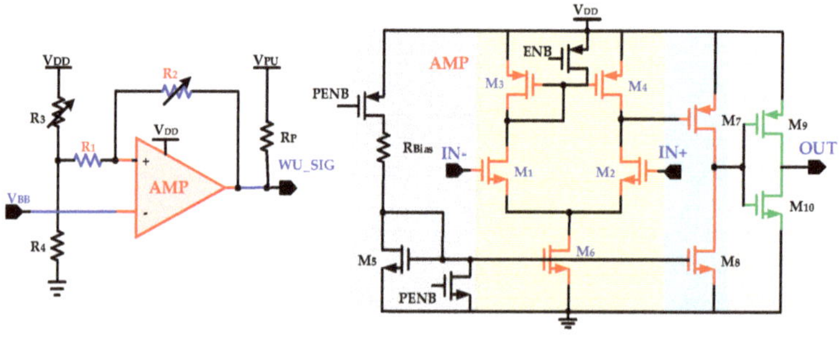

Figure 12. Hysteresis comparator with positive feedback and amplifier.

6. Experimental Results

The presented WuRx is integrated in a DSRC transceiver for ETCS applications. It is fabricated with a 130 nm CMOS process. Figure 13 shows the microphotograph of WuRx and magnified IDC layout. The WuRx occupied chip area is 532 × 910 μm^2 of which IDC takes only 94 × 82 μm^2. The WuRx is measured extensively to ensure its reliability and accuracy. The experimental lab environment is captured in Figure 14a and the measurement board with the fabricated DSRC transceiver chip is depicted in Figure 14b. The board is powered up from the Agilent® DC Power Supply with 5 V and the on-chip low dropout regulator (LDO) generates 0.9 V for the WuRx circuit, which is measured at the output pin with a digital multi-meter (DMM). The lower supply voltage is used to minimize the power consumption and all blocks, and correct operation is verified at a supply voltage of 0.9 V. The OOK baseband wake-up signal is generated from the Tektronix® AFG3101 Function Generator and modulated at 5.8 GHz with the Agilent® E4438C Signal Generator. This modulated RF signal is fed at a *RF_IN* SMA input connector on the board and after passing through external pi-matching network, package pin, and die PAD, it enters the WuRx circuit. The comparator output *WU_SIG* and wake-up interrupt *WK_INT* are plotted on a Tektronix® DSA71254C Digital Serial Analyzer. Different parameters and configurations are programmed through SPI and the graphical user interface (GUI) running on computer.

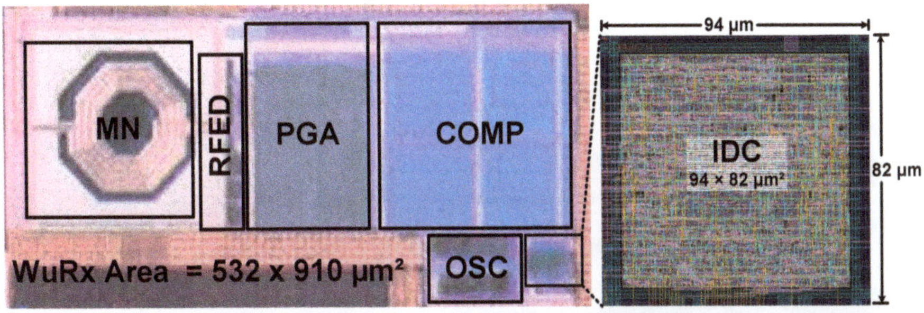

Figure 13. WuRx chip microphotograph and magnified IDC layout.

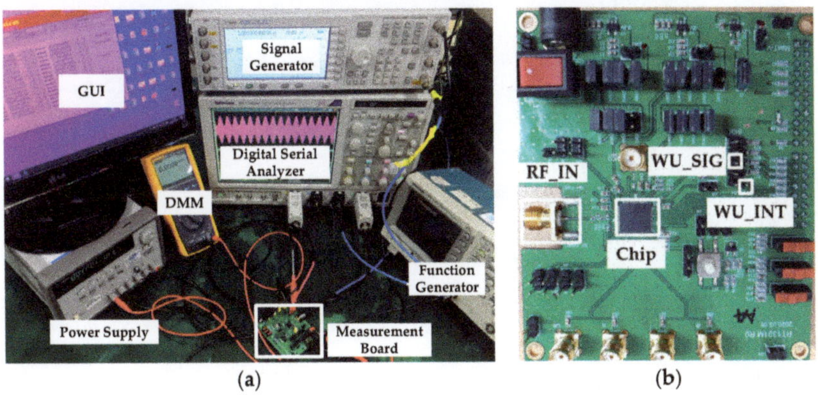

Figure 14. WuRx measurement: (**a**) experimental lab setup; (**b**) measurement board with chip.

The IDC performance is summarized in Table 1. The proposed digital controller is fully synthesizable. With an area of 0.007 mm^2 and a 34.62 nW power consumption, it not only ensures WuRx reliability and accuracy but also replaces complex and power hungry analog blocks such as BPF and ADC. The configurability, operating modes, digital hysteresis, and self-hibernation features prove its sublimity. The power consumption with and without self-hibernation for IDC and OSC and its effect on overall WuRx power performance is summarized in Table 2. Since the DSRC communication between RSE and OBU lasts for a very short interval and the OBU is in sleep mode most of the time, self-hibernation by voltage and frequency scaling has a significant positive impact on battery performance. The performance comparison of the proposed WuRx with the existing designs is listed in Table 3. The wake-up circuits in [8,24] integrate the complex BPF and [9] use the frequency detector (FD) circuit as its interface output stage without ensuring reliability and filtering of non-wake-up signals. The proposed WuRx architecture incorporates a fully synthesizable intelligent controller, which is not only area and power efficient but it also ensures unwanted signals filtering, guarantees WuRx reliability, and improves battery performance. The measurement results report an almost identical sensitivity of −46 dBm and a power consumption of only 2.48 µW.

Table 1. IDC performance summary.

Parameter	Value
CMOS process	130 nm
Occupied area	0.0077 mm^2
Gate count	809
Supply voltage	0.9 V
Current consumption [1]	38.47/9.7 nA
Power consumption [1]	34.62/8.73 nW
Wake-up frequency	1–140 kHz
Configurable architecture	Yes
Reliability and accuracy	Digital hysteresis, BIST, WDT
Operating Modes	WUM (WU-I, WU-M), WOM

[1] Without and with self-hibernation.

Table 2. Power consumption summary with and without self-hibernation.

Block	Without Self-Hibernation		With Self-Hibernation	
	Current	Power	Current	Power
Intelligent digital controller	38.47 nA	34.62 nW	9.7 nA	8.73 nW
RC oscillator	214 nA	192.6 nA	107 nA	96.3 nW
Total WuRx	2.75 µA	2.48 µW	2.62 µA	2.36 µW

Table 3. WuRx performance comparison.

Parameter	[8]	[9]	[24]	This Work
CMOS process (nm)	130	180	130	130
Wake-up frequency (kHz)	14	7~42	14	1–140 [1]
Operating frequency (GHz)	5.8	5.8	5.8	5.8
Sensitivity (dBm)	−44	−47	−45	−46
Power consumption (µW)	36	3.8	45	2.48 *
Chip area (mm^2)	-	-	-	0.484
Interface	BPF	FD	BPF	IDC

[1] Configurable for any specific range. * Without self-hibernation.

The WuRx accuracy and reliability is verified by applying various RF-modulated valid and invalid signals with different amplitudes and frequencies at the *RF_IN* input. When the input signal is valid, meaning its amplitude is greater than the sensitivity and its frequency is in a configured hysteresis range, the WuRx gives out confirmed wake-up interrupt signal. In the measurement results in Figure 15a, initially random, invalid OOK sequence, modulated at 5.8 GHz, with a sensitivity of −46 dBm is applied at the *RF_IN* input. The baseband signal is successfully recovered by RFED and the digital *WU_SIG* from COMP is fed to IDC. As it is clear from the results, the IDC identifies this invalid sequence and does not generate a confirmation signal. Later, the valid wake-up signal is ensured and verified by IDC. If IDC is not used at the comparator output, then WuRx reliability degrades as the main power hungry receiver is turned on, even with a non-wake-up signal. The consecutive burst of valid wake-up signals, as shown in Figure 15b, with exactly 14 clock cycles at 14 kHz is applied for proving proposed WuRx robustness, accuracy, and reliability. For each time, *WU_INT* is generated for approximately 1 ms after confirming five clock cycles (WU_N = 5) and IDC returns to its ideal state for the next wake-up signal sensing.

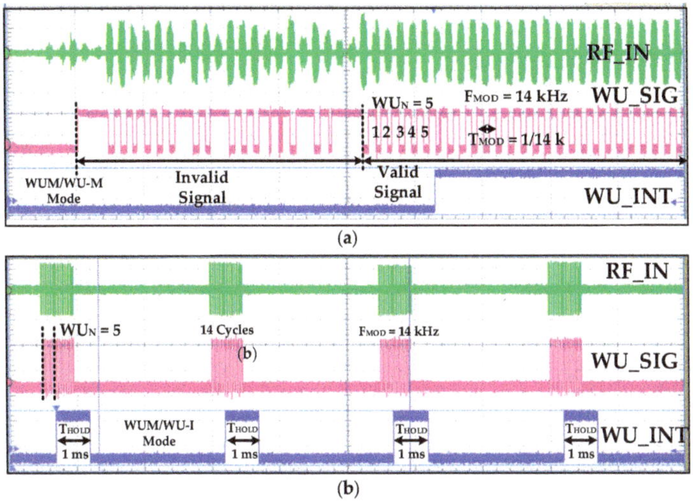

Figure 15. WuRx measurement result: (**a**) initially, invalid random sequence and then valid wake-up signal; (**b**) wake-up valid signal burst for robustness testing.

Figure 16 shows different WUM measurement results for various scenarios. In this measurement, WU_N is set to 5 and digital hysteresis, watch-dog timer, and T_{HOLD} are configured to 11~18 kHz, 142.8 µs, and 1 ms, respectively. The OSC is configured for wake-up and self-hibernation frequencies of 140 kHz and 14 kHz, respectively. Figure 16a,b show WUM with WU-I and WU-M configurations, respectively, in which the *WU_SIG* frequency is 14 kHz. The *WU_INT* is generated after sensing and confirming five successive *WU_SIG* clock cycles. In normal WUM, IDC and OSC current consumption from a 0.9 V supply is 38.47 nA and 214 nA, which reduces to 9.7 nA and 107 nA in self-hibernation, respectively. The self-test is measured in Figure 16c in which the *st_sig* signal of 14 kHz is generated by STPG.

Figure 17 summarizes WuRx measurement results with false, poor, invalid signals. Figure 17a,b shows results when the *WU_SIG* frequency is out of the configured hysteresis range (11 kHz~18 kHz in this case) and identifies false wake-up signals. Poor and false wake-up and noise signals are also perfectly identified and *WU_INT* is not generated. The signal with a valid frequency but insufficient number of cycles (less than WU_N = 5) is identified and filtered accurately by IDC, as shown in Figure 17c. Similarly, noise pulses and glitches in the RF signal are converted to a baseband digital signal and sensed and removed by IDC without generating interrupt and prohibits turning on the power hungry main receiver. If IDC is not integrated, then all these invalid signals are identified as wake-up signals, and as a consequence, power on transceiver falsely and degrading battery performance.

The WOM measurement result is depicted in Figure 18 in which T_{WOI} and T_{WOS} intervals are set for 65 ms and 0.65 s by configuring N_{WOI} = 91 and N_{WOS} = 91,000 according to (8) and (9), respectively.

The measured tuning range of OSC is captured in Figure 19. The OSC capacitor C is implemented as binary weighted capacitor bank which is controlled from an 8-bit *OSC_CTRL* signal from IDC. The measured $f_{CLK.MIN}$ and $f_{CLK.MAX}$ frequencies are 12.16 kHz and 362.37 kHz when *OSC_CTRL* values are all high and all low, respectively, with a total frequency range Δf_{CLK} of 350.21 kHz. The spectrum also shows the WUM frequency which is configured as approximately 140 kHz. In self-hibernation mode, the OSC frequency f_{SH} is configured to about 14 kHz. The WOM clock frequency depends on the configured parameters for T_{WOI} and T_{WOS} intervals. At f_{WU} of 140 kHz, it draws 214 nA current from 0.9 V supply which is reduced to almost half in self-hibernation mode.

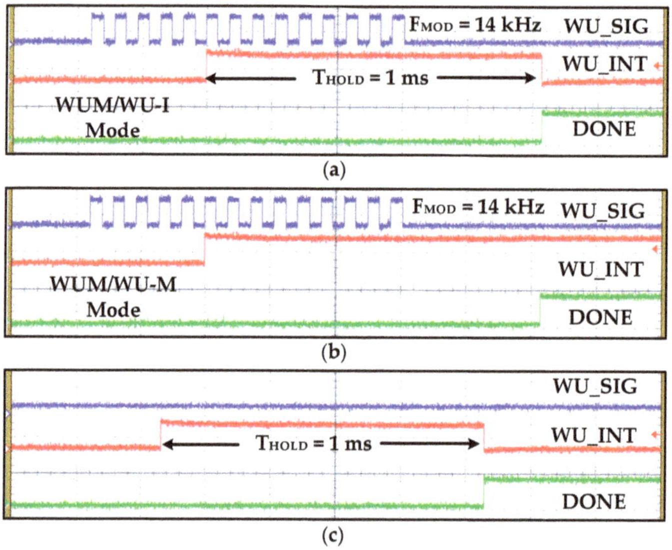

Figure 16. WuRx measurement results for different IDC wake-up modes with valid signals: (**a**) wake-up interrupt (WU-I); (**b**) wake-up monitoring (WU-M); (**c**) self-test with WU-I mode.

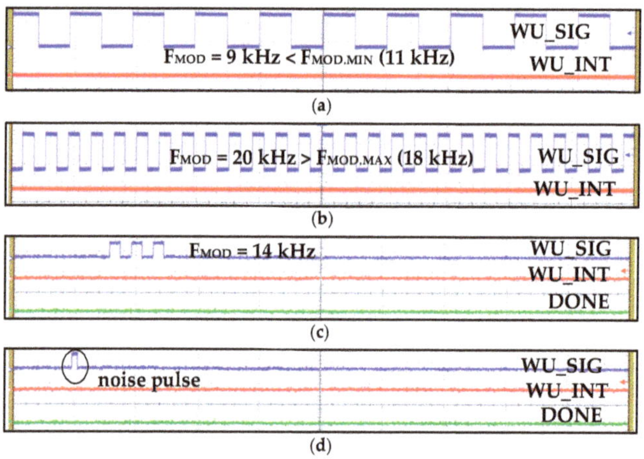

Figure 17. WuRx measurement results for invalid signals and IDC identification and filtering: (**a**) invalid wake-up signal with modulation frequency of 9 KHz, which is less than the minimum hysteresis configured limit of 11 kHz; (**b**) invalid wake-up signal with modulation frequency of 20 KHz, which is greater than the maximum hysteresis configured limit of 18 kHz; (**c**) poor wake-up signal with valid modulation frequency but less number of cycles than the configured value of 5; (**d**) invalid signal of noise pulse.

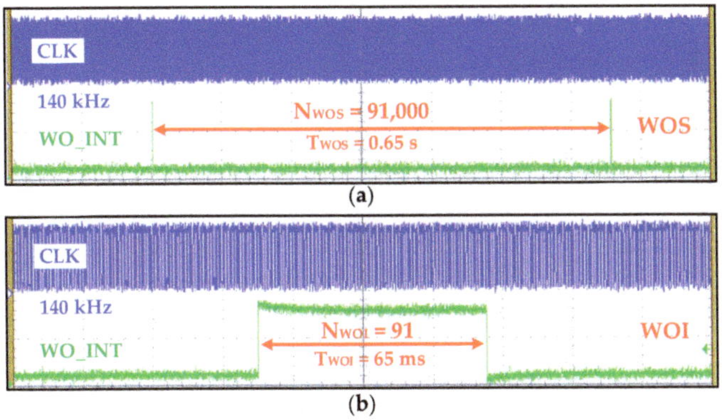

Figure 18. Wake-on mode measurement result: (**a**) Wake-on sleep of 0.65 s duration when N_{WOS} is configured with value of 91,000; (**b**) wake-on interrupt of 65 ms when NWOI is configured as 91.

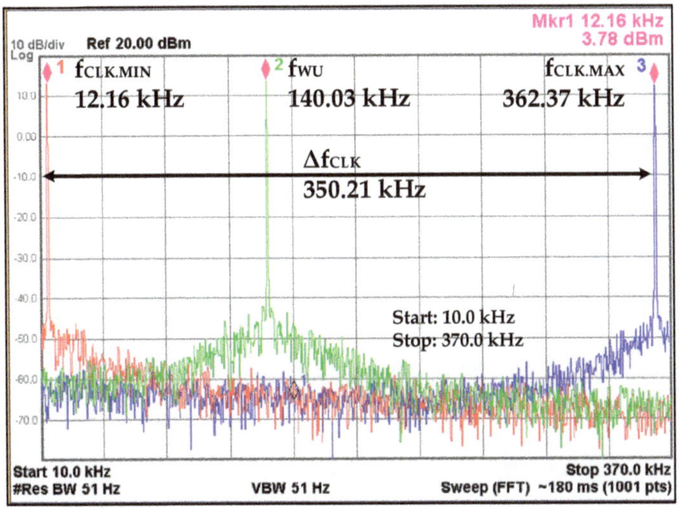

Figure 19. Oscillator (OSC) frequency range and measurement result.

Figure 20 shows the measured reflection co-efficient, $|S_{11}|$, for the proposed WuRx. The measured value of $|S_{11}|$ at 5.8 GHz is about −25.622 dB, which shows the excellent matching. Moreover, $|S_{11}|$ values at 5.75 GHz and 5.85 GHz are −17.138 dB and −12.876 dB, respectively.

Figure 21 summarizes the detailed IDC post place and route (P&R) simulation results using the NC-Verilog® tool. The wake-up interrupt mode simulation result is shown in Figure 21a in which *WU_SIG* with different frequencies is applied. It is clear from the simulation results that when the wake-up signal is either less or greater than the configured hysteresis range (11 kHz~18 kHz), it is identified and filtered out without generating interrupt at *WU_INT*. The self-test simulation with one of the configurations is shown in Figure 21b. The STPG generates a variety of valid and invalid signals for ensuring the functional accuracy of IDC. Instead of a baseband digital *WU_SIG* signal, the test wake-up signal *st_sig* is generated internally by STPG. The IDC accurately generates interrupt *WU_INT* after identifying and verifying the signal, as shown in Figure 21b. The self-test enhances the reliability of IDC itself. The WOM simulation is depicted in Figure 21c in which T_{WOI} and T_{WOS} intervals are set for 20.8 ms and 0.65 s by configuring N_{WOI} = 2912 and N_{WOS} = 91,000 according to (8) and (9), respectively.

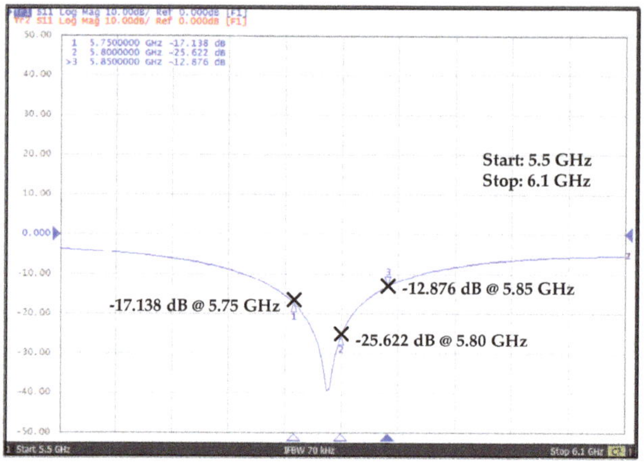

Figure 20. RF external on board impedance matching measurement result.

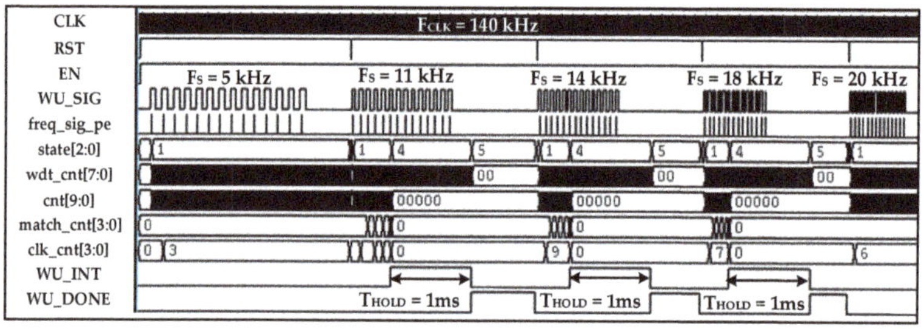

(a)

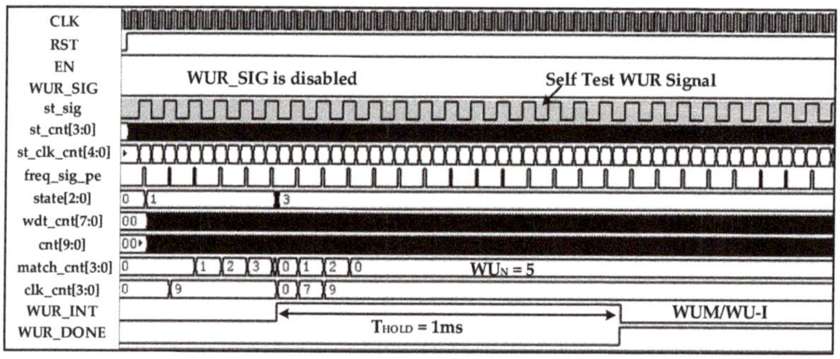

(b)

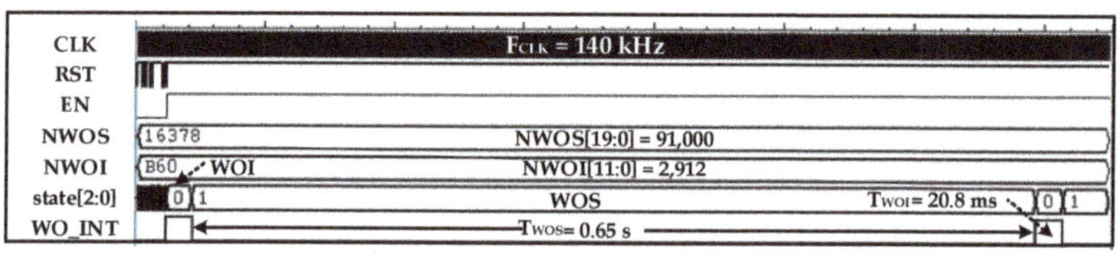

(c)

Figure 21. The IDC post place and route (P&R) simulation result: (**a**) WU/WU-I with different wake-up frequencies; (**b**) self-test with WU_I; (**c**) wake-on mode with interrupt and sleep durations of 20.8 ms and 0.65 s.

7. Conclusions

A highly reliable RF WuRx is presented for ETC systems in this article. For improving WuRx reliability and enhancing battery performance, the IDC is proposed as final stage. The IDC also acts as filter and replaces complex and power demanding analog blocks such as BPF, ADC, and FD. With the proposed configurable digital controller, high reliability and accuracy are achieved by sensing and ensuring a successive, configurable number of wake-up signal cycles before enabling power hungry RF transceiver. The presented self-hibernation technique reduces IDC and RC oscillator current consumption during the non-wake-up period and improves battery life. The digital hysteresis accommodates wake-up signal frequency variation and enhances WuRx accuracy. To avoid uncertain conditions during poor and false wake-up, a watch-dog timer for IDC self-recovery is integrated. During wake-up, the digital controller requires 34.62 nW power. In self-hibernation mode, its current reduces from 38.47 nA to 9.7 nA. It is fully synthesizable and needs 809 gates for its implementation in a 130 nm CMOS process with an area of $94 \times 82\ \mu m^2$. The WuRx measured power consumption is 2.48 μW, has −46 dBm sensitivity, and a 0.484 mm^2 chip area. The extensive measurement and verification make the proposed WuRx an ideal solution for a highly reliable DSRC wake-up circuit.

Author Contributions: K.-Y.L. guided and directed the authors for this research. I.A. proposed, designed, simulated, and implemented the overall architecture and wrote the paper. M.A., M.R.U.R., D.K., and H.Y. contributed to the synthesis and place and route (P & R) simulation. They also contributed to the design of the top layout of the chip. S.J.K. helped writing the paper, designing the testing board, and during the measurements. Y.P. and S.-S.Y. gave advice about implementation issues and reviewed the paper before submission. All authors have read and agreed to the published version of the manuscript.

Funding: This research was supported by the National Research Foundation of Korea (NRF) grant funded by the Korean government (MSIP) (2014R1A5A1011478).

Conflicts of Interest: The authors declare no conflict of interest.

References

1. Kwon, K.; Choi, J.; Choi, J.; Hwang, Y.; Lee, K.; Ko, J. A 5.8 GHz integrated CMOS dedicated short range communication transceiver for the Korea/Japan electronic toll collection system. *IEEE Trans. Microw. Theory Tech.* **2010**, *58*, 2751–2763. [CrossRef]
2. Moazzeni, S.; Sawanm, M.; Cowan, G.E.R. An Ultra-Low-Power Energy-Efficient Dual-Mode Wake-Up Receiver. *IEEE Trans. Circuits Syst. I Reg. Pap.* **2015**, *62*, 517–526. [CrossRef]
3. Polonelli, T.; Huy, T.L.; Lizzi, L.; Ferrero, F.; Magno, M. A wake-up receiver with ad-hoc antenna co-design for wearable applications. In Proceedings of the IEEE Sensors Applications Symposium, Catania, Italy, 20–22 April 2016.
4. Blanckenstein, J.; Klaue, J.; Karl, H. A Survey of Low-Power Transceivers and Their Applications. *IEEE Circuits Syst. Mag.* **2015**, *15*, 6–17. [CrossRef]
5. Research Institute of Highway Ministry. *Electronic Toll Collection-Dedicated Short Range Communication-Part 1: Physical Layer*; China Nat. Standard GB/T 20851.1-2007; Research Institute of Highway Ministry: Beijing, China, 2007.
6. Liu, H.; Qu, X.; Cao, L.; Liu, R.; Zhang, Y.; Zhang, M.; Li, X.; Wang, W.; Lu, C. A 5.8 GHz DSRC digitally controlled CMOS RF-SoC transceiver for China ETC. In Proceedings of the IEEE Asia and South Pacific Design Automation Conference, Jeju, Korea, 22–25 January 2018.
7. Sasho, N.; Minami, K.; Fujita, H.; Takahashi, T.; Iimura, K.; Abe, M.; Yasuda, A. Single-chip 5.8 GHz DSRC transceiver with dual-mode of ASK and Pi/4-QPSK. In Proceedings of the IEEE Radio and Wireless Symposium, Orlando, FL, USA, 22–24 January 2008.
8. Choi, J.; Lee, I.Y.; Lee, K.; Yun, S.O.; Kim, J.; Ko, J.; Yoon, G.; Lee, S.G. A 5.8-GHz DSRC transceiver with a 10-μA interference-aware wake-up receiver for the Chinese ETCS. *IEEE Trans. Microw. Theory Tech.* **2014**, *62*, 3146–3160. [CrossRef]
9. Zhu, W.; Zhang, D.; Gao, T.; Liu, F.; Yang, H. A 2.1 μA wake-up circuit for Chinese ETC system. *J. Electron. China* **2013**, *30*, 308–312.

10. Rekhi, A.; Arbabian, A. A 14.5 mm^2 8 nW −59.7 dBm-sensitivity ultrasonic wake-up receiver for power-, area-, and interference-constrained applications. In Proceedings of the IEEE International Solid-State Circuits Conference, San Francisco, CA, USA, 11–15 February 2018.
11. Masuda, T.; Ohhata, K.; Shiramizu, N.; Hanazawa, S.; Kudoh, M.; Tanba, Y.; Takeuchi, Y.; Shimamoto, H.; Nagashima, T.; Washio, K. Single-chip 5.8 GHz ETC transceiver IC with PLL and demodulation circuits using SiGe HBT/CMOS. In Proceedings of the IEEE International Solid-State Circuits Conference, San Francisco, CA, USA, 7–11 February 2002.
12. Marinkovic, S.J.; Popovici, E.M. Nano-power wireless wake-up receiver with serial peripheral interface. *IEEE J. Sel. Areas Commun.* **2011**, *29*, 1641–1647. [CrossRef]
13. Kolinko, P.; Larson, L.E. Passive RF receiver design for wireless sensor networks. In Proceedings of the IEEE/MTT-S International Microwave Symposium, Honolulu, HI, USA, 3–8 June 2007.
14. Pletcher, N.M.; Gambini, S.; Rabaey, J. A52 μW wake-up receiver with −72 dBm sensitivity using an uncertain-IF architecture. *IEEE J. Solid State Circuits* **2009**, *44*, 269–280. [CrossRef]
15. Shin, S.; Yun, S.; Cho, S.; Kim, J.; Kang, M.; Oh, W.; Kang, S.M. 0.18 μm CMOS integrated chipset for 5.8GHz DSRC systems with +10 dBm output power. In Proceedings of the IEEE International Symposium on Circuits and Systems, Seattle, WA, USA, 18–21 March 2008.
16. Drago, S.; Leenaerts, D.M.W.; Sebastiano, F.; Breems, L.J.; Makinwa, K.A.A.; Nauta, B. A 2.4 GHz 830 pJ/bit duty-cycled wake-up receiver with −82 dBm sensitivity for crystal-less wireless sensor nodes. In Proceedings of the IEEE International Solid-State Circuits Conference, San Francisco, CA, USA, 7–11 February 2010.
17. Pletcher, N.; Gambini, S.; Rabaey, J. A 65 μW, 1.9 GHz RF to digital baseband wakeup receiver for wireless sensor nodes. In Proceedings of the IEEE Custom Integrated Circuits Conference, San Jose, CA, USA, 16–19 September 2007.
18. Cheng, K.; Liu, X.; Je, M. A 2.4/5.8 GHz 10 μW wake-up receiver with −65/−50 dBm sensitivity using direct active rf detection. In Proceedings of the IEEE Asian Solid State Circuits Conference, Kobe, Japan, 12–14 November 2012.
19. Huang, X.; Rampu, S.; Wang, X.; Dolmans, G.; Groot, H.D. A 2.4 GHz/915 MHz 51μW wake-up receiver with offset and noise suppression. In Proceedings of the IEEE International Solid-State Circuits Conference, San Francisco, CA, USA, 7–11 February 2010.
20. Kamalinejad, P.; Keikhosravy, K.; Magno, M.; Mirabbasi, S.; Leung, V.C.M.; Benini, L. A high-sensitivity fully passive wake-up radio front-end for wireless sensor nodes. In Proceedings of the IEEE International Conference on Consumer Electronics, Las Vegas, NV, USA, 10–13 January 2014.
21. Khan, D.; Oh, S.J.; Shehzad, K.; Basim, M.; Verma, D.; Pu, Y.G.; Lee, M.; Hwang, K.C.; Yang, Y.; Lee, K.Y. An Efficient Reconfigurable RF-DC Converter with Wide Input Power Range for RF Energy Harvesting. *IEEE Access* **2020**, *8*, 79310–79318. [CrossRef]
22. Khan, D.; Oh, S.J.; Shehzad, K.; Verma, D.; Khan, Z.H.N.; Pu, Y.G.; Lee, M.; Hwang, K.C.; Yang, Y.; Lee, K.Y. A CMOS RF Energy Harvester With 47% Peak Efficiency Using Internal Threshold Voltage Compensation. *IEEE Microw. Wireless Compon. Lett.* **2019**, *29*, 415–417. [CrossRef]
23. Demirkol, I.; Ersoy, C.; Onur, E. Wake-up receivers for wireless sensor networks: Benefits and challenges. *IEEE Trans. Wireless Commun.* **2009**, *16*, 88–96. [CrossRef]
24. Choi, J.; Lee, K.; Yun, S.; Lee, S.; Ko, J. An interference-aware 5.8 GHz wake-up radio for ETCS. In Proceedings of the IEEE International Solid-State Circuits Conference, San Francisco, CA, USA, 19–23 February 2012.
25. Ali, I.; Rikhan, B.S.; Kim, D.-G.; Lee, D.-S.; Rehman, M.R.U.; Abbasizadeh, H.; Asif, M.; Lee, M.; Hwang, K.C.; Yang, Y.; et al. Design of a Low-Power, Small-Area AEC-Q100-Compliant SENT Transmitter in Signal Conditioning IC for Automotive Pressure and Temperature Complex Sensors in 180 Nm CMOS Technology. *Sensors* **2018**, *18*, 1555. [CrossRef] [PubMed]
26. Chabini, N.; Chabini, I.; Aboulhamid, E.M.; Savaria, Y. Methods for minimizing dynamic power consumption in synchronous designs with multiple supply voltages. *IEEE Trans. Comput. Aided Des. Integr. Circuits Syst.* **2003**, *22*, 346–351. [CrossRef]
27. Joonhyung, L.; Kwangmook, L.; Koonsik, C. Ultra low power RC oscillator for system wake-up using highly precise auto-calibration technique. In Proceedings of the IEEE European Solid-State Circuits Conference ESSCIRC, Seville, Spain, 14–16 September 2010.

28. Ali, I.; Abbasizadeh, H.; Rehman, M.R.U.; Asif, M.; Oh, S.J.; Pu, Y.G.; Lee, M.; Hwang, K.C.; Yang, Y.; Lee, K.-Y. An Ultra-Low Power, Adaptive All-Digital Frequency-Locked Loop with Gain Estimation and Constant Current DCO. *IEEE Access* **2020**, *8*, 97215–97230. [CrossRef]
29. Abbasizadeh, H.; Ali, I.; Rikan, B.S.; Lee, D.S.; Pu, Y.G.; Yoo, S.S.; Lee, M.; Hwang, K.C.; Yang, Y.; Lee, K.-Y. 260-μW DCO with Constant Current over PVT Variations Using FLL and Adjustable LDO. *IEEE Trans. Circuits Syst. II Exp. Briefs* **2018**, *65*, 739–743. [CrossRef]
30. Cheng, K.; Chen, S. An Ultralow-Power Wake-Up Receiver Based on Direct Active RF Detection. *IEEE Trans. Circuits Syst. I Reg. Pap.* **2017**, *64*, 1661–1672. [CrossRef]
31. Furth, P.M.; Tsen, Y.C.; Kulkarni, V.B.; Raju, T.K.P.H. On the design of low-power CMOS comparators with programmable hysteresis. In Proceedings of the IEEE International Midwest Symposium on Circuits and Systems, Seattle, WA, USA, 1–4 August 2010.

sensors

MDPI

Article

A Highly Accurate, Polynomial-Based Digital Temperature Compensation for Piezoresistive Pressure Sensor in 180 nm CMOS Technology

Imran Ali, Muhammad Asif, Khuram Shehzad, Muhammad Riaz Ur Rehman, Dong Gyu Kim, Behnam Samadpoor Rikan, YoungGun Pu, Sang Sun Yoo and Kang-Yoon Lee *

College of Information and Communication Engineering, Sungkyunkwan University (SKKU), Suwon 16419, Korea; imran.ali@skku.edu (I.A.); m.asif@skku.edu (M.A.); khuram1698@skku.edu (K.S.); riaz@skku.edu (M.R.U.R.); rlarlarbrb@skku.edu (D.G.K.); behnam@skku.edu (B.S.R.); hara1015@skku.edu (Y.P.); rapter@kaist.ac.kr (S.S.Y.)

* Correspondence: klee@skku.edu; Tel.: +82-31-299-4954

Received: 14 August 2020; Accepted: 11 September 2020; Published: 14 September 2020

Abstract: Recently, piezoresistive-type (PRT) pressure sensors have been gaining attention in variety of applications due to their simplicity, low cost, miniature size and ruggedness. The electrical behavior of a pressure sensor is highly dependent on the temperature gradient which seriously degrades its reliability and reduces measurement accuracy. In this paper, polynomial-based adaptive digital temperature compensation is presented for automotive piezoresistive pressure sensor applications. The non-linear temperature dependency of a pressure sensor is accurately compensated for by incorporating opposite characteristics of the pressure sensor as a function of temperature. The compensation polynomial is fully implemented in a digital system and a scaling technique is introduced to enhance its accuracy. The resource sharing technique is adopted for minimizing controller area and power consumption. The negative temperature coefficient (NTC) instead of proportional to absolute temperature (PTAT) or complementary to absolute temperature (CTAT) is used as the temperature-sensing element since it offers the best temperature characteristics for grade 0 ambient temperature operating range according to the automotive electronics council (AEC) test qualification ACE-Q100. The shared structure approach uses an existing analog signal conditioning path, composed of a programmable gain amplifier (PGA) and an analog-to-digital converter (ADC). For improving the accuracy over wide range of temperature, a high-resolution sigma-delta ADC is integrated. The measured temperature compensation accuracy is within ±0.068% with full scale when temperature varies from −40 °C to 150 °C according to ACE-Q100. It takes 37 μs to compute the temperature compensation with a clock frequency of 10 MHz. The proposed technique is integrated in an automotive pressure sensor signal conditioning chip using a 180 nm complementary metal–oxide–semiconductor (CMOS) process.

Keywords: temperature compensation; digital controller; piezoresistive; pressure sensor; negative temperature coefficient; ACE-Q100; CMOS

1. Introduction

Presently, research on pressure sensors and transducers has been gaining significant attention. These sensors are being adapted widely in variety of applications such as the automotive industry, biomedical systems, petrochemicals, energy and electric power systems, aerospace, process control and humidity sensing systems [1–4]. The silicon piezoresistive-type (PRT) pressure sensor is widely used due to its simplicity, low cost, small size and robustness. In diverse harsh environment applications, the temperature increases up to 150 °C and consistency in sensor accuracy and performance are expected. For automotive applications, according to AEC-Q100 grade 0, the ambient temperature range is −40 °C to 150 °C [5]. The PRT sensor exhibits non-linear temperature dependency and its output voltage is influenced by temperature significantly. This complex temperature-dependent nature adversely affects the accuracy, reliability, precision and performance of piezoresistive sensors. The PRT sensor input-out and temperature dependent characteristics are elaborated in Figure 1 [6,7]. Other than non-linear behavior, gain and offset errors, depicted in Figure 1a, the PRT sensor output is highly dependent on temperature. At constant input pressure, the PRT sensor output voltage has complex relationship under dynamic temperature environment as shown in Figure 1b. Therefore, real time temperature compensation is mandatory for accurate and reliable measurement results of a sensor. In the harsh automotive environment, the temperature dependent pressure sensor variation must be encountered for accurate and reliable operation. The main drawback of current piezoresistive pressure sensors is the drop of output voltage with increase in the operating temperature which severely reduces the measurement accuracy. With the rise in temperature, the crystalline silicon electrical resistance increases and its piezoresistive coefficient decreases. The PRT sensor temperature sensitivity consists of temperature coefficient sensitivity (TCS) and temperature coefficient offset (TCO) [8]. The negative temperature coefficient of the piezoresistive coefficient is the main cause of TCS. The residual stress on packaging or membrane effects and mismatch values of resistors affects TCO. In the past, different techniques have been introduced to compensate for temperature variations in pressure sensors. Passive and active efforts were made to overcome undesired temperature effect on pressure output. In passive methods, additional resistors are utilized for temperature compensation in half bridge or full bridge arrangements within the sensor instead of signal conditioning circuit. Typically, the TCS and TCO are canceled by utilizing temperature-dependent series and trimmed parallel resistors. In active temperature compensation methodologies, additional temperature sensor is incorporated inside sensor chip. The TCS and TCO are compensated for by additional value from temperature sensor [8]. Based on the implementation method, these are categorized into hardware, software and hybrid approaches [9].

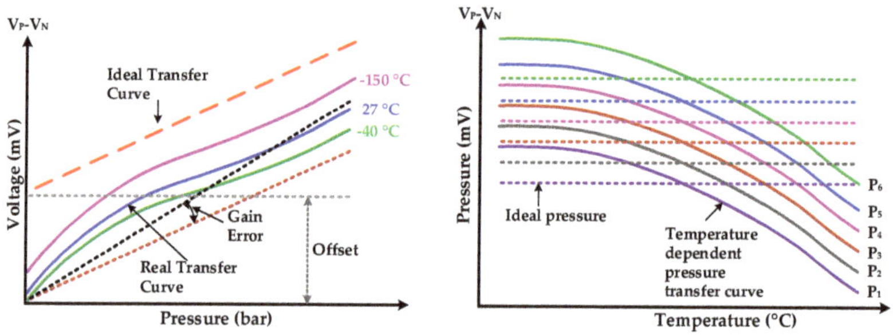

(**a**) (**b**)

Figure 1. Piezoresistive-type (PRT) sensor transfer characteristics: (**a**) PRT input-output characteristics; (**b**) temperature-dependent pressure transfer curves at different constant pressures.

The passive compensation techniques were also adopted to eliminate pressure sensor output voltage drop with the increase in temperature [8,10,11]. For a piezoresistive pressure sensor, a built-in passive temperature compensation technique is introduced in [9]. An extra polysilicon resistor with negative temperature coefficient of resistivity (TCR) is employed inside a sensor-fabricated patch instead of the calibration process. In [11], a similar passive resistor temperature compensation method is presented in which the system parameters are manipulated by using differential equations. These passive techniques reduce TCS but TCO is not eliminated.

In the literature, several active methods using temperature element such as proportional to absolute temperature (PTAT), an analog-to-digital converter (ADC) and lookup tables are proposed [7,12,13]. A signal-conditioning integrated circuit (IC) is presented for piezoresistive pressure sensor in [7,13] in which temperature compensation incorporates on-chip PTAT used as the temperature element. The flash ADC converts the temperature analog signal to digital which is then used to pick a compensation factor from a look-up table. A similar approach is adopted in analog front end IC for an automotive capacitive pressure sensor [12]. A band gap reference (BGR) is used as temperature-sensing element and lookup approach is introduced for compensating temperature effect in pressure value. Such approaches are not accurate due to low-resolution ADC and the lookup table inside IC is integrated based on simulation results which causes errors due to PVT variation after fabrication.

Several software-based techniques using either conventional mathematical computation or artificial neural network (ANN) algorithms have been investigated to compensate for temperature effect on pressure sensor accuracy [14–18]. In [14], machine learning part is implemented on LabVIEW® system for algorithm training and compensation part is designed on a microcontroller for a piezoresistive pressure sensor. The algorithm is trained first on the software system and then trained parameters are loaded into the microcontroller for real-time temperature compensation. However, the temperature compensation is valid only for the temperature range of −40 °C to 85 °C. Two techniques for capacitor pressure sensor modeling are reported in [15,16] which are based on a functional link ANN and back propagation neural network, respectively. These modeling based temperature compensation exhibits better accuracy of 1% FS only in a temperature range of −20 °C to 70 °C. An intelligent scheme using back propagation is proposed in [16] which achieves approximately 98% error reduction when applied to pressure ranges from 0 bar to 1 bar and temperature ranges from 25 °C to 80 °C. A feed-forward neural network is implemented on the CMOS analog application specific integrated circuit (ASIC) [17] for temperature compensation of a piezoresistive pressure sensor. With the presented technique, the error was reduced to 0.1% for compensated sensor in the temperature range of 0 °C to 70 °C. However, ANN-based proposed approaches are complex, requiring machine learning systems with large data set and exhibits low accuracy. Due to large computation and memory size, these methodologies are not appropriate for on-chip integration. These approaches do not clarify in neural networks the configuration and performance. A hybrid approach consisting of hardware and software for temperature compensation in pressure sensors is reported in [18]. It used a small processor as hardware compensation and a cubic B-spline based curve fitting algorithm in software. It is an off-chip compensation which results in an increased complexity. It is also unstable for batch compensation.

In this paper, a polynomial-based highly accurate temperature compensation technique is introduced. The compensation polynomial of a PRT pressure senor is proposed which is implemented in fully digital fashion. The negative temperature coefficient (NTC) sensor is used as a temperature-sensing element and high-resolution sigma-delta ADC (SD-ADC) is integrated. The NTC sensor, connected in Wheatstone bridge configuration has very high sensitivity of −3% to −6% per °C and it demonstrates comparatively very steep resistance-temperature slope and typically suitable for −55 °C to 200 °C temperature range. For automotive applications, according to AEC-Q100 grade 0, as shown in Table 1, it exhibits high accuracy for the ambient temperature range of −40 °C to 150 °C.

Table 1. AEC-Q100 compliant operating temperature grades.

Grade	Ambient Operating Temperature Range
Grade 0	−40 °C to +150 °C
Grade 1	−40 °C to +125 °C
Grade 2	−40 °C to +105 °C
Grade 3	−40 °C to +85 °C
Grade 4	0 °C to +70 °C

The rest of the paper is organized as follows: Section 2 presents architecture of pressure and temperature sensor interface chip top architecture. The detailed design of the proposed temperature compensation technique is described in Section 3. The temperature compensation digital controller is discussed in Section 4. The SD-ADC design is included in Section 5. Section 6 describes a programmable gain amplifier (PGA) design. Section 7 describes the experimental results and analysis. Lastly, the paper is concluded in Section 8.

2. Proposed Pressure Sensor Interface Architecture with Temperature Compensation

Typical signal conditioning integrated circuits nowadays usually perform analog and digital processing for improving automotive PRT sensor linearity, offset and gain errors [7,19]. Figure 2 shows the block diagram for a proposed PRT sensor interface IC with the presented polynomial-based digital temperature compensation. For automotive PRT pressure sensor, the highly reliable and accurate digital temperature compensation is composed of an off-chip NTC sensor, analog multiplexer (MUX), PGA, SD-ADC and polynomial-based configurable digital temperature-compensation controller (TCC). The proposed design may be integrated in any pressure sensor exhibiting this architecture with additional NTC sensor, MUX and TCC. The shared structure is introduced for pressure and temperature information processing. The main controller (MC) selects one of the sensor path for taking current pressure or temperature information from MUX. The sensor signal is amplified by PGA for increasing its range to a proper voltage level. The amplified signal is then fed to SD-ADC for digital conversion. In the proposed architecture, PGA and SD-ADC are shared for both PRT and NTC sensors and thus reduce cost and power consumption significantly. The pressure processing for non-linearities is performed in a pressure processing unit (PPU) while the real-time temperature compensation is achieved in the proposed polynomial-based TCC. The final pressure code with temperature compensation is delivered to an electronic control unit (ECU) interface (EI). The EI is either a digital to analog (DAC) converter with driving buffer amplifier (DA) [7] or it is a digital serial interface such as single edge nibble transmission (SENT) [19]. In the proposed pressure sensor interface chip, SENT is incorporated for its interface with the ECU and the final output signal SOUT is an asynchronous digital signal. Since, the pressure and temperature signals have very low frequencies of a few kHz, therefore, a low speed, high resolution SD-ADC is used for precise digitization of analog signals. The digital processing is more robust and reliable compared to analog processing [20,21]. Also, digital compensation processing is much easier and simpler than in analog techniques. Therefore, the digital temperature compensation and processing approach is adapted in the proposed system.

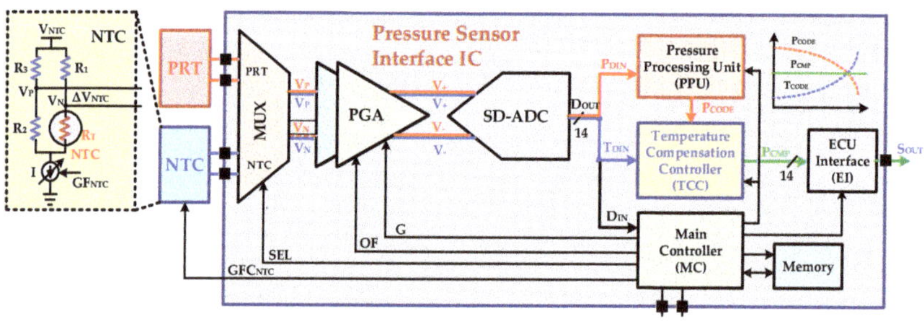

Figure 2. Automotive PRT pressure and temperature sensors signal interface integrated circuit (IC) with proposed polynomial based digital temperature compensation.

3. Proposed Temperature Compensation

The main drawback of current piezoresistive pressure sensors is the drop of output voltage with the increase in the operating temperature which severely reduces the measurement accuracy. Ensuring the accurate operation of the sensor with temperature variation is critical to satisfy the temperature characteristics of the AEC-Q100 [5]. A novel polynomial-based technique is introduced in a digital way to compensate temperature variations. The concept of temperature-dependent PRT characteristics and its compensation polynomial is depicted in Figure 3. The output voltage of the pressure sensor is not a linear function of temperature and the input-output of a pressure sensor have a complex polynomial relationship. Thus the opposite of the polynomial is an ideal solution to compensate temperature variation accurately with zero error for the full range of temperature.

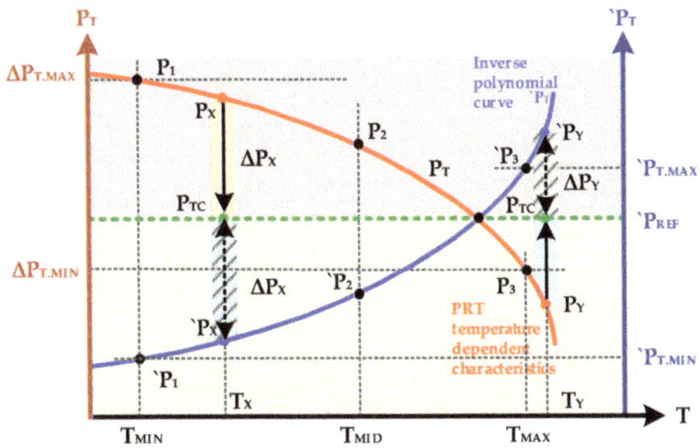

Figure 3. Concept of pressure and its compensation polynomials.

The NTC thermistor is connected in a Wheatstone bridge configuration to detect output voltage variation ΔV_{NTC} as a function of temperature. The NTC gauge factor calibration keeps the NTC output voltage to a certain range. When the NTC sensor is connected to PGA through MUX, after gain, and offset calibration of PGA for NTC sensor signal, the temperature information is converted to digital by a 14-bit SD-ADC. The TCC is a finite state machine (FSM) based configurable digital controller which integrates compensation polynomial characteristics of pressure sensor as function of temperature. Based on the characteristics of a PRT sensor, the coefficients of compensation polynomial may be very small with fractional parts. The scaling technique is introduced to eliminate errors due to fractional part which results in more accurate temperature compensation.

Figure 4 shows the flowchart for preparing temperature compensation parameters. The different parameters like the NTC sensor gauge factor, PGA gain and offset for NTC sensor, and compensation polynomial coefficients are computed before starting temperature compensation. The proposed temperature compensation parameters are achieved with the following steps:

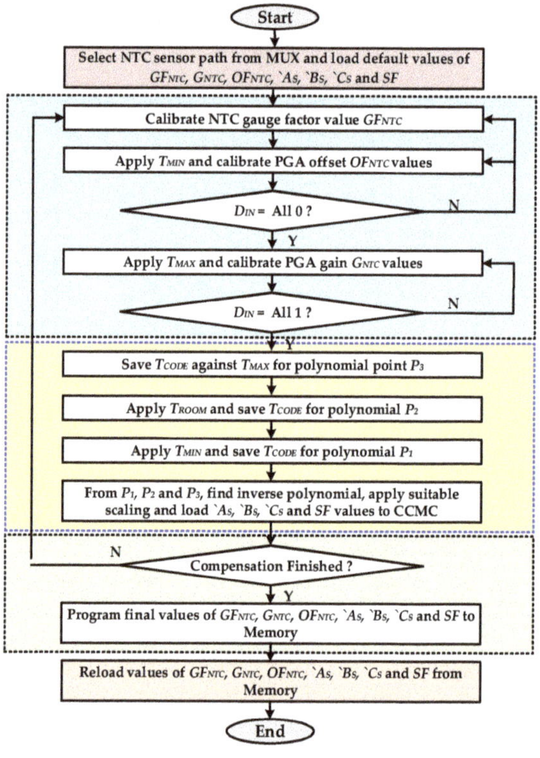

Figure 4. Temperature compensation parameter preparation flow chart.

1 The MC selects the NTC sensor path from MUX and applies default values from memory for the gauge factor GF_{NTC}, PGA offset OF_{NTC} and gain G_{NTC} for NTC sensor.

2 The NTC gauge factor is set to its central value and the PGA offset and gain is determined automatically by the MC for the NTC sensor. For offset cancelation, the minimum temperature is applied and the PGA is tuned to make ADC output nearest to zero. Then the highest temperature is applied and PGA gain is tuned to get the highest possible value of ADC output.

3 To find out the temperature compensation parameters, fixed pressure is applied at the PRT sensor input. The temperature of chip is changed from minimum T_{MIN} to maximum T_{MAX}. Due to non-linear temperature-dependent PRT characteristics, the sensor output voltages decreases when temperature is swept from T_{MIN} to T_{MAX} with even fix pressure at its input. Three values of the pressure code from ADC output in digital format are achieved when the temperature values are −40 °C (minimum), 25 °C (mid) and 150 °C (maximum), respectively. These values give the three points P_1, P_2 and P_3 for the complex temperature-dependent pressure characteristics of PRT sensor as shown in Figure 4. The second-degree polynomial representing this relationship is give as in Equation (1):

$$P_T = aT^2 + bT + c \tag{1}$$

where a, b and c are the coefficients of polynomial, T is temperature and P_T is the temperature-dependent pressure value. This polynomial is valid if temperature is swept at different constant input pressure.

4 Compensation characteristic is computed from the temperature-dependent pressure characteristics with three polynomial points $'P_1$,$'P_2$ and $'P_3$ and is given in Equation (2) as follows:

$$P_T = AT^2 + BT + C \tag{2}$$

5 Since the coefficients A, B and C may have very small values depending on the curve shape for different sensors, so for digital implement with enhanced accuracy of the polynomial, the scaling technique is introduced. Both sides of the equation are multiplied by a suitable number 2^{SF} so that the smallest coefficient has significant integer value, where SF is scaling factor.

$$2^{SF}.P(T) = 2^{SF}.(AT^2 + BT + C) \tag{3}$$

$$P_S = 2^{SF}AT^2 + 2^{SF}BT + 2^{SF}C \tag{4}$$

$$P_S = A_ST^2 + B_ST + C_S \tag{5}$$

The final compensation of the polynomial in Equation (5) is designed and implemented in TCC. During normal operation after temperature compensation and pressure calibration, the compensating value is determined by downscaling the result of Equation (5) by the same factor of 2^{SF} as follows in Equation (6):

$$P = \frac{P_S}{2^{SF}} \tag{6}$$

Since, in downscaling step, the division is involved. The scaling value is selected in the form of the power of 2. This technique eliminates the necessity for a binary divider and downscaling is accomplished by hardware friendly right shift operation.

6 The memory is programmed with compensated parameters of gauge facture GF_{NTC}, PGA offset OF_{NTC}, PGA gain G_{NTC} and compensation polynomial coefficients A_S, B_S, C and scaling factor SF. After reset, these parameters are automatically loaded from memory and are used during temperature compensation.

During normal operation, when the IC is reset or powered on, the polynomial and temperature-compensation parameters are loaded from memory to TCC and MC. Since, the NTC and PRT share the same PGA and SD-ADC, therefore at a time one path is selected by MC. The temperature change rate is not so high and the temperature path selection is less frequent compared to pressure sensing duration and most of the time, the PRT path is selected. The final temperature-compensated pressure code P_{CMP} is given as follows in Equation (7):

$$P_{CMP} = P_{CODE} \pm \Delta P \tag{7}$$

where, P_{CODE} is uncompensated pressure code after PPU processing and ΔP is pressure variation due to temperature which is compensated for by the proposed design. During normal chip operation, if some pressure having digital code of P_X at temperature T_X is applied then it means due to temperature variation the sensor output is more than the real value. This addition pressure variation need to be subtracted from P_X to obtain P_{CMP} as shown in Figure 3 and is given as follows in Equation (8):

$$P_{CMP} = P_X - \Delta P \tag{8}$$

In this case, when 'P_T is less than a reference value 'P_{REF}, the ΔP is the difference of 'P_{REF} and 'P_T and is described in Equation (9) as follows:

$$\Delta P = P_{REF} - P_X \tag{9}$$

Hence, the final compensated value in this case is computed as follows in Equation (10):

$$P_{CMP} = P_X - (P_{REF} - P_X) \tag{10}$$

Similarly, if 'P_T value at current temperature T_Y is greater than 'P_{REF}, then it shows the pressure code is less than the actual value. As is clear from Figure 3, in this case pressure variation of ΔP_Y is mandatory to add in P_Y to obtain temperature-compensated P_{CMP} as explained in Equation (11):

$$P_{CMP} = P_Y + (P_Y - P_{REF}) \tag{11}$$

Finally, if the 'P_T is equal to 'P_{REF} then it means, there pressure code represents the actual value and no compensation is needed as clear from conceptual diagram elaborated in Figure 3. In general, all three P_{CMP} computation scenarios are summarized as follows in Equation (12):

$$P_{CMP} = \begin{cases} P_{CODE} - (P_{REF} - P_T) & if\ P_T < P_{REF} \\ P_{CODE} & if\ P_T = P_{REF} \\ P_{CODE} + (P_T - P_{REF}) & if\ P_T > P_{REF} \end{cases} \tag{12}$$

The reference 'P_{REF} is configurable and its position is adjustable during compensation polynomial design.

4. Temperature-Compensation Controller (TCC)

The main core of the proposed polynomial-based temperature compensation is designed as a configurable digital controller. Since, recent research focuses on the digital solutions rather than analog circuits due to simplicity, scalability, noise immunity along with less power consumption and reduced area requirement, and therefore the proposed compensation polynomial is designed as fully digital circuit. Figure 5 shows the block level architecture of the TCC which is mainly composed of a polynomial FSM controller (PFC), final compensation unit (FCU) and combinational binary multiplier (CBM). When enabled, the PFC computes the polynomial for configured parameters based on the current temperature value. Resource sharing is adopted and a single CBM block is reused for several timely managed multiplication operations to reduce occupied area and power consumption. The configurable TCC architecture is very flexible and easily scalable for computing any higher degree polynomial computation at the cost of additional clock cycles. The PFC computes polynomial value for current input temperature code T_{DIN} when the NTC sensor path is selected. The A_S, B_S and C are configurable polynomial coefficients. The coefficients A_S and B_S are scaled by factor of 2^{SF}. The PFC is designed with a finite state machine control unit and related datapath which mainly computes polynomial in several clock cycles. Single-cycle polynomial computation architecture is also possible at the cost of more hardware for parallel computing. In current design, sequential architecture is adopted with minimum possible hardware utilization because the temperature variation is not abrupt and polynomial evaluation is possible in a very short time interval.

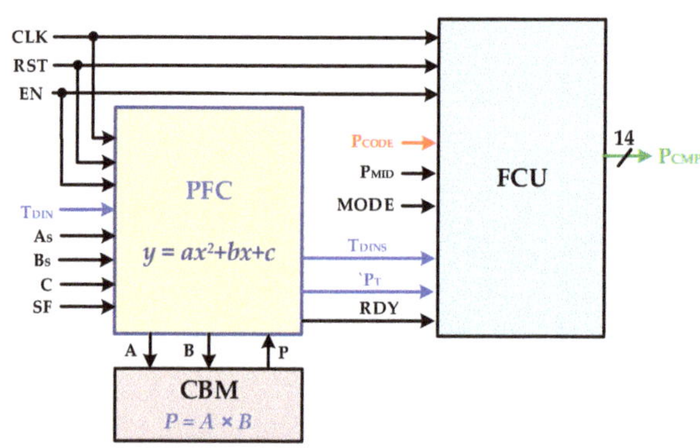

Figure 5. Proposed temperature-compensation controller (TCC) architecture.

The PFC flow diagram is depicted in Figure 6. The controller remains in power-up state for configured duration T_{PU} after turning on or being reset. This allows the other blocks of the chip to be settled down after soft start and also prohibits TCC contributing to the inrush current. The polynomial parameters such as A_S, B_S, C and scaling factor SF are saved in the memory after the compensation procedure and are loaded in TCC registers on power-up. When the signal conditioning chip is in normal operation then PFC waits for the enable signal from MC. The MC first selects the NTC path and enable TCC. The current temperature digital code T_{DIN} is sampled in internal register and polynomial manipulation starts. In the first phase, AT^2 is computed. For this first T^2 is calculated by CBM and its result is saved; then in the second multiplication A_S and T^2 are multiplied and final A_ST^2 is saved in internal register. Then in next phase, BsT is computed and saved in a separate register. The coefficient C is unscaled to reduce its size unlike other scaled coefficients A_S and B_S and, therefore, it is scaled to factor 2^{SF}. The final polynomial value $'P_S$ is calculated and down scaled to original value $'P$ by just right sift operation. The scaling value is chosen as power of 2. This eliminates the requirement of the binary divider and division is achieved by a simple right shift operation. This technique reduces significant area and cost. When compensation value $'P$ based on current temperature and parameters is ready, then FCU computes the final compensated code P_{CMP} as according to Equation (12). During the pressure path, the current pressure digital value P_{DIN} is processed by PPU and its value P_{CODE} remains hold during NTC path. When $'P$ is computed, then it is used for pressure compensation. If T_{DIN} is less than P_{MID}, it means pressure value is higher than the ideal value. In this case, P_{CODE} needs to be reduced sufficiently so that it becomes equivalent to P_{MID}. Similarly, P_{CODE} needs to be increased by an amount if T_{DIN} is greater than P_{MID}. When T_{DIN} code is same as P_{MID} then $'P$ will also be equal to P_{IMID} which means the pressure code is already equal to P_{MID} and does not need to be compensated for. In the proposed structure, compensation polynomial of both positive and negative slope is designed and selectable from the main controller.

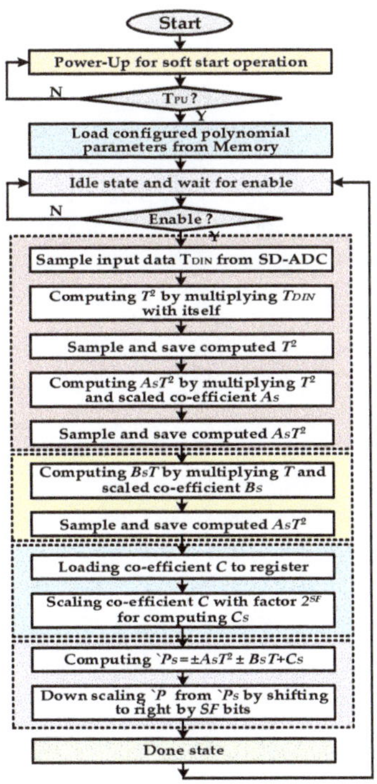

Figure 6. Polynomial finite state machine (FSM) controller flow chart.

5. Sigma-Delta Analog-to-Digital Converter

The ADCs are implemented on system-on-chips (SoCs) so that converted digital data may be used for further digital processing. In the proposed architecture, the signal frequencies to the ADC are in the low range and precise digitized signals are required, which leads to the design of a low speed and high resolution ADC. In the previous designs, successive approximation register (SAR) ADC has been used for this purpose due to its low power consumption. This ADC structure is not suitable for high-accuracy measurements as it has a limitation in terms of resolution [22–25]. A reconfigurable second-order SD-ADC is designed for automotive PRT sensor. In the proposed temperature-compensation architecture, the existing ADC is shared for digitalizing the NTC temperature signal and, therefore, n additional converter is not required. In SD-ADC, different techniques at system and circuit levels have been implemented to address the design challenges. Figure 7a represents a simplified top-block diagram of the SD-ADC. The proposed ADC consists of a second-order sigma-delta modulator (SDM) and reconfigurable decimation filter (RDF). The chopper stabilization technique is used in each SDM integrator stage to reduce the influence of low-frequency noise and offset error. The placement of each cell is also optimized to obtain the required specification of the SD-ADC. A non-overlap clock generator circuit to overcome delay differences is also implemented in the proposed design. In order to ensure better stability performance and low area, a second-order discrete-time (DT) SDM with a cascaded integrator feedback (CIFB) structure is realized. The block diagram of the second order SDM is shown in Figure 7b. The modulator consists of two switched capacitor integrators whose output is fed back to the integrators with the coefficients b_1 and b_2. The values of these coefficients have been derived and fixed through Simulink® modeling which is done in MATLAB®. Figure 7c shows the schematic level implementation of the second order SDM. The coefficients $a1$ and a_2 are implemented as the ratio of two capacitances C_{S1}/C_{I1} and C_{S2}/C_{I2}, respectively. In the modulator, top and bottom reference voltages, V_{REFH} and V_{REFL}, define the feedback coefficients b_1 and b_2 with the optimized values of 0.25 and 1.0, respectively. A gain-boosted topology is employed for amplifier used in the integrator stages of SDM. The folded-cascode with P-type and N-type structures are used in gain-boosted structure. Moreover, the chopper stabilization technique is used to minimize the influence of *1/f* noise at lower frequencies. Moreover, a common mode feedback (CMFB) is applied to all amplifiers to keep the biasing level. The structure uses a single bit quantizer, consisting of a dynamic type comparator followed by two set/reset (SR) latch cells, which determine the outputs of the comparator. For small area and low power, CIC filter structure is used as a decimation filter for SDM [26]. Figure 7d shows that a digital controller is used along with CIC filter to make it reconfigurable and hence it is used for different date rates and input signal bandwidth. The decimation factor is configurable among 32, 64, 128, 256, 512, 1024 and 2048 depending upon the required output data rate and input signal.

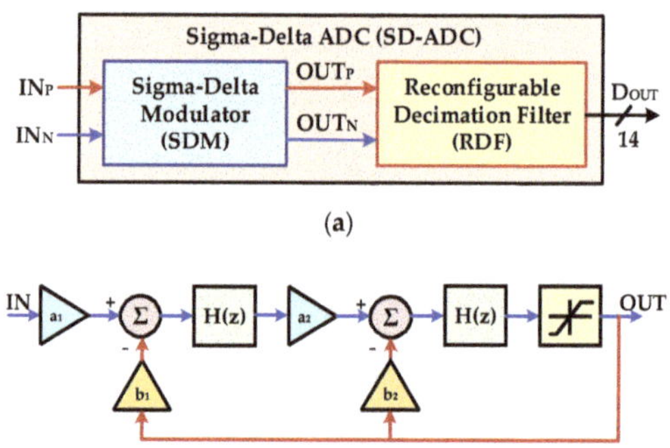

(a)

(b)

Figure 7. *Cont.*

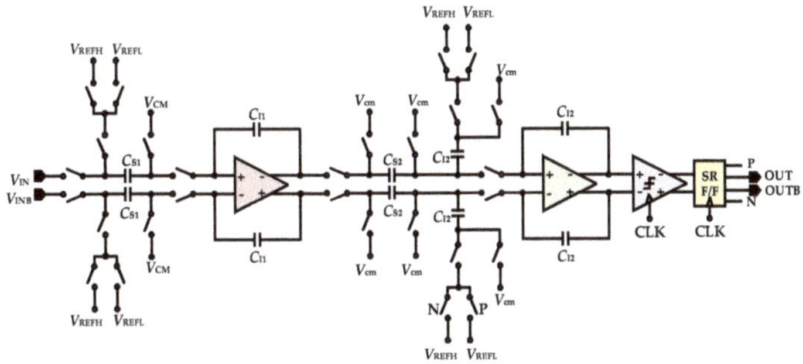

(c)

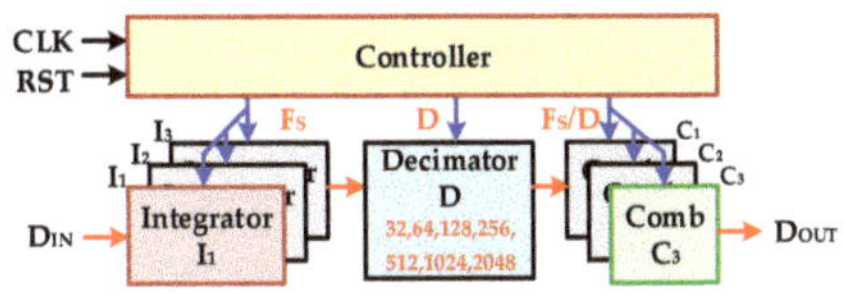

(d)

Figure 7. Sigma-Delta analog-to-digital converter (ADC): (**a**) block diagram with sigma-delta modulator (SDM) and reconfigurable decimation filter (RDF); (**b**) second order SD-ADC block diagram; (**c**) second order SD-ADC circuit diagram; (**d**) reconfigurable decimation filter (RDF) architecture.

6. Programmable Gain Amplifier with Offset Compensation and Single to Differential Circuits

The programmable gain amplifier with offset compensation circuit (OCC) [7] and single to differential (STD) is shown in Figure 8. The PGA is designed with three amplifiers and an additional offset compensation block. To meet the performance requirements, such as high immunity towards noise and a large signal-to-noise ratio, a differential signal is needed to drive the ADC. A single to differential amplifier is designed with two amplifiers to convert a single-ended PGA output signal V_{PGA} to a differential signal pairs V_+ and V_- as shown in Figure 8. The PGA amplifies the difference between input signals VP and VN by gain of APGA as described in Equation (13) as follows:

$$V_{PGA} - V_{OF} = A_{PGA}(V_P - V_N) \tag{13}$$

The PGA gain, A_{PGA} is depended and controlled by ratio of resistors R_1 and R_4 and is give as follows in Equation (14).

$$V_{PGA} = V_{OF} - \Delta RI(1 + \frac{2R_3}{R_4})\frac{R_2}{R_1} \tag{14}$$

where, I is current passing through NTC or PRT sensor and V_{OF} is OCC offset voltage value. With the designed resistor values, the PGA gain is controllable from 4.88 V/V to 25.83 V/V with step size of 0.655 V/V. The designed bandwidth is 2 MHz at maximum PGA gain of 25.83 dB. In the pressure sensor interface IC, the PGA facilitates the digital conversion process for SD-ADC by the amplifying voltage signal obtained from PRT or NTC sensors and enhances resolution. The output voltage for each sensor differs slightly depending on the pressure and temperature. This voltage difference is compensated for by controlling the gain. The offset voltage V_{OF} is applied from OCC. For generating constant offset voltage for PGA, the compensation circuit in low drop-out structure is used. The V_{OF} depends on resistors ratio and is obtained by Equation (15) as follows:

$$V_{OF} = (1 + \frac{R_{OF1}}{R_{OF2}})V_{BGR} \tag{15}$$

where V_{BGR} is BGR output voltage. The PGA should not exceed the input range of the ADC at the pressure range. Therefore, PGA gain and offset are controlled by resistors R_{OF1}, R_1, and R_4 from MC These resistors are implemented as resistor backs which are controllable form MC. The final calibrated resistors codes are stored in memory by the MC digital controller. During normal operation, the calibrated resistor values are fetched from memory and are applied before reading signal values from NTC or PRT sensors.

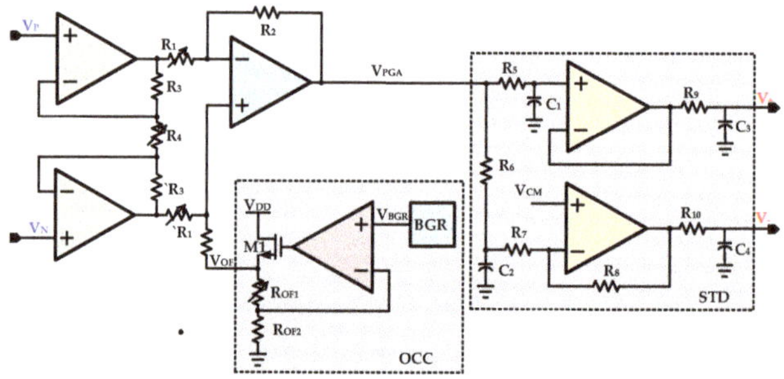

Figure 8. Programmable gain amplifier (PGA) with offset compensation (OCC) and single to differential (STD) circuits.

7. Experimental Results

The proposed polynomial-based digital temperature compensation is integrated in piezoresistive type pressure sensor signal conditioning IC for automotive applications. The design is fabricated with a 1P6M 180 nm CMOS process. The chip microphotograph is depicted in Figure 9 in which TCC, SD-ADC and PGA are highlighted. The existing analog front end (AFE) and ADC used for the pressure path are shared and only with the integration of TCC, the proposed technique is implemented. The TCC is fully synthesizable and occupies only 465 × 180 μm^2 area. The performance of the temperature-compensation controller is summarized in Table 2. The fully scalable and configurable TCC architecture requires only 1386 K logic gates for its full implementation and consumes only 1375 μW for its full operation. Figure 10 shows the pressure sensor module with the integrated chip including the proposed digital compensation. Inside metal cover, PRT, NTC and pressure sensor interface chip are stuffed on a flexible printed circuit board (PCB). The NTC sensor is connected in Wheatstone bridge configuration. The NTC sensor has very high sensitivity of −3% to −6% per °C compared to PTAT or a resistance temperature detector (RTD). It demonstrates comparatively very steep resistance-temperature slope and typically suitable for the −55 °C to 200 °C temperature range. In the pressure sensor interface IC, the NTC sensor is used for temperature information and temperature calibration.

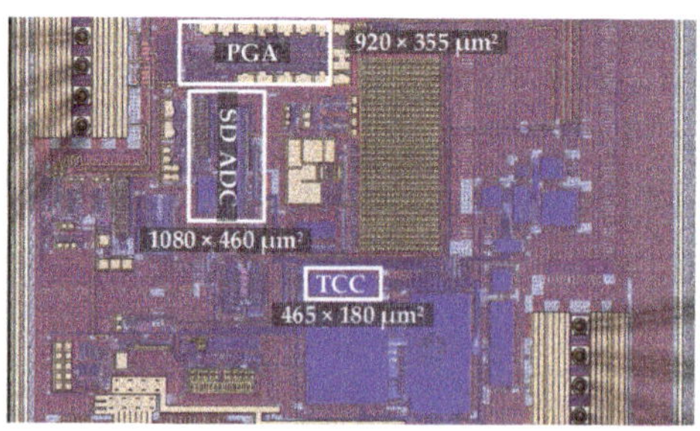

Figure 9. Chip microphotograph.

Table 2. TCC performance summary.

Parameter	Value
CMOS process	180 nm
Occupied area	0.0837 mm^2
Gate count	1.386 K
Supply voltage	1.8 V
Current consumption	764 nA
Power consumption	1.375 μW
Clock Frequency	10 MHz
Polynomial	2nd Order
Scalable	Yes
Configurable architecture	Yes

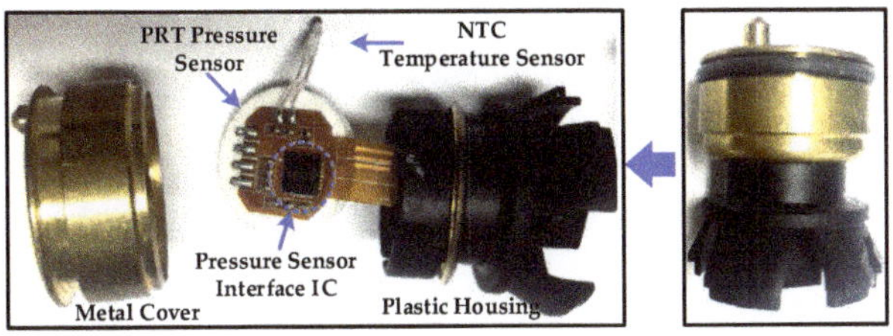

Figure 10. PRT pressure sensor module with PRT and negative temperature coefficient (NTC) sensors and proposed temperature compensation.

The performance comparisons of the proposed temperature compensation with the exiting methods are summarized in Table 3. Most of the prior works adopted PTAT and BGR as the temperature sensor element. Their sensitivity is very low compared to NTC. The NTC exhibits very high sensitivity in the required temperature range according to AEC-Q100 grade 0. In prior works, the lookup table (LUT) which is based on simulation analysis is used which results in poor performance. Furthermore, low-resolution ADC in previous works also limits the accuracy. In the proposed work, very high-sensitivity NTC temperature element and a 14-bit ADC are used for polynomial-based digital on the fly compensation.

Table 3. Temperature-compensation performance comparison.

Parameter	[7]	[12]	[13]	[27]	This Work
CMOS process (μm)	0.35	0.35	0.18	-	0.18
System clock (MHz)	4	8.96	4	-	10
Power consumption (mW) [1]	23.5	25	11.8–64.8	20	22.5
Pressure sensor type	PRT	Capacitive	PRT	SOS [2]	PRT
Temperature range (°C)	−40–+150	−30–120	−40–+85	−20–+140	−40–+150
Temperature method	LUT [3]	LUT [3]	Digital	Software	PDTC [4]
Temperature sensor	PTAT	BGR	PTAT	RTD [5]	NTC
ADC type	Flash	Flash	Charge balancing	Sigma-Delta	Sigma-Delta
ADC resolution	4	4-bit	16	24	14
Deviation (%) FS [6]	0.5	1.0	0.1	0.3	0.068

[1] Chip current and supply voltage. [2] Silicon-on-Sapphire. [3] Lookup table-based temperature compensation. [4] Polynomial-based digital temperature compensation. [5] Resistance temperature detector. [6] Temperature compensation deviation from ideal value for full scale (FS).

The accuracy is the ratio of deviation of output N_{MEAS} from ideal value N_{IDEAL} to full-scale output value N_{FOS}, and is given as follows in Equation (16):

$$Accuracy = Min\left\{1 - \frac{N_{MEAS} - N_{IDEAL}}{N_{FSO}}\right\} \times 100(\%) \quad (16)$$

After PGA gain and offset compensation, the minimum and maximum measured digital temperature codes are 68 and 16,285 when chip temperature is −40 °C and 150 °C, respectively, at fixed pressure. With this measured range, the full-scale value is 16,217. The maximum measured deviation of digital code from ideal value is 11. With these values, from Equation (16) the accuracy is 99.9321%. Hence, the output temperature compensation accuracy is within ±0.06783% with full scale.

Figure 11 explains the experimental environment for measuring temperature-compensation performance. The pressure is applied from nitrogen gas cylinder and different temperature conditions are measured with temperature chamber. The digital serial output is received on a computer for analysis. Different temperature-compensation measurement results are depicted in Figure 12.

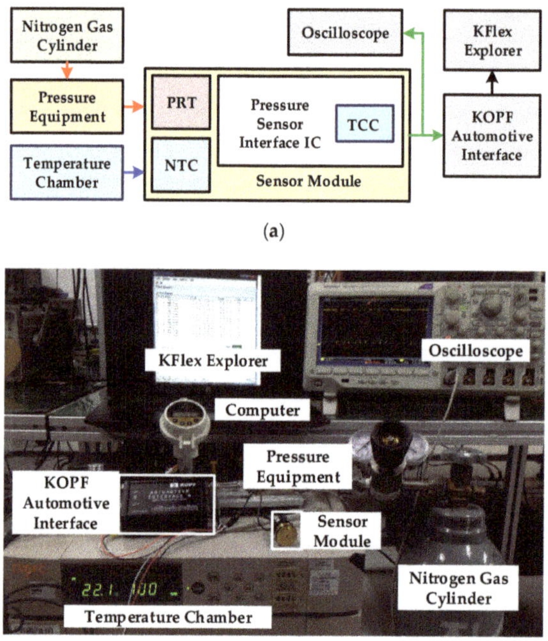

(**a**)

(**b**)

Figure 11. Experiment environment: (**a**) measurement block diagram; (**b**) experimental lab setup.

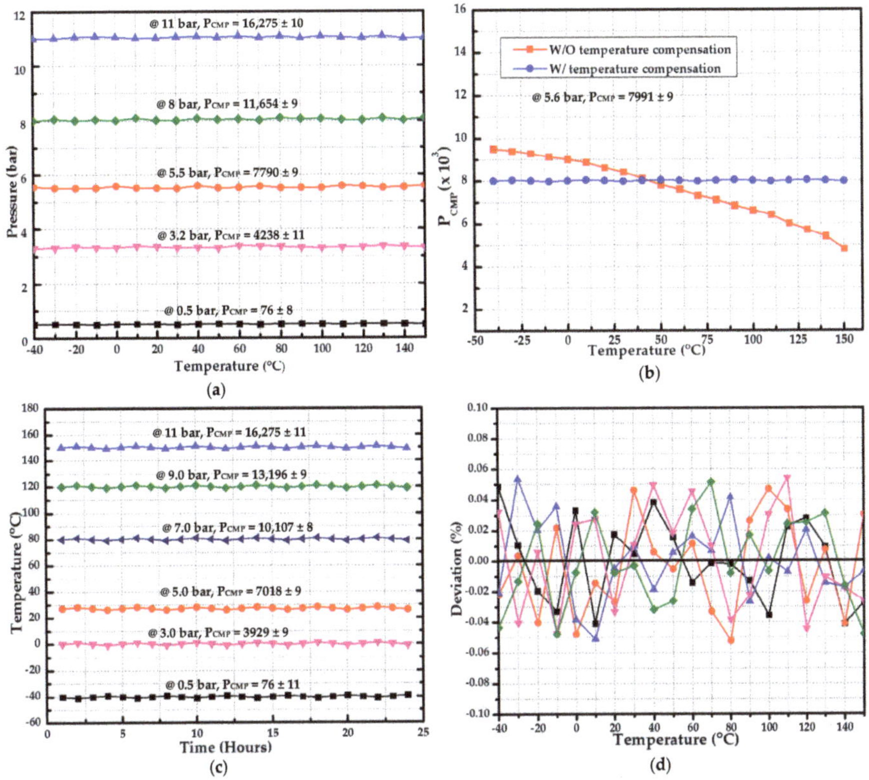

Figure 12. Temperature-compensation measurement results: (**a**) temperature compensation results at different input pressure; (**b**) with and without temperature compensation at 5.6 bar input pressure; (**c**) 24-h measurement at different fixed pressure and constant temperature; (**d**) percentage output deviation from ideal value at different pressures with full temperature range.

In Figure 12a, various fixed pressures are applied and the temperature is changed from –40 °C to 150 °C. The pressure range is 0.5 bar to 11.0 bar. The maximum deviation from ideal value is 11 which results in 0.06783% accuracy including ADC noise. The measurement with and without temperature compensation is depicted in Figure 12b in which fixed 5.6 bar pressure is applied. The ideal digital output is 7991 whereas the maximum deviation of ±9 is reported. The fixed pressure is applied at fixed temperature for 24 h and proposed temperature compensation performance is analyzed. The results are summarized in Figure 12c. As is clear from results, the digital output is almost constant with variation in four least significant bits. Figure 12d explains the percentage output deviation from ideal value for various fixed pressures while sweeping temperature. It also includes ADC noise with input DC value.

Figure 13 shows the measured fast Fourier transform (FFT) spectrum of SD-ADC with an effective number of bits (ENOB) of 13.22 bits and signal-to-noise and distortion ratio (SNDR) level of 81.37 dB. The measurement is done with 0.61 kHz input signal frequency (f_{IN}), input signal level of 300 mV and an OSR 1024 operating at a sampling frequency (f_S) of 2.5 MHz.

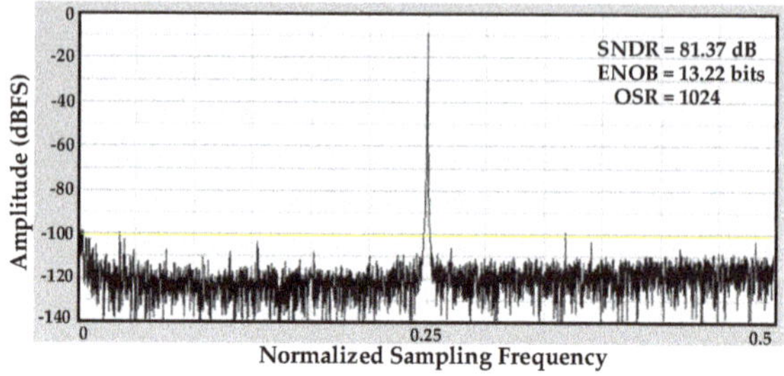

Figure 13. Measured sigma-delta analog-to-digital converter (SD-ADC) fast Fourier transform (FFT) spectrum.

The detailed post-place and route (P&R) simulation results with a NC-Verilog® simulator of TCC are elaborated in Figure 14. The polynomial computation for a single input of temperature value is shown in Figure 14a. The polynomial coefficient parameters A_S, B_S, C and SF are configured to 75, 205,018, 1 and 24, respectively. The P_{REF} and P_{CODE} are 700 and 5242, respectively. In this simulation, the temperature T_{DIN} is 3029 when it is enabled from main controller. After enabling, TCC takes 37 clock cycles to compute compensating value 'P_T of 79 which is adjusted to pressure code to compute compensated pressure P_{CMP} of 4621. The full temperature-seep simulation results are shown in Figure 14b. It is clear from results that the 'P_T is perfect compensation curve of temperature-dependent pressure sensor characteristics represented as P_{CODE}.

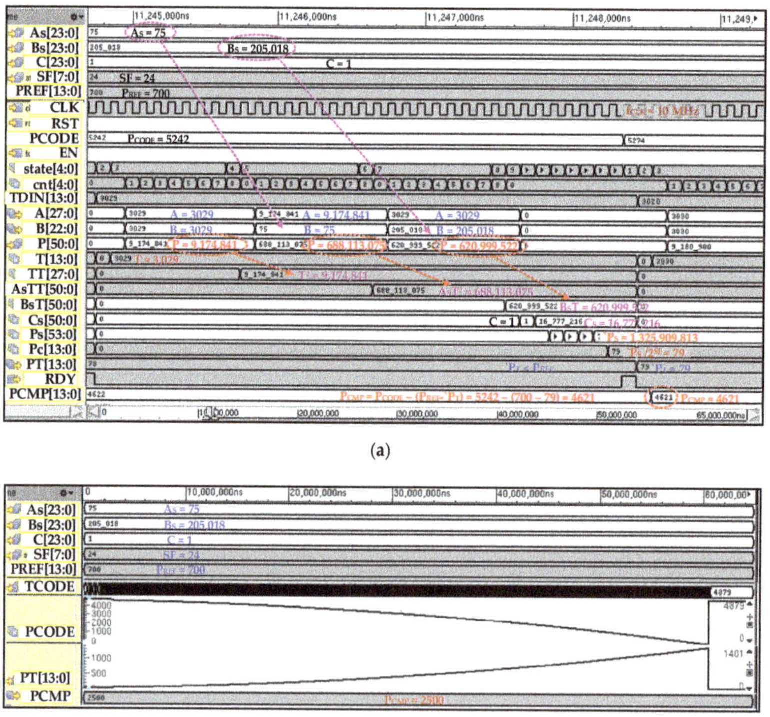

Figure 14. TCC simulation results: (**a**) polynomial computation and temperature compensation for single iteration; (**b**) full temperature sweep from −40 C to 150 C with constant input pressure.

The PGA simulation results including STD for entire configurable gain range is depicted in Figure 15. The S2D differential output voltage ranges from 114.6 mV to 607 mV when PGA input voltage difference is 23.0 mV. The PGA gain is controllable from 4.88 V/V to 25.83 V/V with step size of 0.655 V/V. The gain is controlled from MC with G <4:0> signal. The configurable PGA gain is acceptable for the SD-ADC with 300 mV of peak-to-peak input.

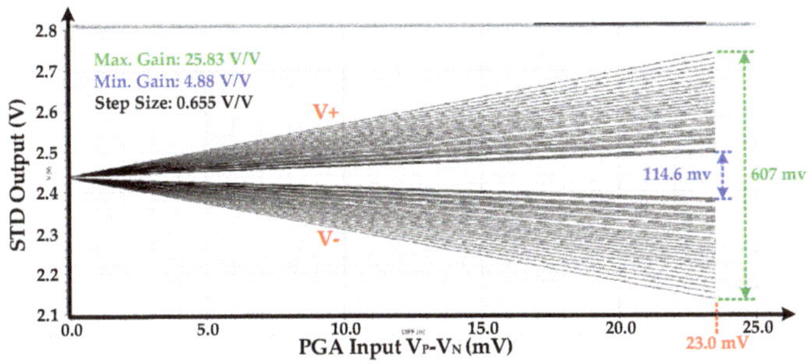

Figure 15. Programmable gain amplifier (PGA) simulation results at different gain including STD.

8. Conclusions

A highly accurate, polynomial-based adaptive digital temperature compensation is presented for automotive piezoresistive pressure sensor applications. By integrating compensation characteristics of the pressure sensor as function of temperature, non-ideal temperature dependency of the pressure sensor is accurately compensated. The compensation polynomial is fully implemented in a digital form with a scaling technique introduced to enhance its accuracy. For area and power efficient design, a resource-sharing technique is adopted. The NTC instead of PTAT or CTAT is used as a temperature-sensing element as it offers the best temperature characteristics for ACE-Q100 grade 0 ambient temperature operating range. A high-resolution 14-bit SD-ADC is proposed for improving accuracy over a wide temperature range. When temperature varies from −40 °C to 150 °C according to ACE-Q100, temperature compensation accuracy reported is 99.93% and it is within ±0.068% at full scale. It takes 37 µs to compute the temperature compensation with 10 MHz of clock frequency. The proposed technique is integrated in an automotive pressure sensor signal conditioning IC using a 1P6M 180 nm CMOS process.

Author Contributions: K.-Y.L. guided and directed the authors for this work. I.A. proposed, designed, simulated and implemented the overall architecture and wrote paper. M.A., K.S., M.R.U.R. and D.G.K. contributed to the synthesis and place and route (P&R). They also contributed to integrating the design in the top layout of the chip. K.S. and B.S.R. helped in designing SD-ADC and paper writing. D.G.K. contributed in designing the PGA, testing board and performing measurement. Y.P. and S.S.Y. gave advice about implementation issues and reviewed the paper before submission. All authors have read and agreed to the published version of the manuscript.

Funding: This work was supported by Institute of Information & communications Technology Planning & Evaluation (IITP) grant funded by the Korea government (MSIT) (No.2020-0-00261, Development of low power/low delay/self-power suppliable RF simultaneous information and power transfer system and stretchable electronic epineurium for wireless nerve bypass implementation).

Conflicts of Interest: The authors declare no conflict of interest.

References

1. Czarnocki, W.; Ding, X.; Schuster, J.P.; Roeckner, B. Programmable CMOS Integrated pressure sensor. *SAE Trans.* **1999**, *108*, 1929–1933.
2. Chiou, J.A. Pressure sensors in automotive applications and future challenges. *Proc. Micro-Electro-Mech. Syst. Symp. Asme Int. Mech. Eng. Congr. Expo.* **1999**, *1*, 525–530.
3. DeHennis, A.D.; Wise, K.D. A wireless microsystem for the remote sensing of pressure, temperature, and relative humidity. *J. Microelectromech. Syst.* **2005**, *14*, 12–22. [CrossRef]

4. Wang, Y.; Maeda, M.R.; Fukunaga, M.; Watanabe, K. A smart thermal environment monitor based on IEEE 1451.2 standard for global networking. *IEEE Trans. Instrum. Meas.* **2005**, *54*, 1321–1326. [CrossRef]
5. Automotive Electronics Council. *Failure Mechanism Based Stress Test Qualification for Integrated Circuits*; AEC-Q100-Rev. H; Automotive Electronics Council: Luton, UK, 2014.
6. Nie, M.; Gao, Y. The analytical calibration model of temperature effects on a silicon piezoresistive pressure sensor. *AIP Adv.* **2017**, *7*, 035120. [CrossRef]
7. Lee, D.S.; Cho, S.H.; Ryu, H.; Pu, Y.; Yoo, S.; Lee, M.; Hwang, K.C.; Yang, Y.; Lee, K.-Y. A Highly linear, AEC-Q100 compliant signal conditioning ic for automotive piezo-resistive pressure sensors. *IEEE Trans. Ind. Electron.* **2018**, *65*, 7363–7373. [CrossRef]
8. Aryafar, M.; Hamedi, M.; Ganjeh, M. A novel temperature compensated piezoresistive pressure sensor. *Measurement* **2015**, *63*, 25–29. [CrossRef]
9. Perraud, E. Theoretical model of performance of a silicon piezoresistive pressure sensor. *Sens. Actuators A Phys.* **1996**, *57*, 245–252. [CrossRef]
10. Yao, Z.; Liang, T.; Jia, P.; Hong, Y.; Qi, L.; Lei, C.; Zhang, B.; Li, W.; Zhang, D.; Xiong, J. Passive resistor temperature compensation for a high-temperature piezoresistive pressure sensor. *Sensors* **2016**, *16*, 1142. [CrossRef]
11. Yao, Z.; Liang, T.; Jia, P.; Hong, Y.; Qi, L.; Lei, C.; Zhang, B.; Xiong, J. A high-temperature piezoresistive pressure sensor with an integrated signal-conditioning circuit. *Sensors* **2016**, *16*, 913. [CrossRef]
12. Lee, D.S.; Tiwari, H.D.; Kim, S.Y.; Lee, J.; Park, H.G.; Pu, Y.G.; Seo, M.; Lee, K.-Y. A highly linear, small area analog front end with gain and offset compensation for automotive capacitive pressure sensors in 0.35 μm CMOS. *IEEE Sens. J.* **2015**, *15*, 1967–1976. [CrossRef]
13. Mailand, M.; Getzlaff, S.; Richter, R.; Apel, S.; Breitmeyer, D.; Meyer, U.; Bohme, E.; Schreiber, R.; Tows, K.; Gunther, D. A standard low-noise sensor interface for 0.66nJ/step-energy-efficient, mobile barometric-pressure-based altitude sensing. In Proceedings of the IEEE International Conference on Sensors, Baltimore, MD, USA, 3–6 November 2013; pp. 1–4.
14. Zhou, G.; Zhao, Y.; Guo, F.; Xu, W. A Smart high accuracy silicon piezoresistive pressure sensor temperature compensation system. *Sensors* **2014**, *14*, 12174–12190. [CrossRef] [PubMed]
15. Patra, J.C.; Kot, A.C.; Panda, G. An intelligent pressure sensor using neural networks. *IEEE Trans. Instrum. Meas.* **2000**, *49*, 829–834. [CrossRef]
16. Patra, J.C.; van den Bos, A. Modeling of an intelligent pressure sensor using functional link artificial neural networks. *ISA Trans.* **2000**, *39*, 15–27. [CrossRef]
17. Pramanik, C.; Islam, T.; Saha, H. Temperature compensation of piezoresistive micro-machined porous silicon pressure sensor by ann. *Microelectron. Reliab.* **2006**, *46*, 343–351. [CrossRef]
18. Futane, N.; Chowdhury, S.R.; Chowdhury, C.R.; Saha, H. Ann based cmos asic design for improved temperature-drift compensation of piezoresistive micro-machined high resolution pressure sensor. *Microelectron. Reliab.* **2010**, *50*, 282–291. [CrossRef]
19. Ali, I.; Rikhan, B.S.; Kim, D.-G.; Lee, D.-S.; Rehman, M.R.U.; Abbasizadeh, H.; Asif, M.; Lee, M.; Hwang, K.C.; Yang, Y.; et al. Design of a low-power, small-area AEC-Q100-compliant SENT transmitter in signal conditioning IC for automotive pressure and temperature complex sensors in 180 Nm CMOS Technology. *Sensors* **2018**, *18*, 1555. [CrossRef]
20. Ali, I.; Abbasizadeh, H.; Rehman, M.R.U.; Asif, M.; Oh, S.J.; Pu, Y.G.; Lee, M.; Hwang, K.C.; Yang, Y.; Lee, K.-Y. An ultra-low power, adaptive all-digital frequency-locked loop with gain estimation and constant current DCO. *IEEE Access* **2020**, *8*, 97215–97230. [CrossRef]
21. Ali, I.; Asif, M.; Rehman, M.R.U.; Khan, D.; Yingge, H.; Kim, S.J.; Pu, Y.; Yoo, S.-S.; Lee, K.-Y. A highly reliable, 5.8 GHz DSRC wake-up receiver with an intelligent digital controller for an ETC system. *Sensors* **2020**, *20*, 4012. [CrossRef]
22. Chen, M.; Liu, Y.T.; Xiao, J.B.; Chen, J. A CMOS detection chip for amperometric sensors with chopper stabilized incremental ΔΣ ADC. *J. Semicond.* **2016**, *37*, 065004.
23. Verma, D.; Shehzad, K.; Khan, D.; Ain, Q.A.; Kim, S.J.; Lee, D.; Pu, Y.; Lee, M.; Hwang, K.C.; Yang, Y.; et al. A Design of 8 fJ/conversion-step 10-bit 8MS/s low power asynchronous SAR ADC for IEEE 802.15.1 IoT sensor based applications. *IEEE Access* **2020**, *8*, 85869–85879. [CrossRef]

24. Kim, S.; Na, S.-I.; Yang, Y.; Kim, S. A 2-MHz BW 82-dB DR continuous-time delta-sigma modulator with a capacitor-based voltage DAC for ELD compensation. *IEEE Trans. Very Large Scale Integr. (VLSI) Syst.* **2018**, *26*, 1999–2006. [CrossRef]
25. Shehzad, K.; Verma, D.; Khan, D.; Ain, Q.U.; Basim, M.; Kim, S.J.; Pu, Y.; Hwang, K.C.; Yang, Y.; Lee, K.-Y. Design of a low power 10-b 8-MS/s asynchronous SAR ADC with on-chip reference voltage generator. *Electronics* **2020**, *9*, 872. [CrossRef]
26. Tao, S.; Rusu, A. A comparative design study of continuous-time incremental sigma-delta ADC architectures. *Int. J. Circuit Theory* **2016**, *44*, 2147–2163. [CrossRef]
27. Guo, Z.; Lu, C.; Wang, Y.; Liu, D.; Huang, M.; Li, X. Design and experimental research of a temperature compensation system for silicon-on-sapphire pressure sensors. *IEEE Sens. J.* **2017**, *17*, 709–715. [CrossRef]

sensors

MDPI

Review

A Review of Microelectronic Systems and Circuit Techniques for Electrical Neural Recording Aimed at Closed-Loop Epilepsy Control

Reza Ranjandish [1] **and Alexandre Schmid** [2,*]

[1] Department of Information Technology and Electrical Engineering, ETH Zürich, CH-8092 Zürich, Switzerland; rranjandish@ethz.ch
[2] Institute of Electrical Engineering, EPF Lausanne, CH-1015 Lausanne, Switzerland
* Correspondence: alexandre.schmid@epfl.ch

Received: 29 July 2020; Accepted: 2 October 2020; Published: 8 October 2020

Abstract: Closed-loop implantable electronics offer a new trend in therapeutic systems aimed at controlling some neurological diseases such as epilepsy. Seizures are detected and electrical stimulation applied to the brain or groups of nerves. To this aim, the signal recording chain must be very carefully designed so as to operate in low-power and low-latency, while enhancing the probability of correct event detection. This paper reviews the electrical characteristics of the target brain signals pertaining to epilepsy detection. Commercial systems are presented and discussed. Finally, the major blocks of the signal acquisition chain are presented with a focus on the circuit architecture and a careful attention to solutions to issues related to data acquisition from multi-channel arrays of cortical sensors.

Keywords: epilepsy; seizure; multichannel neural recording; feature extraction; closed-loop neurostimulator; low-power; low-noise amplifier; implantable medical device

---

1. Introduction

Over a long period of time of being constrained to experiment, bioelectricity has matured into the fundaments of bio-electronic interfaces, paving the way to new therapeutic systems in recording or stimulation modes of operation. Implantable electronic medical devices (IEMDs) have emerged in recent years, based on the success of early medical applications of bioelectricity, including for instance external electro-cardiograms (ECG), recording or external heart defibrillators (stimulation), as well as based on the advancements of low-power integrated microelectronics.

Commercial IEMD products are multiple and include, for instance, heart monitoring systems such as the CardioMEMS HF System provided by Abbott Laboratories for heart failure detection [1], heart pacemakers such as the Momentum X4 provided by Boston Scientific that applies electric pulses to the heart muscles to regulate its contractions [2]. Additional successful IEMDs include cochlear implants aimed at restoring hearing capability, pain control devices, deep-brain implantable systems aimed at controlling and eluding tremors due to Parkinson's disease, and recently, retina implants aimed at restoring lost vision by electrical stimulation of remaining healthy retina tissues.

This paper reviews microelectronic techniques aimed at epilepsy control in a closed loop. Epilepsy is briefly discussed from an engineering perspective in Section 1.1, i.e., with a focus on the conditions and parameters that are specific and relevant to electrical neuromodulation. The architecture of epilepsy-control implantable systems is presented in Section 2. Classical electrodes are presented along with a discussion of electrical recording and stimulation of the brain. Commercial systems aimed at epilepsy control are reviewed, as implantable and external systems, as open and closed-loop systems and discussed in terms of their operational principles, advantages and drawbacks in Section 3.

Finally, analog front-end amplifiers are reviewed in terms of classical topologies aimed at single, and multi-channel recording from the brain in Section 4. The major blocks in the recording chain that are specific to epilepsy control are reviewed and discussed. A detailed review of the stimulation chain is deliberately not considered in this review.

1.1. Engineering Overview of Epilepsy, Seizures and Treatment

Neurological disorders are defined as diseases that affect the nervous system, including the central and peripheral nervous systems. Among multiple different neurological disorders, several diseases are prevalent such as amyotrophic lateral sclerosis, brain tumors, epilepsy, Parkinson, etc. Epilepsy consists of the recurrence of a phenomenon called a seizure. A seizure occurs when the brain produces a brief abnormal and uncontrollable electrical discharge. Clinically, according to the International League Against Epilepsy (ILAE) official report [3], epilepsy is considered to be a disease of the brain defined by any following conditions:

- At least two unprovoked (or reflex) seizures occurring >24 h apart.
- one unprovoked (or reflex) seizure and a probability of further seizures similar to the general recurrence risk (at least 60%) after two unprovoked seizures, occurring over the next 10 years.
- diagnosis of an epilepsy syndrome.

1.1.1. Phases of a Seizure

Seizures often proceed over four consecutive phases; prodromal, auras, ictal and postictal. Some patients experience a prodromal stage which mostly involves emotional signals. This stage may develop over hours or even days before the start of the seizure. The second phase which is called aura occurs immediately before a seizure onset and may last for a few seconds. During aura, patients may experience déjà vu, jamais vu, headaches and other symptoms. The third phase, which is called ictal, covers the period over which a seizure extends. This period is correlated with the electrical seizure activity in the brain. The fourth phase, which is called postictal is a period of recovery from a seizure. The postictal stage is different among the patients and may last from a few minutes to a few hours. The recovery period is determined by the type of seizure (Section 1.1.2) as well as the part(s) of the brain involved in the seizure.

1.1.2. Seizure Classification

Understanding and diagnosis of the seizure type bears a significant implication on the daily life of an epileptic patient. The type of seizure determines whether a patient can safely perform some common daily-life tasks, including driving, and sports. Natural tasks with potential high social impediment are affected by seizures, including walking and communicating. In addition, the type of seizure has a tremendous impact on understanding which medication is suitable for the treatment or which medication may potentially be harmful [4].

Describing the different types of seizures was initiated in the times of Hippocrates. In 1964, Gastaut proposed a new and modern classification of seizures [5]. The traditional classification of seizures is based on anatomy comprising temporal, frontal, parietal, occipital, diencephalic, or brainstem seizures. However, the understanding of the mechanism of seizures has evolved thanks to modern research. In 1981, an ILAE commission classified seizures into partial and generalized-onset, simple and complex partial seizures, as well as various specific generalized types [6,7]. This classification was established from the study of hundreds of video–electroencephalography (EEG) recordings of seizures. This latter classification is the main reference to date, with some terminology revisions [8,9]. One of the newest classifications is the 2017 classification proposed by ILAE [7]. The overview of the new classification is shown in Figure 1. With respect to the 1981 classification, the 2017 classification significantly reduces the number of unclassifiable cases. The combination of motor/non-motor and awareness level features provides better flexibility and detailed seizure description [10]. The new classification is based on several important facts expressed as follows [4]:

- The onset or beginning of a seizure;
- a person's level of awareness during a seizure, and
- whether body movements occur during a seizure.

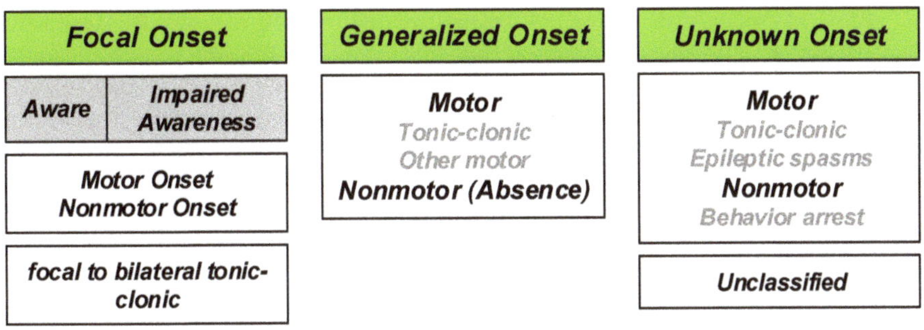

Figure 1. The new classification of seizure types based on the International League Against Epilepsy (ILAE) report (Reprinted with permission of Wiley Periodicals, Inc. © 2017 International League Against Epilepsy) [7].

Based on the 2017 classification, three major groups of seizures are observed, including generalized onset seizures, focal onset seizures and unknown onset seizures. Generalized seizures affect both sides of the brain or group of cells located over both sides of the brain, simultaneously. Focal seizures (Partial seizures in the 1981 classification) start in one area or group of cells in the brain. Unknown onset seizures is the term used when the location of the seizure onset is unknown.

1.1.3. Statistics

According to World Health Organization (WHO) [11], approximately 50 million people worldwide suffer from epilepsy. 80% of them live in low and middle-income countries. Among them 75% do not receive proper treatment. The estimated proportion of the general population with active epilepsy (continuing seizures or with the need for treatment) is between 4 and 10 per 1000 people. However, this number is much higher in low and middle-income countries, i.e., between 7 and 14 per 1000 people.

Each year, 2.4 million people are diagnosed with epilepsy [11]. The estimated proportion of new cases with epilepsy in high-income countries is between 30 and 50 per 100,000 people. This rate doubles in low and middle-income countries.

1.1.4. Epilepsy Treatment

Several studies have explored the reason for the initiation of seizures [12,13]. The importance of finding the reasons and origin lies in the correct diagnosis and treatment of epilepsy. No medical treatment can be safely provided without having knowledge of the different sources that cause epilepsy. The development of modern medical devices and implantable systems participate in improving the understanding of the origins of epilepsy.

The treatment of epilepsy is composed of three levels. The first level is the treatment with medications. Several anti-epileptic drugs (AEDs) are available for epilepsy patients. There are approximately 25 different AEDs suitable for controlling seizures; different AEDs are suitable for different seizure types. The benefits of AEDs consist of the reduction of stopping seizures and risk of accidents. A first-line AED is an AED that is tried first in the therapy. The AEDs added to the first-line

AED are called second-line AEDs. The success rate of the third medication delivered after trying two different medications is approximately 5%.

Epilepsy is defined as a drug-resistant epilepsy when a patient has unsuccessfully tried two different anti-seizure medications. The second level of epilepsy treatment is lobectomy. Lobectomy is the removal of the part of the brain that is responsible for seizure initiation, if the region is uniquely detectable, and if it is not in a sensitive region of the brain. Before lobectomy, surgeons must find the precise location of the brain from which the seizures start. Following modern medical procedures, seizures can be identified and categorized using three Tesla magnetic resonance imaging (3T MRI), video-EEG, single-photon emission computed tomography (SPECT), magnetoencephalography (MEG) scan and positron emission tomography (PET) scan as prerequisites of further treatment steps. A resection surgery would be prescribed in cases of a focal seizure and if a single location can be identified [14]. In a resection surgery, a part of the skull is temporarily removed by craniotomy and the part of the brain which is engaged in the seizure initiation is removed. Laser thermal ablation is another way to stop the seizures, which is a minimally invasive method. This method is performed in the MRI in real time. In this surgery, during an MRI scan, the part of the brain that is engaged in the seizure initiation is heated up and destroyed, while local temperature is monitored using the MRI scanner. This method can only be employed in limited areas of the brain.

In order to precisely detect the part of the brain that is a target of resection surgery or laser thermal ablation, intracranial monitoring is necessary, which is a diagnostic surgery. Intracranial monitoring is carried out in two different ways, including recording using subdural grids and strips or stereo-EEG using SEEG electrodes.

Patients responding to prevalent cures like medications and surgeries are approximately 70%, while approximately 30% of the patients are untreated or poorly treated because of the following reasons:

- Seizures diffuse over an excessively large area;
- seizures occur in sensitive areas of eloquent cortex that may not be surgically treated;
- seizures have multiple foci (multifocal seizures) which are thus difficult to individual localization and in practice impossible to surgically treat;
- surgery may not be tolerable due to specific medical conditions.

Implantable electrical stimulators offer an alternative therapy to untreated or poorly treated patients whose seizures are not controlled using medication or surgery. These stimulator systems either operate in an open-loop or closed-loop mode. IEMDs operating in open-loop stimulation only consist of electrical stimulators. In contrast, closed-loop stimulators record neural signals from the brain and detect the seizure onsets. Upon seizure detection, closed-loop stimulators trigger electrical stimulation. Commercial stimulators are reviewed in Section 3.

Intracranial Pressure (ICP) is reported to influence drug-resistant epilepsy in some cases. In [15], a 23-year old patient was reported as the case study. In this study, an increase in ICP is shown to augment the efficiency of anti-seizure medications. The epileptic patient had undergone a shunt surgery in childhood to decrease the ICP. The patient had uncontrolled seizures in spite of three years of pharmacotherapy. However, the uncontrolled seizures suddenly stopped after shunt removal.

2. Introduction to Epilepsy Control Using Implantable Microelectronic Systems

Several electronic building blocks are essential to design and implement a low-power seizure detection system. These blocks are also some of the main building blocks of an epilepsy control system. Figure 2 shows the general overview of a low-power seizure detection (Figure 2a) and epilepsy control (Figure 2b) system. Both systems have common essential building blocks which are shown with gray boxes including the analog front-end (AFE), analog-to-digital converter (ADC), and seizure onset detector (SON). The Epilepsy control system has an additional essential building block which is the electrical neural stimulator aimed at suppressing or modulating the seizure electrical activity.

In addition, data compression may be applied prior to feeding the data to the SON to lower the power consumption of the systems. Data compression can be done either in analog or digital domains.

In the following, the operation principles of some of these blocks are briefly reviewed, including the neural amplifiers, compressive sensing and feature extractors inside the SON block.

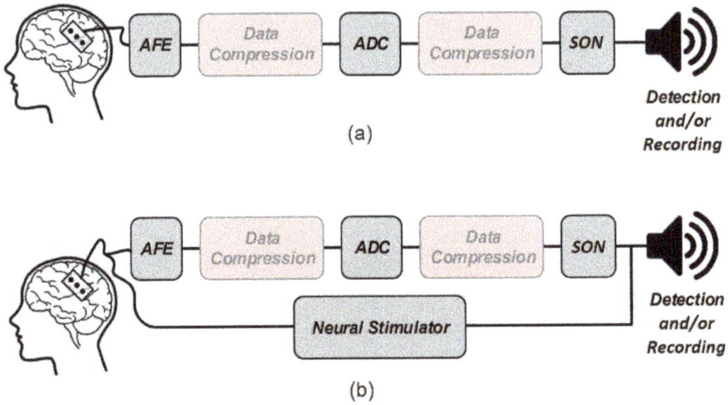

Figure 2. General overview of (**a**) a low-power seizure detection, and (**b**) a low-power seizure control systems.

2.1. *Electrical Stimulation*

The historical developments of electrical stimulators are considered to begin in the European post-Middle Ages, especially from the 16th century when a Dutch scientist, Jan Swammerdam, performed the first experiment on the muscle of a dissected frog [16]. He realized that the severed muscle of a frog is contracted by irritation. The idea of contracting a muscle by a stimuli had an important impact on neuroscience by demonstrating the fact that the observed behavior is based on the stimuli. More than one hundred years after Swammerdam's first experiment, on 6 November 1787, Luigi Galvani realized that a frog muscle can be contracted by placing an iron wire to the muscle and a copper wire to the nerve during a random experiment. He observed that an animal body performs convulsive movements when electricity is applied to it. The work of Galvani inspired Alessandro Volta to invent the voltaic pile in 1799. Using a voltaic pile, Luigi Rolando performed the first cortical stimulation experiment on an animal cortex in 1809. Indeed, Rolando is well-known for his pioneering research on brain localization of function. In 1825, J.-B. Sarlandiere published an extensive study on the benefits of electricity for pain relief by applying electricity to acupuncture needles. Gustav Fritsch and Eduard Hitzing published an article in 1870, showing that the stimulation of some specific part of the cortex leads to muscle contraction in dogs. Robert Bartholow, who was an American physician, was the first to report the findings of a study on electrical stimulation applied to the cerebral cortex of an awake human in 1874. All of these studies led to the design and introduction of different types of modern electrical stimulators and implantable stimulation devices.

Medical devices for electrical stimulation can be considered to belong to two classes, including implantable electrical stimulators and external electrical stimulators. External electrical stimulators are also categorized as transcranial and transcutaneous electrical stimulators. Some of the well-known external electrical stimulation methods include transcranial Alternating Current Stimulation (tACS), transcranial Direct Current Stimulation (tDCS), transcutaneous Trigeminal Nerve Stimulation (tTNS), transcutaneous Vagus Nerve Stimulation (tVNS) and transcutaneous Electric Nerve Stimulation (tENS). Each type of external electrical stimulator is developed, aiming at a specific application and target. For example, tTNS stimulators are currently commercialized for the relief and prevention of headaches. A commercialized tTNS is designed by Cefaly Technology. Implantable stimulators including deep-brain stimulation (DBS), retinal implants, pacemakers, cochlear implants and functional electrical stimulation (FES) are employed to assist or restore the functionality of organs that are not properly functioning. Electrical stimulators operate on the principle of initiating an action potential (AP) upon the transfer of electrical charge into excitable tissue. Electrical stimulators operate in three distinct modes, namely voltage-mode stimulation, current-mode stimulation and charge-mode stimulation.

Safety issues of biological cells impose a strict condition on the electrical pulses, i.e., a biphasic stimulation is necessary to prevent tissue damage and any long-term effects such as pH shift (during the usage of IEMDs) and erosion of the electrodes. Biphasic stimulation consists of a cathodic phase followed by an anodic phase. During the cathodic phase, the cell membrane is depolarized. Then, the anodic phase neutralizes the charge which has been injected during the cathodic phase. For exerting a safe stimulation, the voltage across the electrode must be constraint within a specific window. In addition, to block any direct current passing through the tissue, a large off-chip capacitor, namely a blocking capacitor, is placed in series with the stimulation electrode. This capacitor blocks the flow of any dc current through the tissue in case of semiconductor failure which makes the stimulator fail-safe. However, in multichannel stimulators such as retinal or cochlear implants, a large silicon area cannot be allocated to the large blocking capacitors. Hence, various circuits are proposed in the literature to substitute blocking capacitors with active circuits and to reduce the overall size of the stimulation system.

2.2. *Physiological Signal Recording*

The current healthcare systems are expected to deal with two major issues including chronic diseases and global population aging [17–19]. Early detection, as well as timely treatment of diseases, require monitoring systems that allow physicians to closely monitor the physiological signals of their patients. Acquiring physiological signals requires robust, light-weight and low-power wearable or implantable electronic medical devices. There are several important physiological signals that provide vital information of the human body such as electromyogram signals (EMG), electroencephalogram signals (EEG), electrocardiogram signals (ECG), phonocardiogram signals (PCG), electroretinography signals (ERG) and photoplethysmograms (PPG). Each of these signals has its own electrical characteristics, i.e., they have specific bandwidth and maximum amplitude. The bandwidths of some of these signals are depicted in Figure 3.

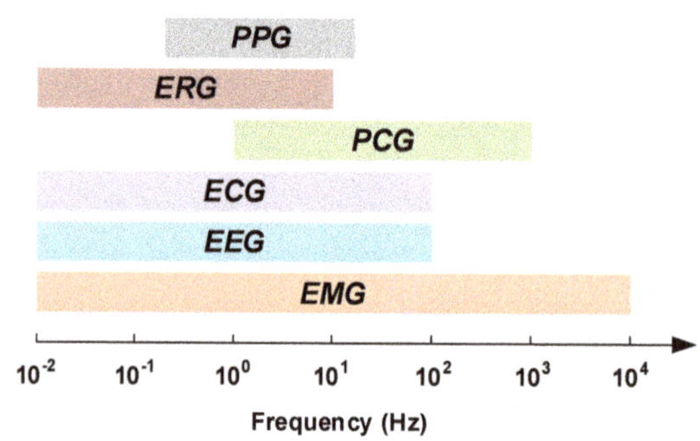

Figure 3. Bandwidths of some vital physiological signals (modified from [20]).

Although epilepsy can be detected from different physiological signals, the most important signal to predict or to detect a seizure onset are neural signals recorded from the brain. The following

Subsection describes the characteristics of different types of neural signals that can be recorded from different types of electrodes.

Neural Signal Recording

Historically, electrophysiology is based on the discovery of Italian scientist Galvani (1737–1798). Galvani realized that the tissues of frog muscles exhibit electrical potential. Studies on the living tissues continued until Hans Berger (1873–1941) discovered the human electroencephalogram (EEG). Berger could record the first human EEG signal using a Siemens double-coil galvanometer in 1924. When Berger's work was confirmed, EEG started to be used in clinical trials. Although vacuum tubes amplifiers were available since 1906, it took long time before they were used in the recording of neural activities. A group of researchers including Fredrick Gibbs (1903–1992), Hallowell Davis (1896–1992) and WG Lennox (1884–1960) with an EEG technician, namely Erna Gibbs (1906–1987), demonstrated EEG signals corresponding to a clinical absence attack [21] in 1935. They showed that the inter-ictal signals corresponding to an absence seizure attack have a specific signature of three spikes per second. Shortly after, in 1936, F. Gibbs demonstrated the importance of EEG in diagnosis and localization of epileptic seizures. Collaboration between F. Gibbs and Albert Grass, a MIT graduate, resulted in the development of EEG recording systems which smoothed the way for Grass Instrument Company. Grass Instrument Company was founded by A. Grass and his wife, Ellen, in 1945 and was acquired by Astro-Med Inc. in 1994, two years after the death of A. Grass. In 1946, the American EEG society (AEEGS) was founded and a year after, in 1947, the first annual meeting of the American EEG society was held in Atlantic City, NJ on 13–15 June [22]. In 1950s, William Grey Walter (1910–1977) developed the first EEG topography machine which can show a map of brain activity. During this decade, Wilder Penfield (1891–1976) and Herbert Jasper (1906–1999), neurosurgeons at the Montreal Neurological Institute, developed electro-corticography (ECoG) as a part of a surgical procedure for treating patients with severe epilepsy. The brain activity is recorded from the cerebral cortex using ECoG. It is shown that recording using ECoG has higher spacial resolution than EEG. Thus, this method is preferred for finding the regions of the cortex that generates epileptic seizures. Until now, recording brain waves has evolved thanks to the improvements in the technology of the electrodes. Nowadays, recording the activity of a single neuron is possible.

Decoding the functional operation of the brain requires recording of the electrical activity of neurons of the central nervous system [23]. Understanding and diagnosing neurological disorders such as epilepsy are based on the recording of the brain electrical activities. Furthermore, neural recording is a major module of brain–machine interfaces and neuroprosthetic technologies that aim at aiding paralyzed patients [24–26]. The demand for technologies that empower the neuroscientist and clinicians to observe the electrical activity of a large population of neurons in the brain has increased in the last decade. Extremely complex circuit solutions are needed to simultaneously monitor a large population of neurons in the brain that exceeds hundreds of cites in some applications [27], Simultaneous recording of the neural activities requires low-noise, low-power and area-efficient amplifiers. Good gain matching over the various channels and low crosstalk over the channels are other criteria.

The neural signals of interest for the recording of the brain activity using implantable devices have a frequency band of 1 Hz to 5 KHz [28]. As shown in Figure 4 [25,28–30], these signals consist of local field potentials (LFPs) and action potentials (APs) which are shown to include biomarkers that are useful for diagnosis and therapy of neurological disorders. LFP signals occupy a frequency band of 1 Hz to 100/200 Hz and a voltage range of 0.5–5 mV. AP signals have a frequency band of 100/200 Hz to 5 kHz with a voltage range of 50–500 µV. In addition, recording electrodes introduce background noise due to their resistance. This noise is thermal noise. Typically, noise integrated in the LFP bandwidth is smaller than 2 μV_{rms} whereas noise integrated into the AP bandwidth is smaller than 5 μV_{rms}. On account of the dynamic range of the LFPs and APs, an ADC with an effective number of bits (ENOB) larger than 12 to 14 bits is required for recording LFPs and an ADC with an ENOB higher than 8 to 10 bits is required for recording APs.

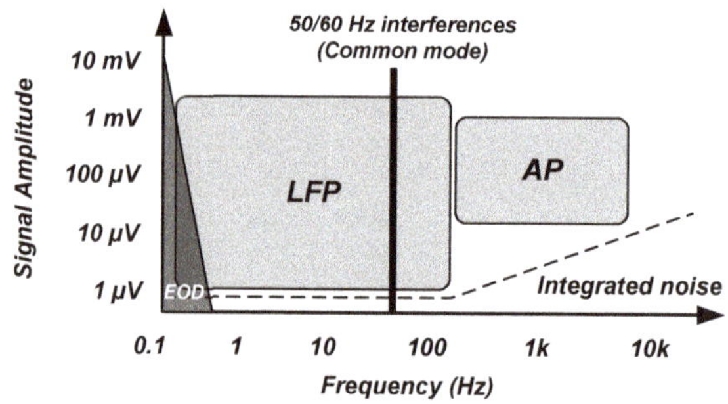

Figure 4. Signals characteristics in neural recording systems.

Signal conditioning is the first step to accommodate the signal before digitizing it to ensure accurate reading of neural activities. As shown in Figure 5, a low-noise amplifier (LNA) is the first stage in the front-end acquisition chain, in which neural signals of small amplitudes are amplified. In this stage, the amplifier should offer a high-pass filtering behavior in order to filter out the large electrode dc offset (EDO). Classically, the gain and bandwidth of the LNA is fixed. Following the LNA, a programmable-gain and bandwidth amplifier (PGA) is used to maximally cover the input range of the ADC that follows in the signal conditioning chain. Analog-to-digital converters (ADCs) are normally used to digitize acquired signals into data prior to its further processing or transmission from the implanted device to the outside of the body. One of the most important features of a neural amplifier is its input impedance. Since most of the electrodes present a 1 kHz impedance (Z_{1kHz}) of less than 200 kΩ [31], the input impedance of the amplifier should be much larger than this value to accurately record the neural signals.

In a conventional sensing system, one amplifier is designed at each sensor site. To ensure matching and to reduce the level of the noise, the input stage of the amplifier consumes the largest portion of the power that is provided which allows it to satisfy matching constraints while reducing the electronic noise level. Different signal-to-noise ratio (SNR) are required to record different types of neural signals, and thus, the amplifiers are tuned for the specific targets of recording. Reducing the amplifier internal noise and its power are a trade-off that must be optimized in careful consideration of the target signals and application.

Recording electrodes are required to transduce neural signals that consist of transmembrane ion exchanges into electrical signals that can be processed by microe-elctronic and information systems. The type of electrode is adapted to the target of the neural recording in the brain. Different types of neural recording electrodes are depicted in Figure 6. A brief description of different types of electrodes is provided in the following:

- Electroencephalography (EEG) Electrode ([32–34]): EEG electrodes are placed on the surface of the scalp. The international 10–20 system is a well-known and internationally recognized distribution of each of the EEG electrodes on the scalp. EEG recording offers several applications including brain-machine interfaces (BMI), polysomnography (PSG) for a sleep study, seizure detection, as well as other medical applications aiming at brain research. EEG recording is not an invasive method. The amplitude and bandwidth of the neural signals recorded by EEG electrodes are significantly smaller than the signals recorded by implantable electrodes due to the filtering behavior of cerebrospinal fluid (CSF), dura, skull and scalp. Furthermore, the fragile EEG signals are more exposed to different sources of artifacts including patient-related artifacts (e.g., movement, sweating, ECG, eye movements) and technical artifacts (50/60 Hz artifact, cable movements, electrode paste-related). The bandwidth of the EEG signals lies in the bandwidth of the LFP signals.

- Intracranial Electroencephalography (iEEG) [35]: Recording the neural signals inside the skull provides better signal quality in terms of signal-to-noise ratio and bandwidth. Intracranial EEG recording can be done using different types of electrodes including epidural electro-corticography (ECoG) electrodes, subdural ECoG electrodes, intracortical electrodes and depth electrodes.

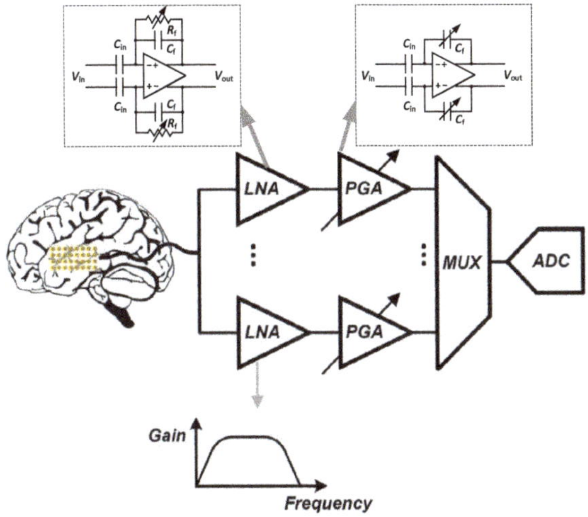

Figure 5. Simplified block diagram of a practical neural recording system.

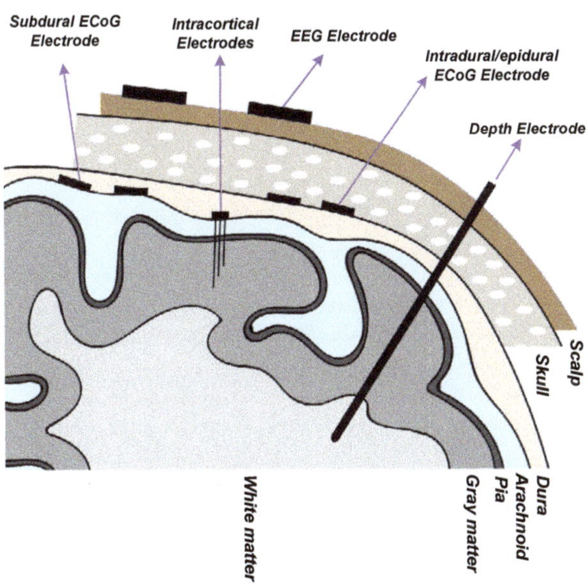

Figure 6. Placement of the different types of electrodes for neural recording.

Epidural ECoG electrodes [36,37]: this type of electrode is placed between the dura and skull. Hence, the dura need not be incised and opened for the placement of the electrodes on the cortex. This type of electrode is implemented in both micro-electrode and macro-electrode arrangements. Epidural electrodes are either strips of electrodes or a multicontact array as shown in Figure 7a. Epidural ECoG electrodes are suitable for both recording and stimulation.

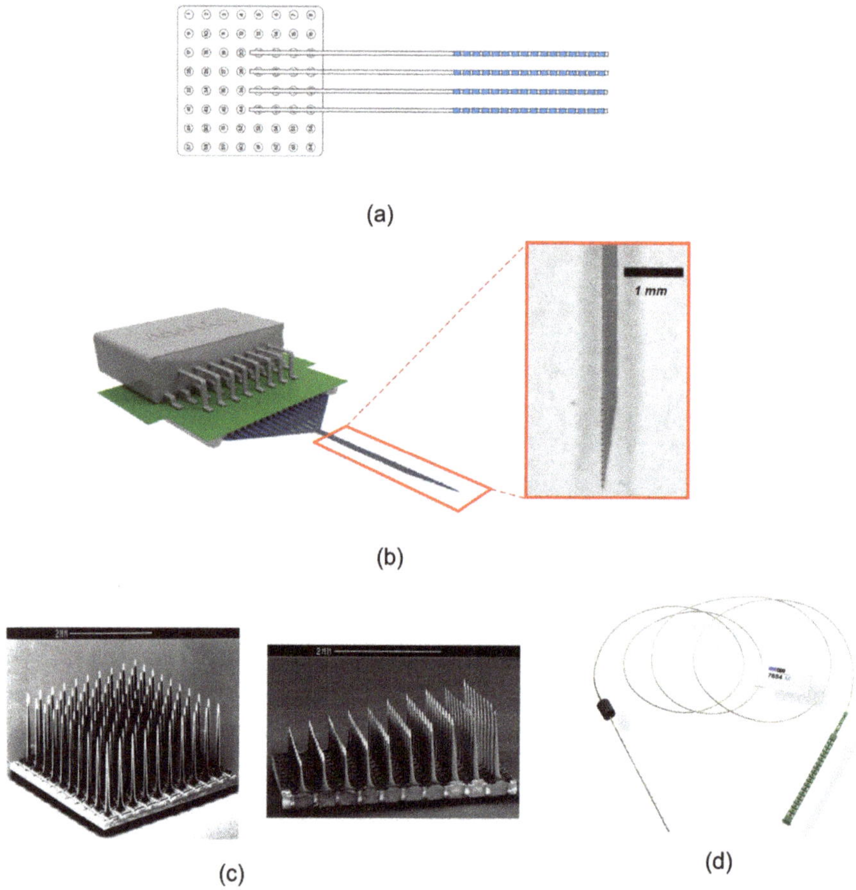

Figure 7. Different types of electrodes for neural recording including (**a**) grid array for intradural or subdural recording, (**b**) laminar intracortical electrodes, (**c**) intracortical micro-electrode array (Reprinted with permission of Wiley Periodicals, Inc., © 2003 The American Laryngological, Rhinological and Otological Society, Inc.), [38]) and (**d**) SEEG electrode (Courtesy DIXI medical).

Subdural ECoG electrodes [39]: This type of electrode is placed between the dura and the surface of the cortex. In this case, the dura must be incised for the placement of the electrodes on the cortex. This type of electrode is implemented in both micro-electrode and macro-electrode arrangements. The design of the Subdural electrodes is identical to Epidural electrodes as shown in Figure 7a. Subdural electrodes record LFP and AP signals, also depending on the size of the electrode. The main advantage of Subdural electrodes is the large coverage of the brain which enables performing a wide range of cognitive studies. Subdural ECoG electrodes are suitable for both recording and stimulation.

Intracortical electrodes [40]: intracortical electrodes are mainly micro-electrodes that are designed to record the signals from different layers of the brain. This type of electrode can perform single-unit recording, where a unique integrated circuit embeds the amplifier(s) or multi-unit recording where several integrated circuits embed multiple amplifiers, depending on the size of the electrode. In general, there are two types of intracortical electrodes including Laminar electrodes (Figure 7b) and micro-electrode arrays. These are also commonly known as neural probes or shank-based electrodes, e.g., [41]. The Utah electrode (Figure 7c) is a micro-electrode array contains up to 96 electrodes which enable high-density multi-channels recording from a large population of neurons, providing valuable data by delivering high spatial resolution within a small area of the brain.

Depth electrodes [42]: depth electrodes are placed at a precise location in the brain using a stereotactic system. Hence, this type of electrodes is also called stereoelectroencephalography (SEEG) electrodes. SEEG electrodes are suitable for recording and stimulation. Figure 7d shows a SEEG electrode manufactured by DIXI Medical [43].

2.3. Additional Blocks of Closed-Loop Epilepsy Control System

Neural signals that originate from several electrodes that are distributed over the surface of the cortex are digitized and can then be processed. Filtering and compressing are two classical digital processing, with which encryption has recently been complemented. In terms of the core functionality of the epilepsy control implantable system, seizures should be detected from the multi-channel recorded data. Algorithms aiming at seizure onset detection and seizure prediction should be accurate in terms of sensitivity and specificity such as to conceive durable and long-lasting IEMDs for seizure prediction and closed-loop stimulation. In addition, algorithms that are used in closed-loop stimulation systems should have a tolerable latency. An ideal seizure detector used in a closed-loop system has a sensitivity of 100% and a specificity of 100% or a false alarm rate (FAR) of zero. The importance of sensitivity is higher than specificity. Indeed, with a perfect sensitivity of 100%, a closed-loop stimulation system can detect and suppress all seizures. Improving the specificity of the seizure detector enables the closed-loop stimulation system to save power by reducing the periods of unnecessary stimulation. Furthermore, the seizures should be detected in advance or with small latency such as to be suppressed by the stimulation. Failing to satisfy the latter constraint, stimulation may not be effective for suppressing the seizures [44].

Delivering power to implanted devices is one of the major challenges in the design of IEMDs. Powering should be carried out over a wireless link, through the living tissues consisting of the scalp or skin layers. If the powering is not efficient, excessive heat due to power loss may damage the tissues around the IEMDs or may cause strange and unwanted sensations to the patients. Powering an IEMD requires applying several technologies including implantable batteries, energy harvesting or wireless power transfer. Two types of batteries are suitable for IEMDs consisting of primary batteries that are non-rechargeable, and secondary batteries that are rechargeable. Medical-grade battery equipped IEMDs are designed to include such technology. Several companies in the world provide medical batteries such as Eaglepicher [45]. Energy harvesting, which is also known as energy scavenging or ambient powering, is a process in which energy is captured from the environment and stored for further usage in small and ultra low-power systems. Energy sources may originate from external sources including solar power, thermal energy, or kinetic energy. Wireless power transfer is also a method that is suitable to power an implantable electronic system. Wireless power transfer (WPT) is also used to recharge the secondary medical-grades batteries in IEMDs.

Implantable electronic medical devices also require a wireless data transceiver. A data transceiver or a data receiver are implemented depending on the type of IEMD that is used for epilepsy control. An IEMD operating in an open-loop stimulation does not necessarily require the presence of a wireless data transmitter; however, a data receiver is needed for setting the stimulation parameters. On the other hand, an IEMD used for intracranial recording or closed-loop stimulation requires a wireless data transceiver to set the internal parameters as well as to send the recorded data to an external base station.

3. Commercial Systems and Products for Epilepsy Control

Commercial devices for seizure alerting and epilepsy control are reviewed in this section. These devices are considered to be partitioned into two categories, namely invasive medical devices, or implantable electronic medical devices, and non-invasive medical devices.

3.1. FDA Approved Implantable Electronic Medical Devices

Commercial IEMDs that are proposed in epilepsy control therapy apply electrical stimulation. These devices may operate as open-loop or closed-loop devices. In the case of open-loop stimulation, electrical stimulation is applied to the brain or group(s) of nerves without detecting any feedback from the body. On the other hand, closed-loop devices record physiological signal(s) and process them to adapt the stimuli. A closed-loop system would only stimulate the brain or a group(s) of nerves upon detection of seizure onset.

Although the quality of epilepsy detection and control of IEMDs is significantly better than it is in non-invasive medical devices, employing IEMDs presents several disadvantages, including the necessary surgery, perioperative risks, as well as side-effects such as hoarseness, throat pain, coughing, dyspnea, paresthesia, to quote the most prominent [46]. The potential of causing these symptoms enters into the decision of implanting and the selection of the device type. Governmental agencies approve devices and systems that can be implanted; these agencies are local to a country or group of countries, and include the U.S. Food and Drug Administration (FDA, USA), the Chinese National Medical Products Administration (NMPA, China; formerly the China Food and Drug Administration, CFDA), the European Economic Area CE Marking (CE marking, Europe).

3.1.1. Vagus Nerve Stimulation Therapy

Vagus nerve stimulation (VNS) therapy was approved by the FDA in July 1997, for the treatment of epilepsy in patients who suffer from drug-resistant epilepsy. The request was presented by Cyberonics Inc. [47] which was subsequently renamed LivaNova. The VNS therapy consists of sending mild electrical pulses to the brain through the vagus nerve (generally the left vagus nerve) using an electrical pulse generator as shown in Figure 8. The stimulation parameters are set within a frequency range of 1 Hz to 30 Hz, a current range of 0 mA (no stimuli) to 3.5 mA and a pulse-width range of 130 μs to 1000 μs [48]. The vagus nerve is one of the longest nerves in the body that originates from the brain stem located on both sides of the brain. The vagus nerve is a part of the parasympathetic nervous system which is responsible for recovery and digestion, in particular. The benefits of VNS therapy may include fewer seizures, shorter seizures, better seizure recovery time, decreased seizure severity, less medication, improved mood and memory and generally improved quality of life of epileptic patients. The major adverse effects of VNS include voice alteration, coughing and shortness of breath as well as headache and neck pain. A study [49] of 60 patients with VNS therapy shows that the most common adverse effects that affect more than 20% of the patients were voice alteration (55%) and headaches (22%). In 1997, the premarket approval application (PMA) from FDA for VNS therapy (PMA number of P970003) indicated that this device is approved for use as a therapy in adults and adolescents over 12 years old. In the new version of this PMA with PMA number P970003/S207, the FDA added the patients within 4–11 years old to the VNS therapy approval. Hence, this made VNS therapy the only candidate of IEMDs that are suitable for patients less than 18 years old.

There are three types of vagus nerve stimulators for epilepsy control proposed by LivaNova including the standard model, the AspireSR and SenTiva. The standard model VNS is the earliest device designed for epilepsy therapy which only offers basic programming features. The most common stimulation pattern using this model consists of 30 s of stimulation every 5 min [50]. In the standard model, a magnet is provided to apply additional stimulation during a seizure. The new model of VNS is a closed-loop VNS based on heart rate signal acquisition and processing. This model, namely the AspireSR (Figure 8a), received the CE mark in Europe in February 2014. A study [51] of 66 seizures from 16 patients using a closed-loop VNS shows that the cardiac-based seizure detection presents more than 80% sensitivity. In addition, the severity of the seizures was significantly reduced after 3–5 days of closed-loop stimulation. In general, this study confirmed that using a cardiac-based closed-loop VNS presents an acceptable sensitivity and specificity for triggering the stimulation. The AspireSR can also apply pre-programmed stimulation throughout the day and night. The third and the newest VNS product is SenTiva shown in Figure 8b. This device offers similar characteristics as the AspireSR with some additional features. For example, SenTiva can detect if a patient is lying after a seizure, revealing a potential loss of consciousness; in addition, it can be programmed to apply a different amount of stimulation at different times of the day. The VNS is FDA-approved for MRI under specific conditions. The estimated battery life for SenTiva is 4.9–10 years [52].

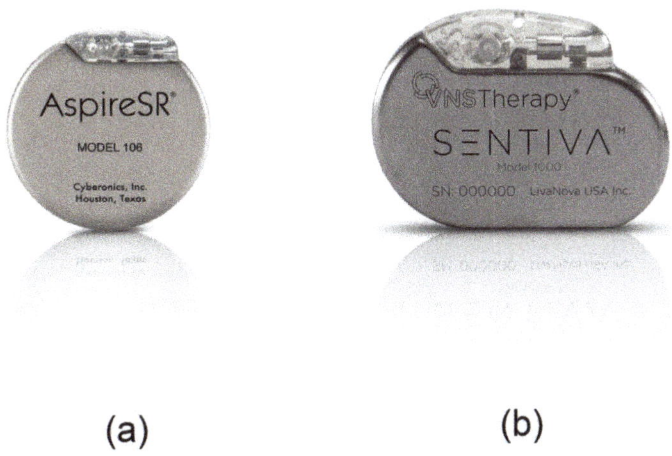

Figure 8. Vagus Nerve Stimulation (VNS) devices by LivaNova [47] including (**a**) AspireSR and (**b**) SenTiva, (Reprinted with permission; copyrighted material of LivaNova).

3.1.2. Responsive Neurostimulation

The responsive neurostimulation (RNS) therapy is a method for epilepsy control that received FDA approval in 2013. The RNS system, which is placed on the skull of the patient as shown in Figure 9, monitors the neural activity of the brain using two leads. Each lead has four electrode contacts that are used for stimulation and recording. Four amplifiers are used to the aim of recording neural signals. Each lead contains four electrodes that can be assigned to one or two out of four amplifiers [53]. The RNS system can store up to 30 min of ECoG activity [53]. The stimulation parameters are set within a frequency range of 1 Hz to 333 Hz, a current range of 1 mA to 12 mA and a pulse-width range of 40 µs to 1000 µs [54]. The current density limit of the stimulation equals 25 $\mu C/cm^2$ in each phase, while the current density is usually much less than the limit [55]. The stimulation can be applied between any two stimulation electrodes or between the electrode and the neurostimulator case. Typical stimulation parameters include a current of 1.5–3 mA, a pulse width of 160 µs, a burst duration of 100–200 ms, and a frequency of 100–200 Hz [54].

The RNS system continuously monitors the brain signals and employs several methods to extract the appropriate feature required for seizure detection. Three algorithms are employed by the RNS system to detect seizures, namely the area, line-length and half-wave algorithms. Physicians can change the algorithms' parameters to obtain an appropriate sensitivity, specificity and latency.

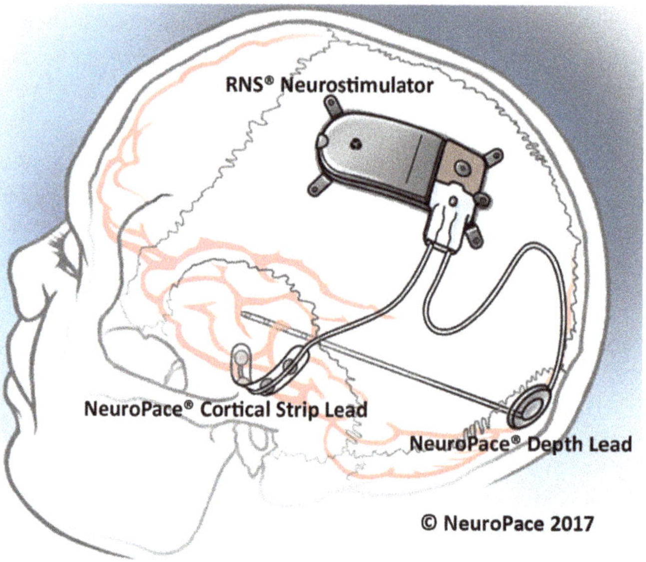

Figure 9. FDA approved responsive neurostimulation (RNS) system for closed-loop epilepsy detection and stimulation (courtesy of NeuroPace, Inc.) [53].

Before implanting the RNS system, the patients should undergo several tests. In order to receive the treatment from the RNS system, the patients should be older than 18 years, and suffer from disabling partial-onset seizure from no more than two foci, and they should be refractory to more than two antiepileptic medications that are properly chosen [53].

In [56], 230 patients with implanted RNS system were studied over time to measure the average decrease in seizures. This study shows that the average decrease in seizures was 44% after the first year, 53% after the second year and up to 66% after 3 to 6 years from implanting the RNS. This trend was observed among the patients who were followed over 7 years and the seizures were decreased by 72%.

3.1.3. DBS

The Medtronic [57] Deep Brain Stimulation (DBS) system for Epilepsy which was approved by the FDA in 2018 is a device that delivers controlled electrical pulses to the brain. The system consists of a pulse generator (IPG) as shown in Figures 10 and 11, that is implanted under the skin of the upper chest, and two leads implanted in the brain. The Medtronic DBS system for epilepsy helps to decrease the frequency of seizures. Unlike the RNS system, the DBS system applies an open-loop stimulation technique. The stimulation parameters are set within a frequency range of 2 Hz to 250 Hz in voltage mode, and 30 Hz to 250 Hz in current mode, a current range of 0 mA to 25.5 mA, a voltage range of 0 V to 10.5 V and a pulse-width range of 60 μs to 450 μs [58]. The Medtronic DBS system for epilepsy is used in conjunction with antiepileptic drugs in individuals 18 years of age or older, with partial onset seizures, with or without secondary generalization, who have not responded to three or more antiepileptic medications. The DBS system is used in patients who have an average of six or more seizures per month. It has not been evaluated in patients with less frequent seizures.

In [59], 157 patients with implanted DBS were studied over time to measure the average decrease in seizures. This study shows that the average decrease in seizures frequency was 56% after the second year of implantation. This study also shows that 54% of the patients had a seizure reduction of at least 50% after the second year of implantation. During this study, 14 patients were reported seizure-free for at least 6 months.

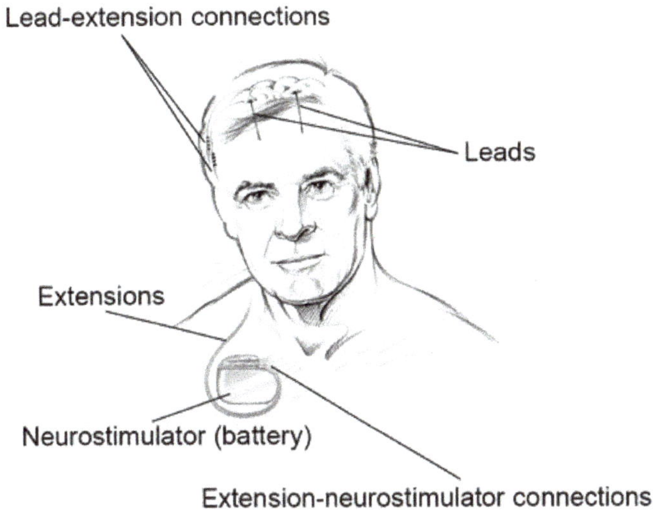

Figure 10. FDA approved deep-brain stimulation (DBS) system for epilepsy (Courtesy Food and Drug Administration (FDA)).

Figure 11. FDA approved DBS system for epilepsy [57] (Image with kind permission of Medtronic).

3.2. Commercialized Non-invasive Medical Devices

Several different commercialized non-invasive medical devices are available in the market. In the following, we mention some of these devices. In general, non-invasive medical devices belong to the two following groups, namely,

- external stimulators, and
- seizure alerting devices.

3.2.1. External Stimulators

In spite of considerable advantages in terms of efficiency and patient comfort and safety, implantable electronic medical devices for epilepsy control have the disadvantage of invasiveness, which may in some cases be intolerable. In contrast to IEMDs, external stimulators are not invasive and thus, no perioperative risks are taken. External stimulators are cheap and easy to use, although they are not as comfortable as IEMDs. Two types of external stimulators have been shown to affect the seizure frequency, namely the transcutaneous Vagus Nerve Stimulation (tVNS) [60] and the External

Trigeminal Nerve Stimulation (eTNS) or transcranial Trigeminal Nerve Stimulation (tTNS). In the following, each type of external stimulation is reviewed and commercialized products are presented.

The effectiveness of tVNS for depressive disorders treatment is reported in [61]. One hundred and twenty cases with mild and moderate depression were studied in a double-blinded randomized clinical trial. This study shows that tVNS has the same effect as VNS in the treatment of depressive disorders. These results encouraged researchers to investigate the effects that tVNS has on epilepsy. In [62], a group of ten patients with drug-resistant epilepsy were studied. tVNS was applied while keeping the dose of the medication. Stimulation was performed using biphasic pulses of pulse-width of 300 μs with an applied average voltage of 25 V. The stimulation frequency was chosen as 10 Hz. This pilot study shows that employing tVNS could reduce the seizure frequency in half of the patients. However, this study has several drawbacks including a lack of randomized stimulation, due to a small number of patients and an inhomogeneous patient group in terms of the type of seizures (focal, generalized epilepsy). This study led to introducing the Nemos system (Figure 12), manufactured by Cerbomed [63], which was taken over by tVNS Technologies GmbH in October 2018. Nemos, which is a transcutaneous device, stimulates the auricular branch of the vagus nerve using a handheld pulse generator. The patients have to apply the stimulation four hours per day. This device received the CE marking in Europe and costs 2499.00 €. The Gammacore is another device introduced by ElectroCore Medical LLC which employs transcutaneous VNS for the treatment of epilepsy. This FDA-approved device which is mainly designed for the treatment of headaches is also suggested as a treatment for epilepsy [64].

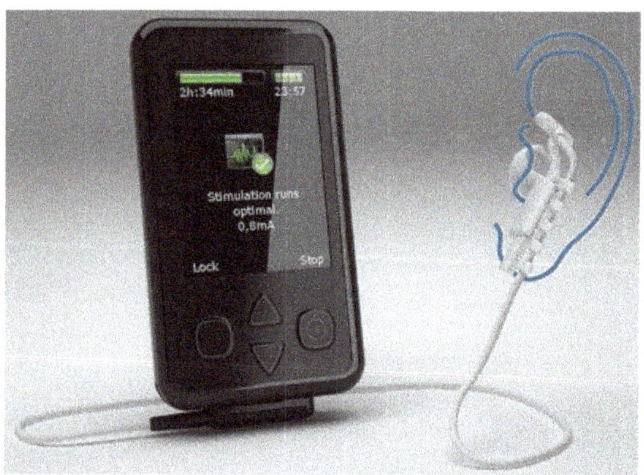

Figure 12. Transcutaneous Vagus Nerve Stimulation device for epilepsy control [63] (Image reprinted with kind permission of tVNS Technologies GmbH).

In 2008, Dr. Christopher DeGiorgio started a project pertaining to the long-term study on the effect of external trigeminal nerve stimulation (eTNS) on epilepsy control [65]. The project started with 50 patients. Out of 50 patients, 35 patients continued the long-term study for one year. In this study, it is shown that the median seizure frequency decreases by 34.8% after one year by employing eTNS. The Monarch eTNS system is an external trigeminal nerve stimulation system proposed by NeuroSigma [66]. This device has European approval for the treatment of epilepsy in adults and children 9 years and older.

3.2.2. Seizure Alerting Devices

Seizure alerting devices are designed to notify the onset of a seizure. These devices help the patients to quickly obtain help from their surroundings. Sudden unexpected death in epilepsy (SUDEP) is a fatal circumstance of epilepsy which often occurs during sleep. Seizure alerting devices help the caregivers and family by notifying the seizure onsets. Some seizure alerting devices are capable of monitoring the breathing of the patients during the sleep; upon detecting any abnormal circumstance, they notify the family or the caregiver of the patients. In addition, alerting the parents of an epileptic child is vital since a prolonged seizure can lead to brain damage, and even death. Hence, seizure alerting systems are important recent devices with a new role in epilepsy handling. Nevertheless, seizure alerting devices are by essence not useful for the patients who are alone, or patients who do not accept to be checked by others. Most of these devices detect the seizures by monitoring the repetitive movements of the patient, which may be ineffective to some types of epilepsy, e.g., tonic-clonic seizures or focal motor seizures. Hence, these devices cannot detect the seizures if the patient does not exercise large movements, e.g., during an absence seizure. Four types of seizure alerting devices are available at the time of writing:

- Watch devices
- Motion devices
- Mattress devices
- Camera devices

Most watch devices used as seizure alerting systems employ an accelerometer to detect abnormal and repetitive movements. Some of these devices have a global positioning system (GPS) device. Hence, if a seizure is detected, the location of the patient is sent by smartphone text messages or email. The Smartwatch Inspyre by Smart Monitor [67] is a smartwatch that detects the repetitive shaking motions and alerts determined contact person(s).

The electrodermal activity (EDA) which is also known as galvanic skin response (GSR) is a biomarker that is shown to be effective in seizure detection. The EDA relates to the electrical characteristic of the skin that changes due to sweating in response to a physiological change in the body. The electrodermal activity is monitored by measuring skin conductance. For example, the conductance can be calculated by applying a low-amplitude constant voltage while measuring the current [68]. In [69], it is shown that using an EDA sensor in addition to the accelerometer used for motion/movement detection increases the quality of the seizure detection, which led to proposing new algorithms for seizure classification reported in [70,71]. This concept is used in the watch device with FDA clearance which was introduced by Empatica Inc. [72]. Embrace2 is the latest watch device for seizure alerting which monitors the EDA, temperature and also employs an accelerometer and gyroscope. This 249 $ device, which is shown in Figure 13 offers more than 48 h of battery life with fast charging capability (30 min). This device sends the data acquired by the sensors to a compatible smartphone using Bluetooth technology. The smartphone processes the data using an application and alerts the family or a caregiver if a seizure is detected. Empatica Inc. also proposes a 1690 $ device namely the E4 for scientific research. The E4 has several sensors including a photoplethysmography (PPG) sensor, an electrodermal activity sensor, an infrared thermopile for reading the skin temperature and a three-axis accelerometer. Raw data delivered by the watch can be viewed in real-time and saved for future use.

Figure 13. Embrace2 by Empatica Inc. [72] for seizure detection (Image reprinted with kind permission of Empatica Inc.).

The EDA is also one of the most important biomarkers of SUDEP, as shown in [73]. It is reported that post-ictal generalized EEG suppression (PGES) appears to be a flat EEG following a seizure which is found in 100% of cases of SUDEP [73]. It is also shown that the duration of PGES correlates with the amplitude of EDA measured from the skin. This founding confirmed that seizure alerting devices such as the Embrace2 can deliver alert if there is a probability of SUDEP, which could save thousands of lives.

The first FDA cleared seizure alerting system that is not based on EEG recording is the Brain Sentinel monitoring and alerting system (known as SPEAC system) by Brain Sentinel, Inc. This device records the surface electromyography (sEMG) signals from the biceps of the patients. Hence, this device can accurately detect the seizures in case of tonic-clonic seizures. In addition to sEMG recording, the system records the audio during each event. This system helps physicians to accurately measure the seizure frequency characteristics.

Mattress seizure alerting devices are placed under a mattress where they can detect vibrations and movements. If a seizure-like movement is detected, an alarm will be triggered. The main difference between mattress devices and watch devices is that mattress devices are not wearable. The MP5-UTB is a mattress seizure-alerting device, designed by Medpage Ltd. The Emfit MM is a mattress device which is designed by Emfit Ltd. and which monitors the movements during sleep. This device was used in a study [74] including 45 patients. In this study, 78 seizures were detected using video-EEG. The Emfit MM system could detect 23 seizures out of the 78 seizures. Most importantly, this device could detect 11 out of 13 tonic-clonic seizures.

Camera devices are alternate non-wearable seizure alerting devices with which the movement of the patient is recorded and processed. The system warns the family or the caregiver of the patient if any unusual movement is detected. SAMi is a camera-based seizure alerting system by Hipass Design LLC. In this system, a remote infrared camera records the patient movements and sends the data to a smartphone for processing. This device records the audio of each event, as well.

4. Neural Recording Circuit Techniques

As a common block to most aforementioned systems, the front-end amplifiers in the recording chain represents one of the most challenging part. Owing to the extremely low amplitude of the recorded brain signals, low-noise electronics are required. Simultaneously, low-power design is necessary to guarantee the autonomy of the implantable and also portable systems. These constraints turn into contradictory design specifications and thus trade-offs must be accepted. While solutions have been provided, new challenges appear to emerge with the increase in the number of electrodes and recording channels. This criteria is in contrast to most other blocks of implantable systems which can be limited even to a single unit. Consequently, recording techniques show significant design challenges, which are explored in the following along with their possible solutions.

4.1. Low-Noise Front-End Amplifier

A neural amplifier is required next to the sensing electrode to amplify the weak neural signals and filter them. A neural amplifier should have five main features as follows:

- high-pass filtering for electrode offset rejection
- appropriate gain for conditioning the signals prior to digitization
- low input-referred noise for sensing weak neural signals
- low-power consumption (for neural amplifiers used in implantable devices)
- compact size (for neural amplifiers used in implantable devices)

In addition to the aforementioned features, a neural amplifier may provide integrated low-pass filter functionality. A low-pass filter is an important part of the analog front-end (AFE) to avoid aliasing. In general, neural amplifiers are considered into two categories, including ac-coupled amplifiers and dc-coupled amplifiers.

The ac-coupled amplifiers use a capacitor placed at the input, in the signal path to block the dc level. Several architectures of ac-coupled neural amplifiers are proposed in literature including capacitive-feedback network neural amplifiers, open-loop network neural amplifiers and chopper amplifiers. Figure 14 shows two common architectures for designing ac-coupled capacitive-feedback neural amplifiers.

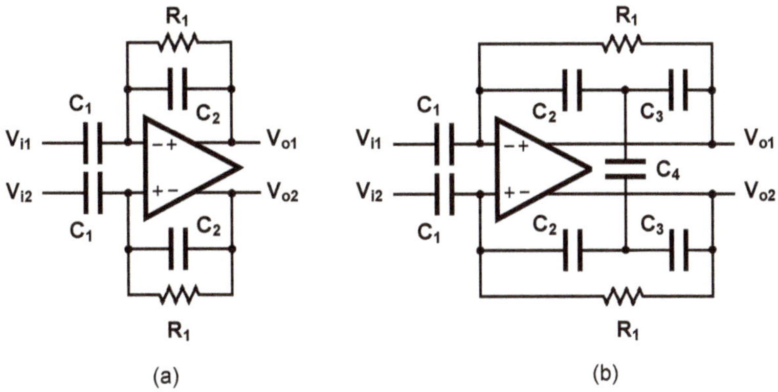

Figure 14. Overview of the ac-coupled capacitive-feedback structure for neural amplifiers. (**a**) Classical feedback topology and (**b**), capacitive T-feedback topology.

Figure 14a presents the general architecture of ac-coupled capacitive-feedback neural amplifiers. In this structure, C_1 is used to block the dc level that is created at the electrode-electrolyte interface. The value of C_1 is typically selected smaller than 20 pF since this capacitor affects the input impedance of the amplifier. The mid-band gain of this amplifier is defined as C_1/C_2. There is a high-pass corner in the frequency response of this amplifier which is determined by C_2 and R_1. For recording low-frequency neural signals, the high-pass corner frequency of the amplifier should be at few MHz to few Hz. Hence, a very large resistance is required to achieve such a corner frequency. In neural amplifiers used in implantable devices, these resistors are implemented as pseudo-resistors to avoid bulky physical resistors. A pseudo-resistor is implemented as a highly-resistive triode-biased MOS transistor as shown in Figure 15a [75]. This type of resistors is highly susceptible to process, voltage and temperature (PVT) variations. In addition, pseudo-resistors are highly non-linear resistors and placing such a resistor between the input and output of an operational transconductance amplifier (OTA) may yield a voltage-dependent resistor. If the voltage swing across this resistor is large, then its value may change significantly with respect to the voltage across it, and thus create a voltage-dependent high-pass pole for the neural amplifier. In addition, the structure shown in Figure 15a is not tunable and only creates one value of resistance. In order to design a neural amplifier with an adjustable high-pass pole, a tunable pseudo-resistor is required as shown in Figure 15b. The voltages across the Gate-Source of the PMOS transistors are set using an NMOS and a current source, such that the equivalent resistance changes [76].

For achieving a high-gain amplification in the structure shown in Figure 14a, C_2 is typically selected very small. Lowering the value of C_2 affects the common-mode rejection ratio (CMRR) of the amplifier as well as the gain precision. Figure 14b shows an architecture that is similar to Figure 14a, yet with a minor topological difference. In Figure 14b the capacitor -feedback network is implemented using a capacitive-T topology. This topology allows the implementation of a low-value of the feedback capacitor using higher values of capacitors [77]. This technique improves the matching between feedback capacitors, thus improves the CMRR of the amplifier. The equivalent feedback capacitor in this architecture is calculated as:

$$C_{efb} = \frac{C_2C_3}{2C_4 + C_2 + C_3}. \quad (1)$$

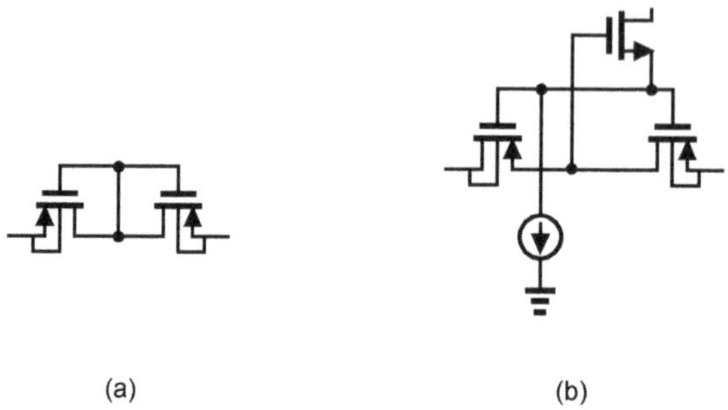

Figure 15. Pseudo-resistor architectures. (**a**) Fixed-value and (**b**) tunable pseudo-resistor.

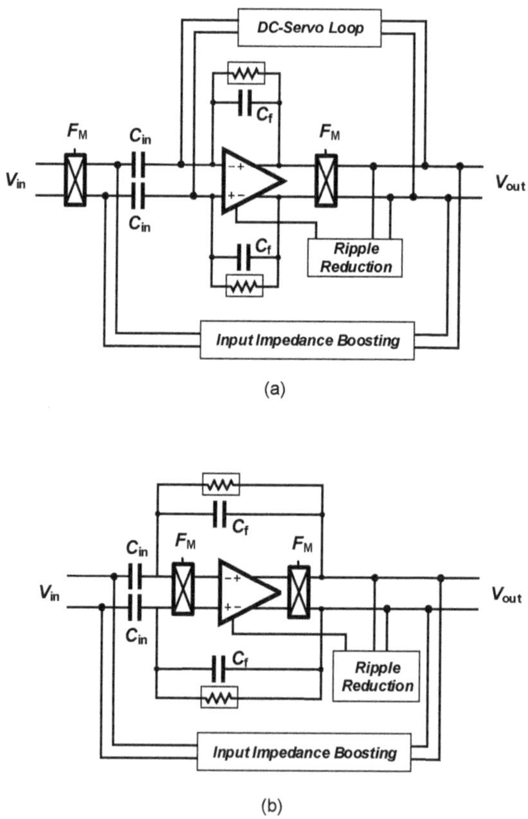

Figure 16. The ac-coupled chopper amplifier architectures. (**a**) Classical topology with servo loop and (**b**) alternated topology without servo loop.

Amplifying LFP signals using a neural amplifier creates significant issues related to Flicker noise, also named $1/f$ noise that lies in the same frequency range. In order to deal with this issue, chopper amplifiers are widely used for sensing LFP signals. However, ac-coupled chopper amplifiers carry over new issues which require new circuit and systems solving techniques. The classical architecture of a chopper amplifier is shown in Figure 16a. Chopper switches are placed before the ac-coupling capacitors. Hence, the dc value of the signal is modulated to a higher frequency and the ac-coupling capacitors can not filter it out. In order to filter the dc component of the input signal, a dc servo-loop is used in this circuit. Furthermore, chopper switches used in conjunction with ac-coupling capacitors reduce the input impedance of the amplifier. An impedance-boosting circuit must be used in the architecture of the chopper amplifier. Another issue of this circuit relates to the ripples introduced into the signal path by the chopper switches. This issue is addressed using a ripple-reduction circuit in the architecture of chopper amplifiers.

An alternate architecture employing a chopper amplifier is shown in Figure 16b. The chopper switches are placed after the ac-coupling capacitors in the signal path. Hence, this architecture does not require a dc-servo loop to block the dc component of the input signal. However, placing the chopper switches after the ac-coupling capacitors results into shaping the OTA thermal noise with $1/f$ characteristic when referred to the input of the neural amplifier [78]. Hence, this architecture requires a very high value of C_1 (in the order of 300–500 pF) to degrade this effect. In [79], ac-coupling capacitors are implemented using off-chip capacitors.

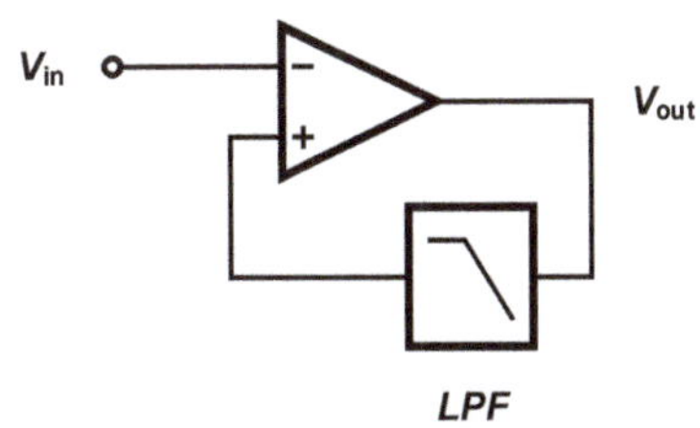

Figure 17. General architecture of dc-coupled amplifiers.

In contrast to ac-coupled amplifiers, dc-coupled amplifiers use a low-pass filter to block the dc component of the input signal, as shown in Figure 17. Several methods are introduced in the literature to implement a dc-coupled amplifier. The low-pass filter used in the structure of a dc-coupled amplifier can be implemented in analog [80], digital [81] or a combination of the domains [82].

Amplifier Sharing Methods

A technology trend is currently observed towards increasing the number of recording sites, which in turn creates severe constraints to the amplifier-related microelectronics. Hence, alternate solutions must be explored to satisfy the constraints in dense multichannel designs in terms of silicon area and power consumption. Techniques supporting sharing one amplifier among multiple channels become attractive as the number of recording channels increases. The sharing technique can be either applied at system-level or circuit-level. In circuit-level techniques, the amplifier is not totally shared and the current that is provided to an input differential pair circuit is reused by the other channels' input pairs. This method, which is called orthogonal current reuse technique, is introduced in [83]. System-level techniques are widely presented in the literature. Two different methods aiming at sharing a single amplifier are introduced in literature, namely the time-division multiplexing (TDM) [84] and frequency-division multiplexing (FDM) [85] techniques which are shown in Figure 18a,b, respectively. Several critical design challenges yield from these sharing techniques, such as crosstalk, settling, accuracy, and filtering. If the amplifier used in TDM is not fast enough, it may introduce large crosstalk between channels. the timing of all signals is very critical in TDM. As discussed in [86], in addition to the crosstalk, timing issues may cause settling errors during the analog-to-digital conversion, in turn causing noise aliasing degrading the noise performance of the circuit.

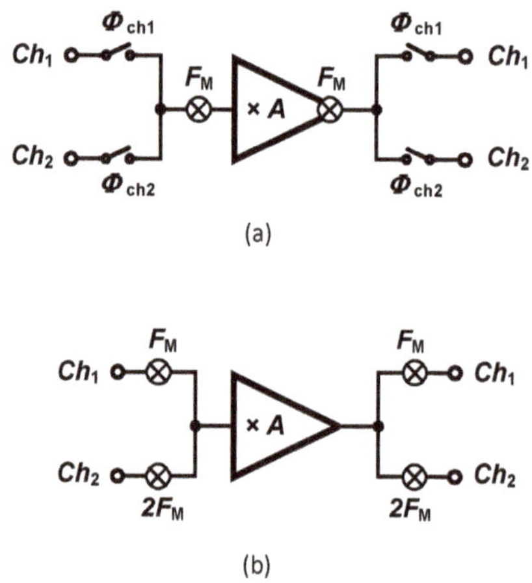

Figure 18. Amplifier sharing techniques: (**a**) time-division multiplexing (TDM) technique proposed in [84] and (**b**) frequency-division multiplexing (FDM) technique proposed in [85].

The first chopper amplifier based on FDM was proposed in [85]. As shown in Figure 18b, inputs are modulated using different chopping frequencies that are orthogonal. In order to guarantee the orthogonality between chopping frequencies, $2^n \times F_M$ is proposed, in which $n = 0, \ldots, N-1$ and N is the number of channels. Rademacher functions are actually applied, which are sequences of orthogonal functions that have two values of ± 1 and are defined by the conditions expressed in Equation (2).

$$\begin{array}{ll} \phi_0(\frac{t}{T}) = 1 & (0 \leq \frac{t}{T} < \frac{1}{2}) \\ \phi_0(\frac{t}{T}) = -1 & (\frac{1}{2} \leq \frac{t}{T} < 1) \\ \phi_0(\frac{t}{T}+1) = \phi_0(\frac{t}{T}) & \\ \phi_n(\frac{t}{T}) = \phi_0(\frac{2^n t}{T}) & n = 1, 2, \ldots \end{array} \quad (2)$$

in which ϕ_n is called the nth Rademacher's function and T is the period of the functions. The waveform of the first four Rademacher's functions are shown in Figure 19. Except ϕ_0, Rademacher's functions can be used as the modulating signal in a chopper amplifier based on FDM. Nevertheless, the technique proposed in [85] presents some major drawbacks which are discussed in the following.

A significant issue regarding the usage of different chopping frequency for different channels relates to the mismatch between the input impedance of the channels. An additional issue of the technique relates to the need for a low-pass filter after the demodulation of each channel. Since the chopper is located after the opamp, high-frequency signals that remain in the demodulated signal (Flicker noise and offset) are not filtered out. Therefore, an additional low-pass filter is required to clean the output signal from high-frequency contents. A third issue of the technique relates to the gain mismatch between channels which in turn relates to the bandwidth of the opamp used in the chopper amplifier.

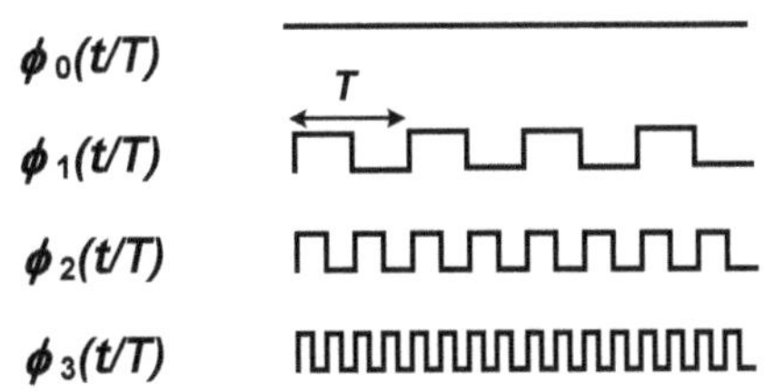

Figure 19. First four Rademacher functions.

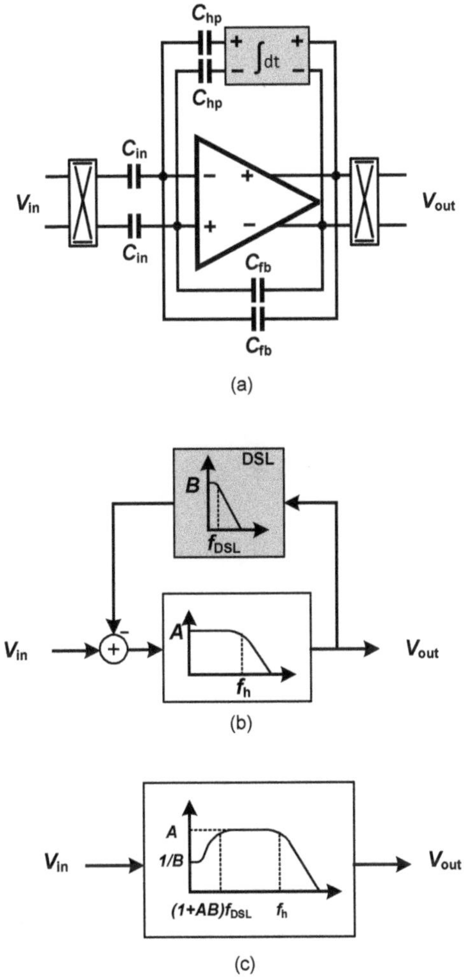

Figure 20. Effect of the dc servo loop on the overall frequency response of the chopper amplifier. (**a**) Topology including a dc-servo loop. (**b**) Block-level frequency responses of the topology including a servo loop. (**c**) Overall frequency response.

The usage of a two-channel chopper amplifier proposed in [85] as a neural amplifier presents a significant drawback. The lack of a dc servo-loop indeed represents a severe issue of two-channel chopper neural amplifiers. A large electrode dc offset can easily saturate the amplifier if it is not filtered. In chopper amplifiers, the dc servos loop (DSL) is employed to implement a high-pass corner in the frequency response of the chopper amplifier, as shown in Figure 20a. Figure 20b,c illustrate the effect of the DSL on the overall frequency response of the amplifier, using control theory. The DSL introduces a high-pass corner at a frequency of $(1 + AB)f_{DSL}$, in which f_{DSL} is the bandwidth of the integrator used in the DSL. At the circuit level, (Figure 20a), the high-pass corner is calculated using Equation (3).

$$f_{hp} = \frac{C_{hp}}{C_{fb}} f_{0DSL}, \tag{3}$$

in which f_{0DSL} is the unity-gain frequency of the integrator in the DSL.

The noise performance of a two-channel orthogonal chopper amplifier is also degraded with respect to the single-channel amplifier. In a single-channel amplifier as shown in Figure 21a, the signal gain from input to the output is equal to $-C_{in}/C_{fb}$ and the noise gain from the amplifier input to the output (V_{out}/V_n) is equal to 1 + C_{in}/C_{fb}. In a two-channel chopper amplifier as shown in Figure 21b, the signal gain from the input to the output is equal to $-C_{in}/C_{fb}$; however, the noise gain from the amplifier input to the output (V_{out}/V_n) is equal to 1 + $2C_{in}/C_{fb}$. Therefore, the input-referred amplifier noise of a two-channel chopper amplifier is approximately two times higher than single-channel chopper amplifier.

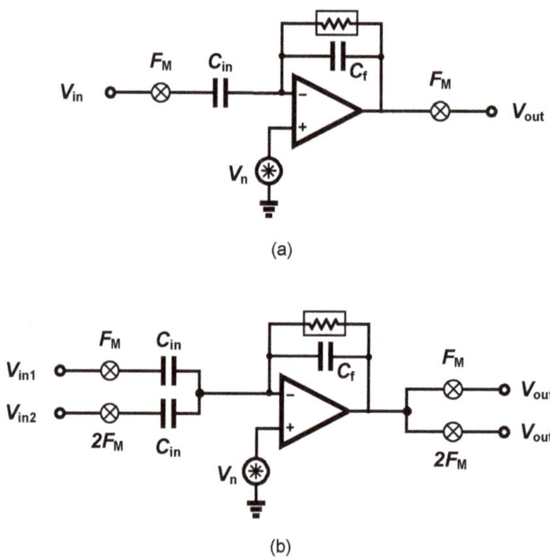

Figure 21. Noise analysis model of (**a**) a single-channel chopper amplifier and (**b**) a two-channel chopper amplifier.

4.2. Data Compression

Data compression is a well-known method to reduce the power consumption of wireless transmitters in implantable electronic medical devices. Using data compression also reduces the power consumption of feature extractors as shown in [87]. Compressive sensing (CS) is a compression scheme presenting relevant characteristics for multi-channel neural recording. This emerging compression method has lower complexity in comparison to established compressing methods. Compressive sensing reduces the number of measurements for a high-dimensional signal with respect to the number of measurements dictated by the Nyquist sampling theorem. Compressive sensing can be applied in three different domains, including analog, digital and multichannel. Analog compressive sensing (ACS) is a method applied to reduce the data rate and digitization power. However, due to the multi-path nature of this technique, a large on-chip silicon area is required. Digital compressive sensing (DCS) is shown to be a superior method over ACS in terms of power consumption [88]. However, this technique requires several accumulation blocks for each channel, which is not tractable in a multi-channel system with a limited silicon area.

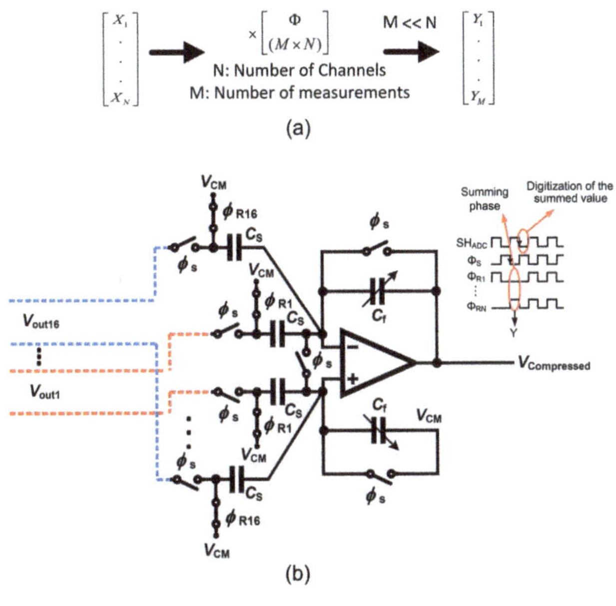

Figure 22. Multichannel compressive sensing (MCS). (**a**) Overview of the MCS concept. (**b**) Structure of the conventional multi-input single-output compressive sensing (MISOCS) block with 16 inputs.

The multi-channel compressive sensing (MCS) technique has been developed as a hardware-suitable compressive sensing that allows a straightforward circuit design as well as an efficient power × area performance [87]. Figure 22a depicts the simplified concept of the MCS operating over N channels. Φ is a measurement matrix that consists of a random sequence of ones and zeros, striclty. Φ is used to project a matrix that stores data from N channel into the compressed domain. The result of the operation is a vector comprising data that is the compressed image of the data recorded from N channels.

Circuit area and power dissipation can be optimized by applying the MCS at the sensing node. The complexity is deferred to the receiver that applies the data reconstruction algorithm from compressed data. The complex data reconstruction operation is performed in the digital domain and involves a significant latency which would slow down any seizure pattern detection or further feature extraction processes. Feature extraction from spatially filtered data based on the MCS is presented in [87] to the aim of detecting seizures. A multi-input single-output compressive sensing (MISOCS) block carries out the projection of data originating from N channels into the compressed domain. The MISOCS block processes the input channels in parallel owing to its implementation as a summing amplifier, which is detailed in the following.

Figure 22b presents the system architecture of a 16-channel ($N = 16$) conventional MISOCS block and the signals that control the corresponding circuit. A compression ratio of 16 ($CR = 16$) is achieved by the system [87]. The 16 inputs to the MISOCS block originate from the front-end neural amplifiers differential outputs. The control signals of the MISOCS block consist of a sampling signal ϕ_s as well as a measurement matrix $\Phi \in \mathbb{R}^{1\times16}$. Sixteen random sampling signals $\phi_{R1}, \ldots, \phi_{R16}$ form the measurement matrix. The conventional MISOCS block operates in two phases. Signal ϕ_s controls the sampling phase during which the output of the individual channels are sampled. ϕ_s at the LOW level marks the summation phase. The value presented on the channels that correspond to a random sampling value equal to one in the measurement matrix are accumulated at the output node of the summing amplifier. This conventional architecture of the MISOCS block is straightforward but suffers from circuit-level intricacies.

First, the random nature of the measurement matrix yields a random number of ones in the random sampling signals [87]. The offset of the summing amplifier depends on the number of ones, thus causing an error. In practical terms, a random signal which has a value between zero and $V_{os}(1+\frac{KC_{in}}{C_f})$ is summed up with the compressed signal, at the output of the summing amplifier. The error is random in nature and derives from the implementation of the method at the circuit level. The random error causes a degradation of the seizure detection efficiency which is due to a degradation of the quality of the signal reconstruction.

In addition, careful consideration must be devoted to controlling the dynamic range of the compressed signal, as it is generated at the output node of the summing amplifier. The dynamic range of the compressed signal at the output of the MISOCS block is calculated in the following, considering the two extreme cases. The first extreme case consists of a full correlation of input signals (C = 1) while the random sampling signals are all at one. This situation yields an output corresponding to a linear sum of all inputs and thus the highest level of the compressed signal. In the highest level worst case, all the N random sampling signals are at one; thus, the output consists of a linear sum of the input resulting in a highest value of the compressed signal that is N times larger than a single channel. The second extreme case consists of a single random sampling signal at one, while all others are at zero. This situation yields the lowest level of the compressed signal which is meaningful, and corresponds to the RMS noise of a single channel analog front-end (AFE).

The dynamic range of the compressed signal can be expressed from the highest and the lowest meaningful levels that are acceptable at the output node of the amplifier. Consequently, the dynamic range of the compressed signal in an N-channel conventional MCS system is N times larger than a single AFE channel. The dynamic range of the compressed signal is thus used in the calculation of the resolution of the ADC that follows the MISOCS block, as expressed in Equation (4).

$$B_{ADC} \approx B_{Channel} + \log_2 N, \tag{4}$$

where $B_{Channel}$ is the required equivalent resolution of the AFEs.

For example, a 16-channel acquisition system with an equivalent bit-resolution of each channel assumed to be equal to 8-bit yields the necessity of a 12-bit ADC to accurately reconstruct the compressed signal.

The mMISOCS block is presented in the following [89] as a modified MISOCS block aiming at overcoming the issues of the conventional architecture that were described above. The mMISOCS architecutre is shown in Figure 23. The circuit topology includes a butterfly switch which is used to invert the output of a channel $V_{out,i}$ when the corresponding random sampling value is at zero; the result is then summed up with other channels output during the summing phase. Hence, in contrast to the conventional MISOCS block, the mMISOCS architecutre generates compressed data that is composed of data originating from all channels, i.e., not a random number of channels with coefficient at one. As a result, the noise level of the compressed data is fixed at $\sqrt{N}$ times the RMS noise of a single AFE. Consequently, the dynamic range of the compressed signal is also fixed at $\sqrt{N}$ times larger than the dynamic range of a single AFE. The ADC resolution is thus expressed in Equation (5).

$$B_{ADC} \approx B_{Channel} + \log_2 \sqrt{N}. \tag{5}$$

As an additional benefit of the mMISOCS topology, the offset signal that appears in the compressed signal for any combination of the values of the sampling signals is fixed because all input signals participate in the output node summation in a positive or negative value. The offset that appears at the output of the summing stage is expressed in Equation (6)

$$V_{os,err} = V_{os}(1 + 16\frac{C_{in}}{C_f}). \tag{6}$$

Consequently, the error is constant and can thus be canceled by processing in the digital domain.

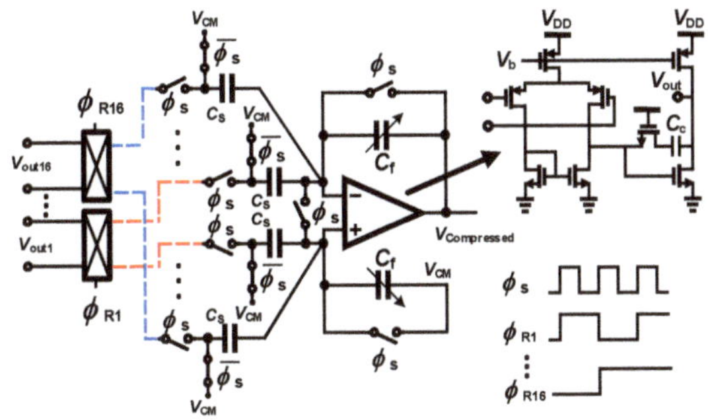

Figure 23. Schematic of the proposed multiple-input single-output compressive sensing block, mMISOCS.

4.3. Feature Extraction

Feature extraction has an important role in the quality of the seizure detection [90]. Several classical features are introduced, which must be implemented in the digital domain, in an implantable epilepsy control device.

Considering a one-dimensional signal representing the electrical activity recorded from the brain, features can be extracted in the time and spectral domains.

- Time-domain feature extraction. The raw signal originating from the signal conditioning chain to the ADC is in turn processed. Usually, the algorithms process the data delivered as successive windows comprising a fixed number of samples. The processed feature score is compared to a threshold yielding a decision.
- Frequency-domain features. Spectral-based feature extractors operate in the digital domain. A fast-Fourier transform is typically applied to the input signal originating from an ADC, and prior to extracting features in the frequency domain. Bandwidths of interests are determined, and the energy is computed within a selected frequency range, e.g., [91]. The process is repeated in time (or time-window) yielding a decision.

Features are typically patient-specific, and thus parameters must be tuned such as to improve the accuracy of the decision. Recently, artificial intelligence and machine learning techniques also including deep-learning methods have been applied to support the feature extraction and classification, Refs. [92–94]. Moreover, the success of advanced AI and deep-learning algorithms in epilepsy detection has opened the way to epilepsy prediction, where interictal signals that are observed between seizures are studied with the aim of extracting reliable markers of a future seizure [95].

The hardware resources that are involved in frequency-domain feature extractors as well as deep-learning and machine-learning methods are important. The nature of the algorithms dictates a large number of memory accesses, multiplications, as well as non-linear operations processed using lookup tables. All of the aforementioned operations are time- and energy-consuming. In addition, the algorithms are developed such as to accurately operate over 32-bit number formats, while reducing the bit representation decreases the reliability. Hence, these techniques are considered not suitable for implantable devices, as excessive consumers of hardware resources and power. Recently, complex deep-learning and machine learning algorithms find adaptations to portable hardware that are dictated by new requirements of internet-of-things. Nevertheless, implantable devices require low-power blocks and feature extraction methods which are best found as the time-domain features presented in the following [96].

- Energy: the energy feature is a popular feature. The average energy of d samples is calculated as expressed in Equation (7).

$$Average\ Energy = \frac{1}{d}\sum_{d} x^2[n]. \tag{7}$$

  A multiple and accumulate block is used to process the data that streams-into. The inputs of the multiplier are identical, yielding the x^2 operation.
- Accumulated energy: the accumulated energy extractor applies the energy criteria over several time-windows.
- Variance and Hjorth variance The variance criteria has extensively been applied in EEG studies. The variance is processed over a window of d samples, and then averaged. The intuitive formulation that directs the hardware implementation is expressed in Equation (8).

$$Variance = \frac{1}{d}\sum_{d} x^2[n] - \mu^2. \tag{8}$$

  The hardware is more complex than the hardware required in the energy extractor, and consists of multipliers and accumulators, subtractors and temporary storage registers.
- Line-length or Coastline: the line-length is a measure of the absolute value of the length between two consecutive data points. Line-length is a feature that increases with low-amplitude while high-frequency signals are presented, as well high-amplitude while low-frequency signals are presented. The line-length feature for d samples is calculated as expressed in Equation (9).

$$Line - length = \frac{1}{d}\sum_{d} |x[n] - x[n-1]| \tag{9}$$

  The hardware is relatively straightforward and consists of multipliers and accumulators, temporary storage registers as well as multiplexers.
- Area: area is a popular feature for seizure detection. The simplicity of the algorithm enables a low-cost and accurate seizure detection. Area is one of the features used in RNS (Section 3.1.2. The area feature for d samples of the signal is calculated as expressed in Equation (10).

$$Area = \frac{1}{d}\sum_{d} |x[n]| \tag{10}$$

- Non-linear autocorrelation: non-linear autocorrelation feature extraction is based on detecting and accumulating the minimum of the maximum of the samples in three consecutive windows, also detecting and accumulating the maximum of the minima of the samples in three consecutive windows, and finally subtracting the latter from the former result, as expressed in Equation (11).

$$\begin{aligned} HV &= min(max(X_{win_i}), max(X_{win_{i+1}}), max(X_{win_{i+2}})) \\ LV &= max(min(X_{win_i}), max(X_{win_{i+1}}), max(X_{win_{i+2}})) \\ Autoc &= \sum_{3} HV - LV. \end{aligned} \tag{11}$$

The hardware requires many resources including a multiplier-accumulators, subtractors, storage resisters as well as several comparators.

5. Discussion

Several challenges must be solved in open and closed-loop IEMDs, from the point of view of microelectronics engineering. Limitations are imposed from the technical and engineering perspectives, and are combined with limitations imposed by medical regulations, methods and ethical concerns. Some of the relevant open issues are summarized in the following.

- Size: the most important challenge of an implantable system is the size. Any implantable medical device (IEMD) is composed of several electrical modules. Some of these modules consist of off-chip components such as wireless powering modules or the wireless data transmitter. The specification of these modules should be defined in a way that the IEMD system presents an acceptable size. The IEMD weight is a related parameter. Increasing the size and the weight of IEMDs also increases the complexity of the surgery. Hence, for the convenience of the patients, IEMDs should have a minimum number of off-chip components in order to present a minimum size and weight.

  A solution for decreasing the size of an implant is to integrate the active circuits as close as possible to the electrode. One method to realize this solution consists of fabricating a silicon-based electrode which allows the active circuits to be implemented on the same silicon or by attaching the active circuitry to the silicon-based electrode using post-CMOS processes.
- Power consumption and temperature elevation: a limitation for temperature raise is imposed by medical regulations for IEMDs. IEMDs temperature should not exceed predefined limits. Generally, the temperature of the outer surface of an implanted device must be limited to 2 °C above body temperature [97]. However, this limit is reported to be 1 °C above body temperature in IEEE standards [98], specially in cortical implants [99]. A device that exceeds this limit should be turned off immediately. Hence, temperature sensors should be considered in the design of IEMDs and stimulators to enable temperature management capabilities of the systems [100].
- Battery powering and rechargeability: IEMDs should offer freedom to patients to proceed in their life with regular activities. This autonomy cannot be provided without using an implanted battery. Moreover, to increase the lifetime of IEMDs, the battery should be rechargeable. Therefore, patients undergo less surgery for the placement and/or removal of the IEMDs. However, the main challenge in the design of rechargeable IEMDs consists of wirelessly and efficiently recharging the implanted battery. Efficiency in the wireless battery charging is very important since this procedure may generate heat and cause skin burning or unpleasant feeling during the battery charging process.
- Biocompatibility: the package and enclosure of IEMDs must be bio-compatible. A biocompatible package serves as a barrier between the electronics and other chemical materials to which a biological system may adversely react. The host response to an implanted IEMD (resulting from tissue trauma during the implantation of an IEMD and the presence of the device in the body [101]) depends on the type of material that is used for the packaging and the enclosure of the IEMD. The importance of the biocompatibility lies in the fact that the systemic toxicity impairs the entire biological system such as the nervous or the immune system [101]. In addition, the reason for a systemic reaction due to the biocompatibility cannot be traced back to its origin since it generally takes place at a location far from the point of contact of an IEMD. Due to all aforementioned issues, biocompatibility has become the most important part of the U.S. FDA approval procedure, even

for Class I devices (lowest risk). Furthermore, biocompatibility is the major part of acquiring an ISO (International Organization for Standardization) standard such as ISO 10993.

- Data storage: in order to increase the accuracy of the seizure detection as well as to provide freedom and autonomy to the patient during the recording period, the implant should store data over a few hours. This feature is important since patients should not have to wear any bulky holder of an external unit (helmet, belt) during some specific activities or during sleeping. Hence, the system should save the recorded data on an implanted memory. If the IEMD is powered by a rechargeable battery, the IEMD should save the recorded data on a non-volatile memory since the IEMD may be turned off by the under-voltage lockout detection circuit.

While many of the aforementioned challenges find solutions at circuit and system levels, we observe a trend aiming at efficiently solving these issues considering the circuit and the system as a whole, i.e., some part of the trade-offs is solved exploiting specific circuit characteristics, while other parts of the trade-offs find solutions exploiting algorithms or taking benefits of system-level improvements.

6. Conclusions

This review paper presents and discusses the major blocks of the signal recording and seizure detection of an implantable system aiming at epilepsy control. While classical techniques pertaining to analog and digital circuit design are used, the paper focuses on specificities related to neural signal recording in multichannel systems. Epileptic signals are discussed from an engineering perspective, i.e., summarizing the major electrical characteristics and their impact on the microelectronics front-ends. The commercial implantable systems aiming at epilepsy control which command over an official approval at the moment of writing are reviewed. External systems which represent non-invasive systems are presented as a new trend in seizure alarming.

Circuit and system techniques are presented that aim at solving contemporary issues related to low-power, low-noise and low-area trade-offs met in modern microelectronics multi-channel neural acquisition systems. Specifically, the amplifier-sharing technique, as well as the compressed sensing technique, are discussed and presented.

Author Contributions: Conceptualization, R.R. and A.S.; circuits design and validation, original draft preparation, R.R.; review and editing, supervision, funding acquisition, A.S. All authors have read and agreed to the published version of the manuscript.

Funding: This research was supported by the SNSF under grant No. 200020-175790.

Conflicts of Interest: The authors declare no conflict of interest.

Abbreviations

The following abbreviations are used in this manuscript:

ACS	Analog Compressive Sensing
ADC	Analog-to-Digital Converter
AED	Anti-Epileptic Drug
AFE	Analog Front-End
BMI	Brain-Machine Interface
CC-LNA	Capacitive-Coupled Low-Noise Amplifier
CMRR	Common-Mode Rejection Ratio
CS	Compressive Sensing
CSF	Cerebrospinal Fluid
DBS	Deep-Brain Stimulation
DCS	Digital Compressive Sensing

DSL	dc Servo Loop
ECG	Electrocardiogram
ECoG	Electrocotricogram
EDA	Electrodermal Activity
EDO	Electrode dc Offset
EEG	Electroencephalogram
EMG	Electromyogram
ENOB	Effective Number of Bits
ERG	Electroretinogram
eTNS	External Trigeminal Nerve Stimulation
FDA	Food and Drug Administration
FDM	Frequency-Division Multiplexing
FE	Feature Extractor
FES	Functional Electrical Stimulation
ICP	Intracranial Pressure
iEEG	Intracranial Electroencephalography
IEMD	Implantable Electronic Medical Device
ILAE	International League Against Epilepsy
IMD	Implantable Medical Device
IPG	Integrated Pulse Generator
LFP	Local Field Potentials
LL	Line-length
LNA	Low-Noise Amplifier
MCS	Multi-channel Compressive Sensing
MEG	Magnetoencephalography
MISOCS	Multi-Input Single-Output Compressive Sensing
mMISOCS	modified Multi-Input Single-Output Compressive Sensing
MRI	Magnetic Resonance Imaging
NEF	Noise Efficiency Factor
OTA	Operational Transconductance Amplifier
PCG	Phonocardiogram
PET	Positron Emission Tomography
PGA	Programmable-Gain Amplifier
PMA	Premarket Approval
PPG	Photoplethysmogram
PSG	Polysomnography
RNS	Responsive Neurostimulation
PVT	Process, Voltage and Temperature
SEEG	Stereo-EEG
sEMG	Surface Electromyography
SNR	Signal-to-Noise Ratio
SPECT	Single-Photon Emission Computed Tomography
SUDEP	Sudden Unexpected Death in EPilepsy
tACS	transcranial Alternating Current Stimulation
tDCS	transcranial Direct Current Stimulation
TDM	Time-Division Multiplexing
tENS	transcutaneous Electric Nerve Stimulation
tTNS	Transcranial Trigeminal Nerve Stimulation
tVNS	transcutaneous Vagus Nerve Stimulation
VLSI	Very Large-Scale Integration
VNS	Vagus Nerve Stimulation
WHO	World Health Organization

References

1. Abbott, Inc., Cardiovascular. CardioMEMS HF. Available online: https://www.cardiovascular.abbott/ (accessed on 11 June 2020).
2. Boston Scientific Corp. Available online: http://www.bostonscientific.com/ (accessed on 11 June 2020).
3. Fisher, R.S.; Acevedo, C.; Arzimanoglou, A.; Bogacz, A.; Cross, J.H.; Elger, C.E.; Engel, J., Jr.; Forsgren, L.; French, J.A.; Glynn, M.; et al. ILAE official report: a practical clinical definition of epilepsy. *Epilepsia* **2014**, *55*, 475–482. [CrossRef] [PubMed]
4. Epilepsy Foundation. Available online: https://www.epilepsy.com/ (accessed on 11 June 2020).
5. Caveness, W.F.; AM LORENTZ, H.J.M.; RADERMECKER, J. A Proposed International Classification of Epileptic Seizures. *Epilepsia* **1964**, *5*, 297–306. [CrossRef]
6. Bancaud, J.; Henriksen, O.; Rubio-Donnadieu, F.; Seino, M.; Dreifuss, F.; Penry, J. Commission on classification and terminology of the international league against epilepsy. Proposal for revised clinical and electroencephalographic classification of epileptic seizures. *Epilepsia* **1981**, *22*, 489–501.
7. Fisher, R.S.; Cross, J.H.; French, J.A.; Higurashi, N.; Hirsch, E.; Jansen, F.E.; Lagae, L.; Moshé, S.L.; Peltola, J.; Roulet Perez, E.; et al. Operational classification of seizure types by the International League Against Epilepsy: Position Paper of the ILAE Commission for Classification and Terminology. *Epilepsia* **2017**, *58*, 522–530. [CrossRef] [PubMed]
8. Korff, C.M.; Scheffer, I.E. Epilepsy classification: A cycle of evolution and revolution. *Curr. Opin. Neurol.* **2013**, *26*, 163–167. [CrossRef]
9. Berg, A.T.; Scheffer, I.E. New concepts in classification of the epilepsies: Entering the 21st century. *Epilepsia* **2011**, *52*, 1058–1062. [CrossRef]
10. Gao, H.; Sander, J.W.; Xiao, Y.; Zhang, Y.; Zhou, D. A comparison between the 1981 and 2017 International League Against Epilepsy classification of seizure types based on an outpatient setting. *Epileptic Disorders* **2018**, *20*, 257–264. [CrossRef]
11. World Health Organization. Available online: https://www.who.int/ (accessed on 11 June 2020).
12. Traub, R.D.; Wong, R. Cellular mechanism of neuronal synchronization in epilepsy. *Science* **1982**, *216*, 745–747. [CrossRef]
13. Gómez-Gonzalo, M.; Losi, G.; Chiavegato, A.; Zonta, M.; Cammarota, M.; Brondi, M.; Vetri, F.; Uva, L.; Pozzan, T.; de Curtis, M.; et al. An excitatory loop with astrocytes contributes to drive neurons to seizure threshold. *PLoS Biol.* **2010**, *8*, e1000352. [CrossRef]
14. Toole, C.; Martinez-Juárez, I.E.; Gaitanis, J.N.; Blum, A.; Sunderam, S.; Ding, L.; DiCecco, J.; Besio, W.G. Source localization of high-frequency activity in tripolar electroencephalography of patients with epilepsy. *Epilepsy Behav.* **2019**, *101*, 106519. [CrossRef]
15. Uchida, D.; Fujimoto, A.; Yamazoe, T.; Yamamoto, T.; Enoki, H. Seizure frequency can be reduced by changing intracranial pressure: A case report in drug-resistant epilepsy. *Epilepsy Behav. Case Rep.* **2018**, *10*, 14–17. [CrossRef] [PubMed]
16. Verkhratsky, A.; Krishtal, O.; Petersen, O. From Galvani to patch clamp: The development of electrophysiology. *Eur. J. Physiol.* **2006**, *453*, 233–247. [CrossRef] [PubMed]
17. Bonato, P. Wearable sensors/systems and their impact on biomedical engineering. *IEEE Eng. Med. Biol. Mag.* **2003**, *22*, 18–20. [CrossRef] [PubMed]
18. Ryoo, D.W.; Kim, Y.S.; Lee, J.W. Wearable systems for service based on physiological signals. In Proceedings of the 2005 IEEE Engineering in Medicine and Biology 27th Annual Conference, Shanghai, China, 17–18 January 2006.
19. Grasso, A.D.; Marano, D.; Palumbo, G.; Pennisi, S. Design methodology of subthreshold three-stage CMOS OTAs suitable for ultra-low-power low-area and high driving capability. *IEEE Trans. Circuits Syst. Regul. Pap.* **2015**, *62*, 1453–1462. [CrossRef]
20. Li, Y.; Poon, C.C.; Zhang, Y.T. Analog integrated circuits design for processing physiological signals. *IEEE Rev. Biomed. Eng.* **2010**, *3*, 93–105. [CrossRef]
21. Gibbs, F.A.; Davis, H. Changes in the human electroencephalogram associated with loss of consciousness. *Am. J. Physiol.* **1935**, *113*, 49–50.
22. American Clinical Neurophysiology Society. Available online: https://www.acns.org/ (accessed on 11 June 2020).

23. Engel, A.K.; Moll, C.K.; Fried, I.; Ojemann, G.A. Invasive recordings from the human brain: Clinical insights and beyond. *Nat. Rev. Neurosci.* **2005**, *6*, 35–47. [CrossRef]
24. Lebedev, M.A.; Nicolelis, M.A. Brain–machine interfaces: Past, present and future. *Trends Neurosci.* **2006**, *29*, 536–546. [CrossRef]
25. Schwartz, A.B.; Cui, X.T.; Weber, D.J.; Moran, D.W. Brain-controlled interfaces: Movement restoration with neural prosthetics. *Neuron* **2006**, *52*, 205–220. [CrossRef]
26. Hochberg, L.R.; Serruya, M.D.; Friehs, G.M.; Mukand, J.A.; Saleh, M.; Caplan, A.H.; Branner, A.; Chen, D.; Penn, R.D.; Donoghue, J.P. Neuronal ensemble control of prosthetic devices by a human with tetraplegia. *Nature* **2006**, *442*, 164–171. [CrossRef]
27. Viswam, V.; Chen, Y.; Shadmani, A.; Dragas, J.; Bounik, R.; Milos, R.; Müller, J.; Hierlemann, A. 2048 action potential recording channels with 2.4 µVrms noise and stimulation artifact suppression. In Proceedings of the 2016 IEEE Biomedical Circuits and Systems Conference (BioCAS), Shanghai, China, 17–19 October 2016.
28. Chandrakumar, H.; Marković, D. A High Dynamic-Range Neural Recording Chopper Amplifier for Simultaneous Neural Recording and Stimulation. *IEEE J. Solid-State Circuits* **2017**, *52*, 645–656. [CrossRef]
29. Burkhard, P.R.; Vingerhoets, F.; Berney, A.; Bogousslavsky, J.; Villemure, J.G.; Ghika, J. Suicide after successful deep brain stimulation for movement disorders. *Neurology* **2004**, *63*, 2170–2172. [CrossRef] [PubMed]
30. Heldman, D.A.; Wang, W.; Chan, S.S.; Moran, D.W. Local field potential spectral tuning in motor cortex during reaching. *IEEE Trans. Neural Syst. Rehabil. Eng.* **2006**, *14*, 180–183. [CrossRef] [PubMed]
31. Ward, M.P.; Rajdev, P.; Ellison, C.; Irazoqui, P.P. Toward a comparison of microelectrodes for acute and chronic recordings. *Brain Res.* **2009**, *1282*, 183–200. [CrossRef] [PubMed]
32. Millett, D.; Coutin-Churchman, P.; Stern, J. Basic Principles of Electroencephalography. *Neurosci. Biobehav. Psychol.* **2015**, *1*, 75–80. [CrossRef]
33. Srinivasan, R.; Nunez, P. Electroencephalography. *Encyclopedia of Human Behavior*, 2nd ed.; Elsevier: Amsterdam, The Netherlands, **2012**; pp. 15–23. [CrossRef]
34. Im, C.; Seo, J. A review of electrodes for the electrical brain signal recording. *Biomed. Eng. Lett.* **2016**, *6*, 104–112. [CrossRef]
35. Parvizi, J.; Kastner, S. Promises and limitations of human intracranial electroencephalography. *Nat. Neurosci.* **2018**, *21*, 474–483. [CrossRef]
36. Shokoueinejad, M.; Park, D.W.; Jung, Y.; Brodnick, S.; Novello, J.; Dingle, A.; Swanson, K.; Baek, D.H.; Suminski, A.; Lake, W.; et al. Progress in the Field of Micro-Electrocorticography. *Micromachines* **2019**, *10*, 62. [CrossRef]
37. Volkova, K.; Lebedev, M.A.; Kaplan, A.; Ossadtchi, A. Decoding Movement From Electrocorticographic Activity: A Review. *Front. Neuroinform.* **2019**, *13*, 74. [CrossRef]
38. Badi, A.; Kertesz, T.; Gurgel, R.; Shelton, C.; Normann, R. Development of a novel eighth-nerve intraneural auditory neuroprosthesis. *Laryngoscope* **2003**, *5*. [CrossRef]
39. Bundy, D.T.; Zellmer, E.; Gaona, C.M.; Sharma, M.; Szrama, N.; Hacker, C.; Freudenburg, Z.V.; Daitch, A.; Moran, D.W.; Leuthardt, E.C. Characterization of the effects of the human dura on macro- and micro-electrocorticographic recordings. *J. Neural Eng.* **2014**, *11*, 016006. [CrossRef]
40. Wang, M.; Guo, L. Intracortical Electrodes. In *Neural Interface Engineering: Linking the Physical World and the Nervous System*; Springer: New York, NY, USA, 2020; pp. 67–94.
41. NeuroNexus. Available online: https://neuronexus.com/ (accessed on 2 September 2020)
42. Cagnan, H.; Denison, T.; McIntyre, C. Emerging technologies for improved deep brain stimulation. *Nat. Biotechnol.* **2019**, *37*, 1024–1033. [CrossRef] [PubMed]
43. DIXI Medical. Available online: http://www.diximedical.com/ (accessed on 11 June 2020).
44. Salam, M.T.; Velazquez, J.L.P.; Genov, R. Seizure suppression efficacy of closed-loop versus open-loop deep brain stimulation in a rodent model of epilepsy. *IEEE Trans. Neural Syst. Rehabil. Eng.* **2016**, *24*, 710–719. [CrossRef] [PubMed]
45. EaglePicher Technologies. Available online: https://www.eaglepicher.com/ (accessed on 11 June 2020).
46. Schachter, S.; Schmidt, D. *Epilepsy: Problem Solving in Clinical Practice*; CRC Press: Boca Raton, FL, USA, 2000; p. 448.
47. Cyberonics Inc. Available online: https://www.cyberonics.com/ (accessed on 11 June 2020).
48. Fogel, B.S.; Greenberg, D.B. *Psychiatric Care of the Medical Patient*; Oxford University Press: Oxford, UK, 2015; p. 505.

49. Sackeim, H.A.; Rush, A.J.; George, M.S.; Marangell, L.B.; Husain, M.M.; Nahas, Z.; Johnson, C.R.; Seidman, S.; Giller, C.; Haines, S.; et al. Vagus nerve stimulation (VNST) for treatment-resistant depression: Efficacy, side effects, and predictors of outcome. *Neuropsychopharmacology* **2001**, *25*, 713–728. [CrossRef]
50. Labiner, D.M.; Ahern, G.L. Vagus nerve stimulation therapy in depression and epilepsy: Therapeutic parameter settings. *Acta Neurol. Scand.* **2007**, *115*, 23–33. [CrossRef] [PubMed]
51. Boon, P.; Vonck, K.; van Rijckevorsel, K.; El Tahry, R.; Elger, C.E.; Mullatti, N.; Schulze-Bonhage, A.; Wagner, L.; Diehl, B.; Hamer, H.; et al. A prospective, multicenter study of cardiac-based seizure detection to activate vagus nerve stimulation. *Seizure* **2015**, *32*, 52–61. [CrossRef]
52. VNSTherapy, LivaNova. An Introduction to VNS Therapy. Available online: https://vnstherapy.co.uk/sites/vnstherapy.com/files/vns-therapy-patient-brochure_1.pdf (accessed on 7 October 2020).
53. *RNS System User Maual*. Available online: https://www.neuropace.com/wp-content/uploads/2015/11/PDMSManual.pdf (accessed on 7 October 2020).
54. Thomas, G.P.; Jobst, B.C. Critical review of the responsive neurostimulator system for epilepsy. *Med. Devices* **2015**, *8*, 405.
55. Sun, F.T.; Morrell, M.J. Closed-loop neurostimulation: the clinical experience. *Neurotherapeutics* **2014**, *11*, 553–563. [CrossRef]
56. Bergey, G.K.; Morrell, M.J.; Mizrahi, E.M.; Goldman, A.; King-Stephens, D.; Nair, D.; Srinivasan, S.; Jobst, B.; Gross, R.E.; Shields, D.C.; et al. Long-term treatment with responsive brain stimulation in adults with refractory partial seizures. *Neurology* **2015**, *84*, 1210–1212. [CrossRef]
57. Medtronic Inc. Available online: https://www.medtronic.com/ (accessed on 11 June 2020).
58. *Medtronic Activa PC, Implant Manual*; Medtronic. 2008. http://manuals.medtronic.com (accessed on 7 October 2020).
59. Carlson, C. Epilepsy Treatment Stimulus Package? Deep Brain Stimulation in Treatment-Resistant Focal Epilepsy. *Epilepsy Curr.* **2010**, *10*, 148–150. [CrossRef]
60. Yap, J.Y.Y.; Keatch, C.; Lambert, E.; Woods, W.; Stoddart, P.R.; Kameneva, T. Critical Review of Transcutaneous Vagus Nerve Stimulation: Challenges for Translation to Clinical Practice. *Front. Neurosci.* **2020**, *14*, 284. [CrossRef] [PubMed]
61. Rong, P.J.; Fang, J.L.; Wang, L.P.; Meng, H.; Liu, J.; Ma, Y.g.; Ben, H.; Li, L.; Liu, R.P.; Huang, Z.X.; et al. Transcutaneous vagus nerve stimulation for the treatment of depression: A study protocol for a double blinded randomized clinical trial. *BMC Complement. Altern. Med.* **2012**, *12*, 255. [CrossRef] [PubMed]
62. Stefan, H.; Kreiselmeyer, G.; Kerling, F.; Kurzbuch, K.; Rauch, C.; Heers, M.; Kasper, B.S.; Hammen, T.; Rzonsa, M.; Pauli, E.; et al. Transcutaneous vagus nerve stimulation (t-VNS) in pharmacoresistant epilepsies: A proof of concept trial. *Epilepsia* **2012**, *53*, e115–e118. [CrossRef] [PubMed]
63. Cerbomed GmbH. Nemos, Vitos devices. Available online: https://cerbomed.com/ (accessed on 11 June 2020).
64. ElectroCore, Inc. Gammacore. Available online: https://www.gammacore.com/ (accessed on 11 June 2020).
65. Soss, J.; Heck, C.; Murray, D.; Markovic, D.; Oviedo, S.; Corrale-Leyva, G.; Gordon, S.; Kealey, C.; DeGiorgio, C. A prospective long-term study of external trigeminal nerve stimulation for drug-resistant epilepsy. *Epilepsy Behav.* **2015**, *42*, 44–47. [CrossRef] [PubMed]
66. NeuroSigma, Inc. Available online: http://www.neurosigma.com/ (accessed on 11 June 2020).
67. SmartMonitor. Available online: https://smart-monitor.com/ (accessed on 11 June 2020).
68. Benedek, M.; Kaernbach, C. A continuous measure of phasic electrodermal activity. *J. Neurosci. Methods* **2010**, *190*, 80–91. [CrossRef]
69. Poh, M.Z.; Loddenkemper, T.; Reinsberger, C.; Swenson, N.C.; Goyal, S.; Madsen, J.R.; Picard, R.W. Autonomic changes with seizures correlate with postictal EEG suppression. *Neurology* **2012**, *78*, 1868–1876. [CrossRef]
70. Onorati, F.; Regalia, G.; Caborni, C.; Picard, R. Improvement of a convulsive seizure detector relying on accelerometer and electrodermal activity collected continuously by a wristband. In Proceedings of the Epilepsy Pipeline Conference, San Francisco, CA, USA, 25–27 February 2016.
71. Regalia, G.; Onorati, F.; Migliorini, M.; Picard, R. An improved wrist-worn convulsive seizure detector based on accelerometry and electrodermal activity sensors, American Epilepsy Society Annual Meeting 2015. Available online: https://www.aesnet.org/meetings_events/annual_meeting_abstracts/view/2327131 (accessed on 11 June 2020).
72. Empatica Inc. Available online: https://www.empatica.com/ (accessed on 11 June 2020).

73. Sarkis, R.A.; Thome-Souza, S.; Poh, M.Z.; Llewellyn, N.; Klehm, J.; Madsen, J.R.; Picard, R.; Pennell, P.B.; Dworetzky, B.A.; Loddenkemper, T.; et al. Autonomic changes following generalized tonic clonic seizures: An analysis of adult and pediatric patients with epilepsy. *Epilepsy Res.* **2015**, *115*, 113–118. [CrossRef]
74. Poppel, K.V.; Fulton, S.P.; McGregor, A.; Ellis, M.; Patters, A.; Wheless, J. Prospective study of the Emfit movement monitor. *J. Child Neurol.* **2013**, *28*, 1434–1436. [CrossRef]
75. Harrison, R.R.; Charles, C. A low-power low-noise CMOS amplifier for neural recording applications. *IEEE J. Solid-State Circuits* **2003**, *38*, 958–965. [CrossRef]
76. Tajalli, A.; Leblebici, Y. A widely-tunable and ultra-low-power MOSFET-C filter operating in subthreshold. In Proceedings of the 2009 IEEE Custom Integrated Circuits Conference, Rome, Italy, 13–16 September 2009.
77. Ng, K.A.; Xu, Y.P. A compact, low input capacitance neural recording amplifier. *IEEE Trans. Biomed. Circuits Syst.* **2013**, *7*, 610–620. [CrossRef]
78. Xu, J.; Yazicioglu, R.F.; Grundlehner, B.; Harpe, P.; Makinwa, K.A.; Van Hoof, C. A 160 *mu*W 8-Channel Active Electrode System for EEG Monitoring. *IEEE Trans. Biomed. Circuits Syst.* **2011**, *5*, 555–567. [CrossRef] [PubMed]
79. Verma, N.; Shoeb, A.; Bohorquez, J.; Dawson, J.; Guttag, J.; Chandrakasan, A.P. A micro-power EEG acquisition SoC with integrated feature extraction processor for a chronic seizure detection system. *IEEE J. Solid-State Circuits* **2010**, *45*, 804–816. [CrossRef]
80. Gosselin, B.; Sawan, M.; Chapman, C.A. A low-power integrated bioamplifier with active low-frequency suppression. *IEEE Trans. Biomed. Circuits Syst.* **2007**, *1*, 184–192. [CrossRef] [PubMed]
81. Yeager, D.; Biederman, W.; Narevsky, N.; Alon, E.; Rabaey, J. A fully-integrated 10.5 μw miniaturized (0.125 mm^2) wireless neural sensor. In Proceedings of the 2012 Symposium on VLSI Circuits (VLSIC), Honolulu, HI, USA, 13–15 June 2012.
82. Yazicioglu, R.F.; Merken, P.; Puers, R.; Van Hoof, C. A 200*mu*W Eight-Channel EEG Acquisition ASIC for Ambulatory EEG Systems. *IEEE J. Solid-State Circuits* **2008**, *43*, 3025–3038. [CrossRef]
83. Johnson, B.; Molnar, A. An orthogonal current-reuse amplifier for multi-channel sensing. *IEEE J. Solid-State Circuits* **2013**, *48*, 1487–1496. [CrossRef]
84. Altaf, M.A.B.; Zhang, C.; Yoo, J. A 16-channel patient-specific seizure onset and termination detection soc with impedance-adaptive transcranial electrical stimulator. *IEEE J. Solid-State Circuits* **2015**, *50*, 2728–2740. [CrossRef]
85. Tsai, Y.L.; Lee, F.W.; Chen, T.Y.; Lin, T.H. 5.3 A 2-channel- 83.2 dB crosstalk 0.061 mm^2 CCIA with an orthogonal frequency chopping technique. In Proceedings of the 2015 IEEE International Solid-State Circuits Conference - (ISSCC) Digest of Technical Papers, San Francisco, CA, USA, 22–26 February 2015.
86. Zhang, C.; Altaf, M.A.B.; Yoo, J. Design and implementation of an on-chip patient-specific closed-loop seizure onset and termination detection system. *IEEE J. Biomed. Health Inf.* **2016**, *20*, 996–1007. [CrossRef]
87. Ranjandish, R.; Schmid, A. A Sub-μW/Channel, 16-Channel Seizure Detection and Signal Acquisition SoC Based on Multichannel Compressive Sensing. *IEEE Trans. Circuits Syst. Express Briefs* **2018**, *65*, 1400–1404. [CrossRef]
88. Chen, F.; Chandrakasan, A.P.; Stojanovic, V.M. Design and analysis of a hardware-efficient compressed sensing architecture for data compression in wireless sensors. *IEEE J. Solid-State Circuits* **2012**, *47*, 744–756. [CrossRef]
89. Ranjandish, R.; Ture, K.; Maloberti, F.; Dehollain, C.; Schmid, A. All Wireless, 16-Channel Epilepsy Control System with Sub-μW/Channel and Closed-Loop Stimulation Using a Switched-Capacitor-Based Active Charge Balancing Method. In Proceedings of the ESSCIRC 2018-IEEE 44th European Solid State Circuits Conference (ESSCIRC), Dresden, Germany, 3–6 September 2018.
90. Sunderam, S. How to Detect and Quantify Epileptic Seizures. In *EPILEPSY the Intersection of Neurosciences, Biology, Mathematics, Engineering, and Physics*; CRC Press: Boca Raton, FL, USA, **2011**; Chapter 10.
91. Akter, M.S.; Islam, M.R.; Iimura, Y.; Sugano, H.; Fukumori, K.; Wang, D.; Tanaka, T.; Cichocki, A. Multiband entropy-based feature-extraction method for automatic identification of epileptic focus based on high-frequency components in interictal iEEG. *Sci. Rep.* **2020**, *10*. [CrossRef]
92. Si, Y. FMachine learning applications for electroencephalograph signals in epilepsy: A quick review. *Acta Epileptol.* **2020**, *2*, 1–7. [CrossRef]
93. Siddiqui, M.; Morales-Menendez, R.; Huang, X.; Hussain, N. Review of epileptic seizure detection using machine learning classifiers. *Brain Inform.* **2020**, *7*, 1–18. [CrossRef] [PubMed]

94. Kim, T.; Nguyen, P.; Pham, N.; Bui, N.; Truong, H.; Ha, S.; Vu, T. Epileptic Seizure Detection and Experimental Treatment: A Review. *Front. Neurol.* **2020**, *11*, 701. [CrossRef] [PubMed]
95. Rasheed, K.; Qayyum, A.; Qadir, J.; Sivathamboo, S.; Kwan, P.; Kuhlmann, L.; O'Brien, T.; Razi, A. Machine Learning for Predicting Epileptic Seizures Using EEG Signals: A Review. *arXiv* **2020**, arXiv:2002.01925.
96. Raghunathan, S.; Gupta, S.K.; Markandeya, H.S.; Roy, K.; Irazoqui, P.P. A hardware-algorithm co-design approach to optimize seizure detection algorithms for implantable applications. *J. Neurosci. Methods* **2010**, *193*, 106–117. [CrossRef]
97. Borton, D.A.; Yin, M.; Aceros, J.; Nurmikko, A. An implantable wireless neural interface for recording cortical circuit dynamics in moving primates. *J. Neural Eng.* **2013**, *10*, 026010. [CrossRef]
98. *IEEE Standard for Safety Levels with Respect to Human Exposure to Radio Frequency Electromagnetic Fields, 3 kHz to 300 GHz*. IEEE Std C95.1-2005 (Revision of IEEE Std C95.1-1991); IEEE: Piscataway, NJ, USA, 2006; pp. 1–238. [CrossRef]
99. Majidzadeh, V.; Schmid, A.; Leblebici, Y. Energy Efficient Low-Noise Neural Recording Amplifier With Enhanced Noise Efficiency Factor. *IEEE Trans. Biomed. Circuits Syst.* **2011**, *5*, 262–271. [CrossRef]
100. Tran, N.; Bai, S.; Yang, J.; Chun, H.; Kavehei, O.; Yang, Y.; Muktamath, V.; Ng, D.; Meffin, H.; Halpern, M.; et al. A complete 256-electrode retinal prosthesis chip. *IEEE J. Solid-State Circuits* **2014**, *49*, 751–765. [CrossRef]
101. Pittsburgh Plastics Manufacturing Inc. 8 Things to Know About Why Biocompatible Materials are Important in a Medical Device. Available online: http://www.pittsburghplastics.com/ (accessed on 11 June 2020).

sensors

Letter

An Ultra-Wide Load Range Voltage Converter Using Proactive Phase Frequency Modulation for IoT Sensors

Saad Arslan [1,2], Syed Asmat Ali Shah [3] and HyungWon Kim [1,*]

1 Department of Electronics Engineering, Chungbuk National University, Chungdae-ro 1, Seowon-gu, Cheongju 28644, Korea; saad@chungbuk.ac.kr
2 Department of Electrical and Computer Engineering, COMSATS University Islamabad, Park Road, Tarlai Kalan, Islamabad 45550, Pakistan
3 Department of Electrical and Computer Engineering, COMSATS University Islamabad, Abbottabad Campus, University Road, Tobe Camp, Abbottabad 22060, Pakistan; asmat_ali@cuiatd.edu.pk
* Correspondence: hwkim@cbnu.ac.kr

Received: 6 October 2020; Accepted: 2 November 2020; Published: 4 November 2020

Abstract: Modern sensor nodes have multiple operating states, which causes a conventional voltage converter to perform poorly over a wide load range of the operating states. This paper proposes a voltage converter whose switching frequency and output voltage are proactively adjusted to maintain high conversion efficiency. This allows the converter to exploit a wider frequency range to cover a wide load range. In addition, the proposed converter uses multiple smaller capacitor banks and employs multiphase operation to provide low output ripple voltage. A distributed topology for non-overlapping signal generation is proposed and used in the converter to minimize the number of wires running from connecting the controller to the converter. The proposed voltage converter has been implemented in a chip using a 0.13 um CMOS process. The measurement results demonstrate the ability to support a wide load range of 10 μA to 10 mA, for switching frequencies ranging from 100 kHz to 200 MHz, while providing an efficiency of above 80%.

Keywords: switched capacitor; voltage converter; wide load range; multiphase operation; variable frequency

---

1. Introduction

Modern sensor nodes perform a set of tasks repeatedly, and thus often exhibit periodic transition between different load states based on a schedule [1–3], as shown in Figure 1. The predictable transition schedule allows the software to proactively reconfigure the voltage converter to supply required amount of current for various load conditions. Under varying load conditions, however, a large Switched Capacitor (SC) converter often provides poor efficiency at smaller loads. It is known that, in recent mobile platforms, around 32% of the overall power is wasted, in step-down conversions [4]. This is because most voltage converters are optimized targeting the average load condition, and thus perform poorly under peak and minimum load conditions.

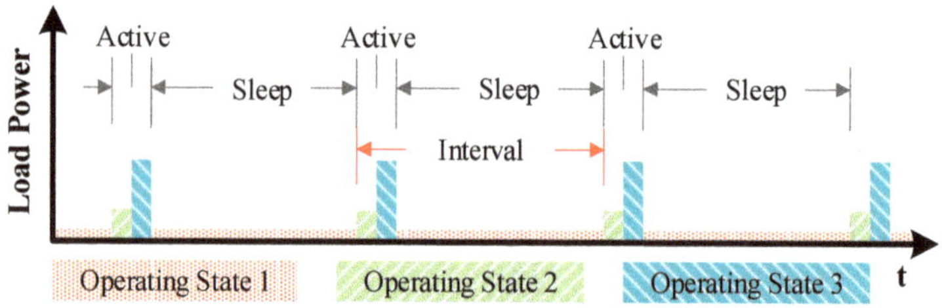

Figure 1. Load behavior with multiple operating states, repeated with a schedule/interval.

Another challenge faced by modern SC converters is the large output ripple voltage, which is often alleviated by having a large output capacitor and/or high switching frequency. These approaches to reducing ripple result in large area overhead and high switching losses, respectively. In [5], a dithered capacitance modulation scheme is presented to reduce ripple at the expense of much higher oscillator frequency than the converters' switching frequency and circuit complexity. A simpler and effective approach to reduce ripple is splitting a converter into smaller units and operating the units at a fixed phase difference [5–7]. A multi-phase converter, due to its low ripple, can offer high conversion efficiency as well. For example, a multiphase operation equally distributes the current surges from the input capacitor, over time. This reduces the ripple voltage $\Delta V = V_{Max} - V_{Min}$ on the input capacitor. It is known that charging a capacitor with small voltage difference results in high energy efficiency in charge transfer process between the supply and capacitor voltage [8].

Under light load conditions, the smaller units can be individually turned-off to reduce losses [9]. However, turning off some of the capacitor banks operating at multiphases causes the output ripple voltage to increase unevenly. Therefore, to fully benefit from a converter using multiphase, all capacitor banks must be operational at all times. A popular technique to reduce switching losses is utilizing Pulse Frequency Modulation (PFM), which reduces the switching of the converter depending on the load [6,10,11]. Most of existing PFM methods are reactive (act in response to an event), which change the switching frequency based on the output voltage feedback. This results in a slower response and failure to utilize a wide frequency range, which degrades conversion efficiency at light loads. Furthermore, PFM methods (especially when combined with multi-phase) generate low frequencies from a high-speed oscillator with a fixed phase difference. Therefore, in PFM methods, the control circuit's power consumption dominates for small load current leading to poor conversion efficiency [10].

Multi-phase voltage converters (comprising of multiple units) need a large number of switches and associated control signals. Therefore, it is infeasible to apply the conventional techniques for non-overlapping signal generation, since they adopt a centralized approach [9,11] and need a large number of wires. Due to the large number of wires, the conventional techniques incur disadvantages such as large area, high power consumption, switching uncertainty and susceptibility to noise. The researchers of [5] suggest reducing the number of phases to reduce the power consumption induced by non-overlapping signals and other switching circuits. A distributed non-overlapping scheme is presented in [6], where an additional NMOS or PMOS is added in the MOSFET driver circuit to delay the turning-on. However, adding an additional NMOS or PMOS in the last stage of the driver would add significant area overhead and slowly turn on the MOSFET (in additional to the delay).

This paper proposes a multi-phase SC converter targeting sensor nodes that operate in predictable schedules of alternating active and sleep periods exhibiting a wide range of load as shown in Figure 1. The converter maintains high conversion efficiency by proactively adjusting its switching frequency, using an integrated oscillator, based on the schedules of load current demanded by the target load. This approach allows the converter to operate with a wider frequency range than existing PFM designs. The proposed distributed non-overlapping signal generation is realized in the MOSFET drivers by delaying the turn-on transition with minimal area overhead and reduced power consumption.

Section 2 presents the proposed converter in more detail. Section 3 discusses the implementation of a test chip to evaluate the proposed converter. Section 4 describes the test setup and demonstrates the converter's performance measurements. Section 5 discuss the measurement results and compares with other works. Finally, concluding remarks are presented in Section 6.

2. Proposed SC Voltage Converter

2.1. Architecture

This paper proposes a voltage converter that offers high conversion efficiency over a wide range of loads and reduced ripple voltage. To deliver a wide load range, the operating frequency of the converter is adjusted by power management software based on the task scheduling prediction. To provide reduced ripple voltage, the SC voltage converter is split into multiple units (capacitor banks) that are operated with a predefined phase difference. Figure 2a illustrates the overall structure of the proposed voltage converter which consists of an oscillator, main controller, delay elements and multiple capacitor banks. Here, each capacitor bank includes a set of Non-Overlapping Signal Drivers (NOSDs), which can minimize the switching losses at very low cost of circuit overhead.

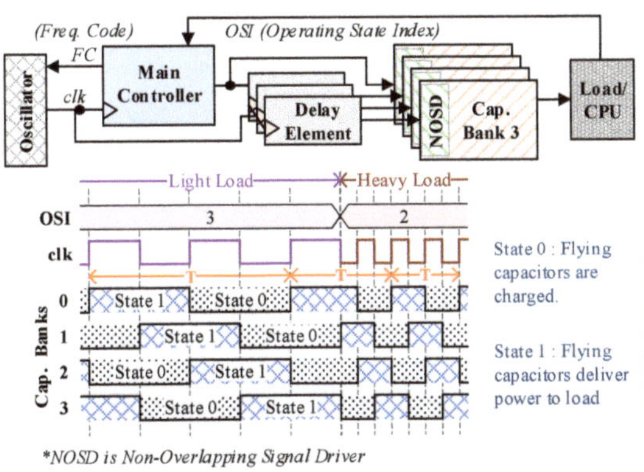

(a)

OSI	VCR	FC (*Freq.*)
1	1 (3/4)	4'b1011 (*100M*)
2	1 (3/4)	4'b0111 (*6.4M*)
3	2 (2/3)	4'b0101 (*1.6M*)
• • •	• • •	• • •
N-1	3 (1/2)	4'b0001 (*0.1M*)
N	4 (1/2)	4'b0001 (*0.1M*)

(b)

VCR	State 0	State 1
1 (3/4)	0x2A94	0x0429
2 (2/3)	0x1694	0x1449
3 (1/2)	0x1694	0x154A
4 (1/3)	0x1491	0x154A
5 (1/4)	0x0831	0x154A
6 (1/1)	0x3318	0x3318

(c)

Figure 2. Proposed Switched Capacitor (SC) converter: (**a**) block diagram and operation; (**b**) an example Operating State Index (OSI) table in the controller; (**c**) Voltage Conversion Ratio (VCR) table specifying control signals for each state.

The SC converter in Figure 2a has four identical capacitor banks, which operate at a phase difference of T/4. Each capacitor bank toggles between two states: state0 (charging) and state1 (delivery). Thus, one period of converter operation comprises of two clock cycles; see Figure 2a. All the capacitor banks share one controller, while each delay element defers the control signal for individual capacitor bank by T/4.

2.2. Operation

Before the SC converter starts a normal operation, its main controller is pre-programmed with the sequence of the operating states of the load. Figure 2b shows an example of such a program for operation states. The program stores a sequence of Operating State Indexes (OSI), which contains the target Voltage Conversion Ratio (VCR) and Frequency Code (FC). The VCR indicates the ratio of the target output voltage to the input voltage. The FC in each entry in the OSI table is programmed to match the target load current for the corresponding operating state. To configure the proper voltage ratio, the main controller stores a VCR table, which is shown in Figure 2c. Each entry of the VCR table provides control signals for the two states of the capacitor banks, e.g., state0 (charging) and state1 (delivery).

When the power management software in the load CPU of sensor nodes is about to change its operating state, it sends its operating state index (OSI) to the controller of the converter. The main controller looks up the OSI table and retrieves VCR and the FC for the corresponding OSI. The VCR is passed to the VCR table to retrieve the control signals for each state of the capacitor bank. The FC is provided to the oscillator to generate the frequency that allows the converter to supply the required current. The main controller maintains its current frequency and VCR, until a new OSI is received from the power management software.

2.3. Oscillator

An adjustable oscillator can easily change the operating frequency of the converter when the load condition changes. In the proposed converter, the integrated oscillator provides sixteen frequencies ranging from 0.1 to 200 MHz, which are selected by the main controller using 4-bit FC. For our voltage converter, the accuracy of the frequency is not critical to maintain the target output voltage, which is proven by the measurement results. Therefore, a simple low-power oscillator suffices our needs.

Figure 3 illustrates the proposed oscillator, which is a ring oscillator consisting of nine current-starved inverters. The current-starved inverters are controlled by a biasing circuit using 4-bit digital control to tune the delay of each inverter. The digital current control has multiple current multiplication branches, one of which is selected based on the FC to provide appropriate current to the biasing circuit. A low-power Schmitt trigger is used at the output to reduce the power consumption of the subsequent buffers.

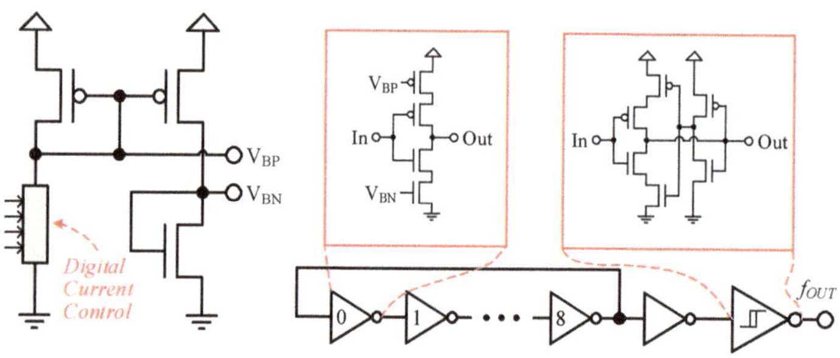

Figure 3. Ring oscillator with current starved inverters and biasing circuit.

2.4. Non-Overlapping Signal Driver (NOSD)

Conventional multiphase converters employ a shared delay controller to generate non-overlapping signals. Such architectures, however, require excessive number of long routing wires as shown in Figure 4a. In contrast, this paper proposes a distributed architecture that generates non-overlapping signals using edge delay circuits at each MOSFET switch, which is illustrated in Figure 4b. The proposed architecture can significantly reduce the number of long routing wires and their associated buffers, consequently reducing the power consumption. Each switch (NMOS and PMOS), in the capacitor banks of the proposed converter, includes a driver that generates non-overlapping control signals by delaying the turn-on transition of the switch, as shown in Figure 4c. The NMOS and PMOS drivers in the NOSDs, respectively, delay the rising and falling edges of the resulting non-overlapping signals. These drivers ensure that no short-circuit path occurs during switching, thus reducing the switching losses. From Figure 4c, it can be observed that the turn-on delay of a switch is approximately equal to the propagation delay of first buffer (t_{p1}) in the drivers. This distributed topology reduces the number of wires connecting the controller with the switches.

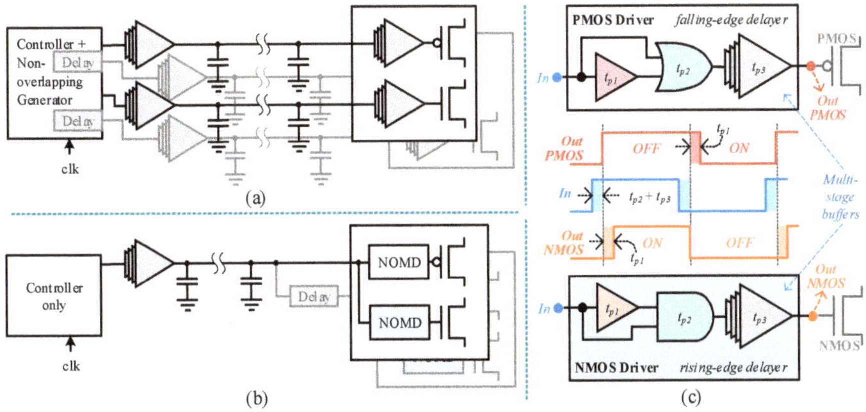

Figure 4. (**a**) Conventional non-overlapping architecture; (**b**) proposed non-overlapping architecture with reduced buffers and wires; (**c**) proposed non-overlapping MOSFET drivers with edge-delaying.

Consider the propagation delays of the gates in drivers as: t_{p1} for the buffer, t_{p2} for OR/AND gates, and t_{p3} for the final multi-stage buffer. The rising- and falling-edge delays of PMOS driver can be represented as:

$$t_{pp-r} = t_{p2} + t_{p3}, \tag{1}$$

$$t_{pp-f} = t_{p1} + t_{p2} + t_{p3}, \tag{2}$$

Similarly, the propagation delays for rising and falling edges through NMOS driver can be represented as:

$$t_{pn-r} = t_{p1} + t_{p2} + t_{p3}, \tag{3}$$

$$t_{pn-f} = t_{p2} + t_{p3}, \tag{4}$$

By using the above equations, we can calculate the non-overlapping periods. Upon rising-edge, NMOS turns-on and PMOS turns-off, the non-overlapping period can be formulated as:

$$t_{nol-r} = t_{pn-r} - t_{pp-r}, \tag{5}$$

$$t_{nol-r} = t_{p1} + t_{p2} + t_{p3} - \left(t_{p2} + t_{p3}\right), \tag{6}$$

$$t_{nol-r} = t_{p1}, \tag{7}$$

Upon falling-edge, PMOS turns-on and NMOS turns-off, the non-overlapping period can be formulated as:

$$t_{nol-f} = t_{pp-f} - t_{pn-f}, \tag{8}$$

$$t_{nol-f} = t_{p1} + t_{p2} + t_{p3} - \left(t_{p2} + t_{p3}\right), \tag{9}$$

$$t_{nol-f} = t_{p1}, \tag{10}$$

It is important that OR- and AND-gate have equal propagation delays to ensure symmetric non-overlapping periods for both the edges.

3. Implementation of Test Chip

To verify the proposed architecture, we have implemented a test chip consisting of four capacitor banks, a reconfigurable test controller, and clock oscillator using 130 nm 1.5 V CMOS process, as shown in Figure 5a. The individual outputs of the four capacitor banks have been brought out of chip for measurement purposes, and they are combined on the PCB. In addition, to ensure accurate output voltage observations, four buffered outputs have been added. For test chip implementation, the oscillator does not include a digital current control circuit. Therefore, an external current Digital-to-Analog Converter (DAC) is used to drive the biasing circuit of the oscillator, as shown in Figure 5b. In addition, some components of the main controller (of Figure 2), such as OSI and VCR tables, have been moved to the off-chip CPU module to have flexibility in testing.

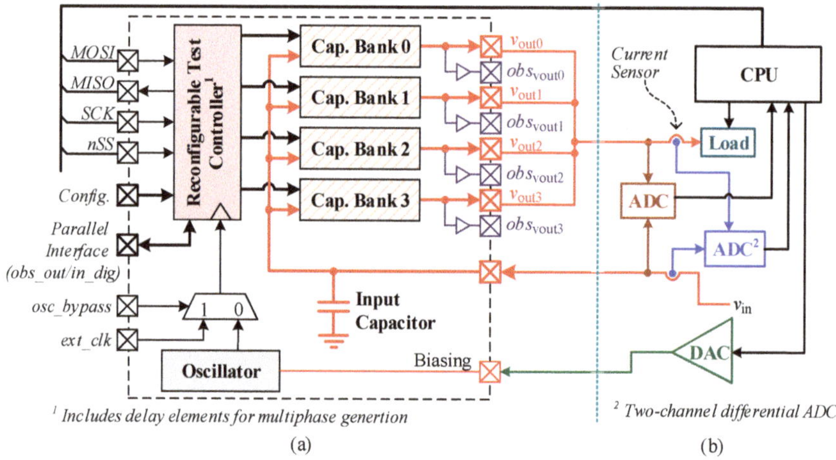

Figure 5. (**a**) Test chip implementation block diagram; (**b**) off-chip components for testing.

3.1. *Reconfigurable Test Controller Implementation*

A test controller is designed with a host interface that emulates the interface to power management software. It operates the four capacitor banks at different phases (or in phase) and can be programmed through its host interface, as shown by the simplified block diagram in Figure 6. In addition to its main functions, we implemented various test and fallback modes in the controller for test purposes. The next entry of the OSI table provides the state information for each capacitor bank, which is stored in VCR state registers (state_reg[0] and state_reg[1]). The VCR state registers' values are used to generate control signals for each attached capacitor bank.

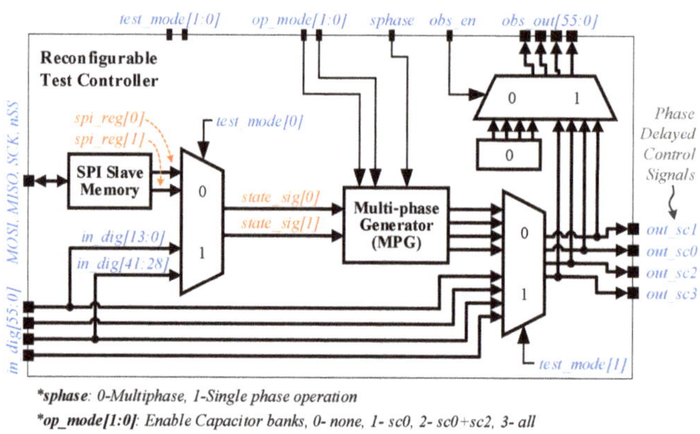

Figure 6. Reconfigurable test controller with interfaces, for a multiphase SC converter.

The multi-phase generator (MPG) block reads the state signals and operates the capacitor banks by supplying phase-delayed signals. The four operation modes of MPG are listed in Figure 6, which allows one, two or all four converters to operate concurrently. It is also possible to operate two or four capacitor banks in phase, by asserting *sphase* input, to compare the performance of single-phase and multi-phase operations.

The test controller has been synthesized using Synopsys design tools targeting 250 MHz. The final implemented layout of the controller has an active area of 590 µm × 15 µm. The aspect ratio of the test controller was chosen to minimize the routing overhead when connecting the controller to a stack of four capacitor banks.

3.2. Capacitor Banks

For the test chip, we implemented four identical series-parallel switched capacitor banks. A circuit block diagram of one bank is shown in Figure 7. Each bank has three flying capacitors which provide five buck voltage conversion ratios (1/4, 1/3, 1/2, 2/3, 3/4). For our target load current range, we chose 80 pF for each flying capacitor and 160 pF for the load decoupling capacitor in each capacitor bank.

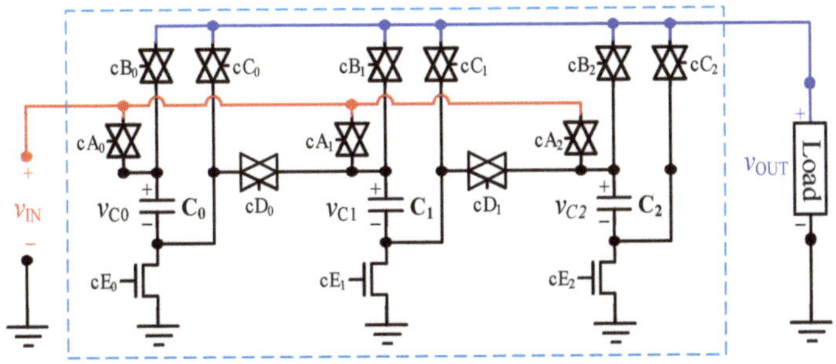

Figure 7. Capacitor bank circuit diagram with three switched capacitor branches.

Figure 8 illustrates an example of non-overlapped switching operation. Figure 8a shows example MOSFET switches corresponding to transmission gates cA0 and cB0 of Figure 7. The NMOS and PMOS drivers are part of the NOSD of a capacitor bank shown in Figure 4c. Figure 8b shows the post-layout simulation result of the circuit in Figure 8a. It reveals that the NMOS and PMOS gates have a non-overlapped period when transitioning between ϕ_1 and ϕ_2, which ensures that no short-circuit path occurs while switching. For the test chip, the non-overlapped period introduced by the NOSD varies from 130 to 200 ps, depending on the potential on the source and drain terminals of the MOSFET being driven.

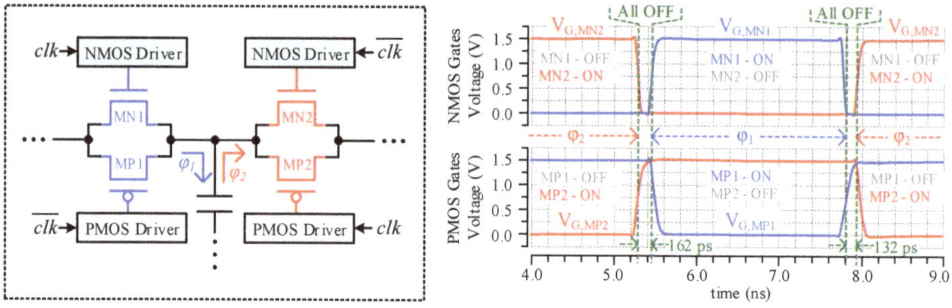

(a) (b)

Figure 8. (**a**) Transmission gate switches controlled by Non-Overlapping Signal Drivers (NOSDs). For example, the blue transmission gate indicates cA_0 and the red transmission gate indicates cB_0 of Figure 7. (**b**) Post-layout simulation of the NOSDs, demonstrating non-overlapping control to mitigate short-circuit losses.

3.3. Overall Test Chip Implementation

Complete layout (including pads) of the overall test chip is shown in Figure 9a, where the converter occupies an area of 0.59 mm^2. The voltage supplies are located on the top left, whereas analog/power outputs (v_{out} and obs_{vout}) are located on the bottom. Decoupling capacitors are added to the voltage supplies of MOSFET drivers, oscillator and reconfigurable test controller, as shown in Figure 9a. The test chip shown in Figure 9a occupies total area of 1.76 mm^2 (1.56 mm × 1.13 mm), including test circuits and pads. For this implementation, only metal–insulator–metal (MIM) capacitors are used for capacitor banks to minimize switching losses [12] caused by the bottom plate capacitance ($C_{bot.,mim}$ = 0.0025 × C). To reduce the overall area, a metal–oxide–semiconductor (MOS)+MIM stack can be safely used as load capacitors, without sacrificing conversion efficiency. Using an MOS+MIM stack as a load capacitor, the area can be reduced by 21% (from 0.59 mm^2 to 0.461 mm^2), compared to an MIM-only load capacitor. Moreover, our analysis also shows that replacing all capacitors by an MOS+MIM stack yields a 54% reduction in area (from 0.59 mm^2 to 0.266 mm^2) at the expense of conversion efficiency.

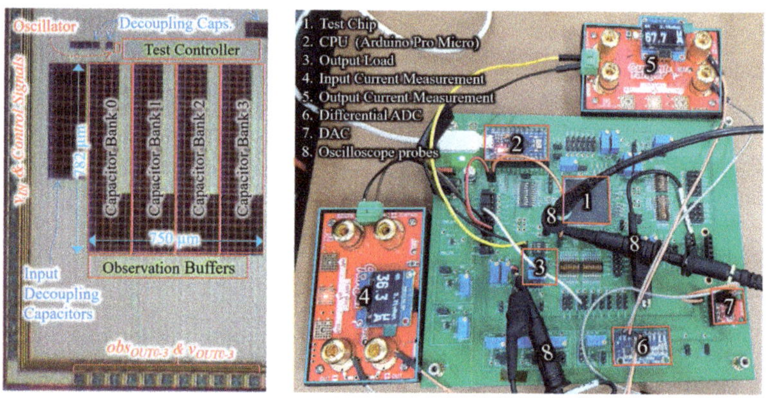

(a) (b)

Figure 9. (**a**) Die microphotograph of the implemented proposed controller with test controller and oscillator; (**b**) measurement setup showing test PCB and off-PCB current measurement devices.

4. Results

The measurement setup to verify the implemented test-chip is shown in Figure 9b. For accurate current measurements at the input and output of the converter, two current sensors [13] and high-resolution Analog-to-Digital Converters (ADCs) are employed. To adjust the frequency of the on-chip oscillator, a 12-bit DAC is used. For the sake of testing, we have split the load and the power management CPU of Figure 2 into separate external components. Two series variable resistors are acting as a load, one of which can be bypassed to toggle between heavy and light load conditions. A CPU module is used to run a power management program and governs the on-chip controller, the external DAC, and the load. The CPU module also reads the current measurements using the external ADC and the voltage measurement using the internal ADC.

The first measurement result is the difference between single-phase and multi-phase operations, which is shown in Figure 10. It can be observed that the ripple voltage of the multi-phase case is significantly reduced compared with the single-phase. Moreover, the average output voltage of the multi-phase case has increased, resulting in improved efficiency.

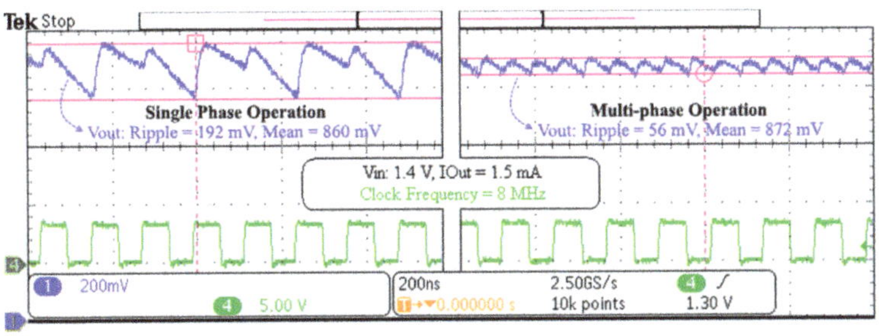

Figure 10. Single-phase vs. multi-phase operation of the implemented converter.

To evaluate the performance of the implemented converter over a wide load range, a set of measurement are performed by reconfiguring the load and the oscillator frequency. The load is configured by adjusting the off-chip resistor, while the converter takes 1.4 V as input. The oscillator frequency is adjusted by the power management CPU via the DAC. The results provided in Figure 11 demonstrate that the converter is able to maintain a high efficiency of above 80% for the targeted load range from 10 μA to 10 mA, by adjusting the switching frequency using the on-chip oscillator. Furthermore, the converter provides a conversion efficiency of 74% at 15 mA, while operating at 150 MHz. In the test chip, where the source and load are off-chip, the conversion efficiency drops above 7.5 mA due to a large voltage drop across the bonding and metal wires. However, for true on-chip implementation, this efficiency will be maintained above 7.5 mA as well. At high frequencies, the power losses (switching) increase while the power delivered slowly saturates due to conduction losses. This results in the degradation of the conversion efficiency at higher frequencies.

To verify the transition operation between heavy (active) and light (sleep) load, the CPU module momentarily changes the load and the frequency. Figure 12 shows the transition operation when the load switches between 68 μA and 10 mA by switching the frequencies between 600 kHz and 100 MHz, respectively. Usually in an active state, a higher voltage is needed to support fast computation in the load (CPUs or SoCs) than the sleep state [14,15]. Therefore, in the test for Figure 12, we configured the voltage conversion ratio to 1/2 for sleep and 3/4 for active period.

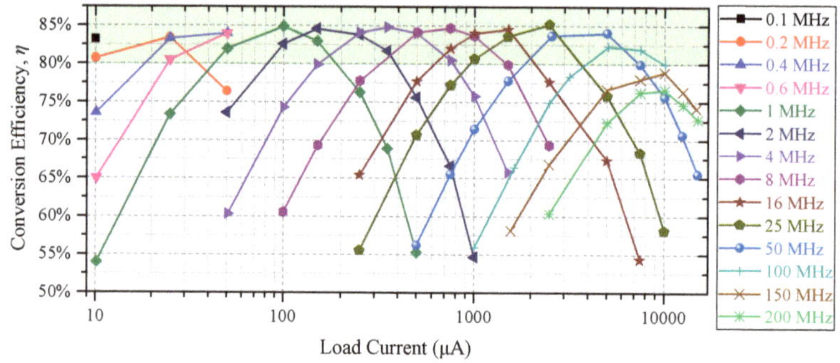

Figure 11. Measurement results demonstrating wide load range performance over various switching frequencies for a voltage conversion ratio of 3/4.

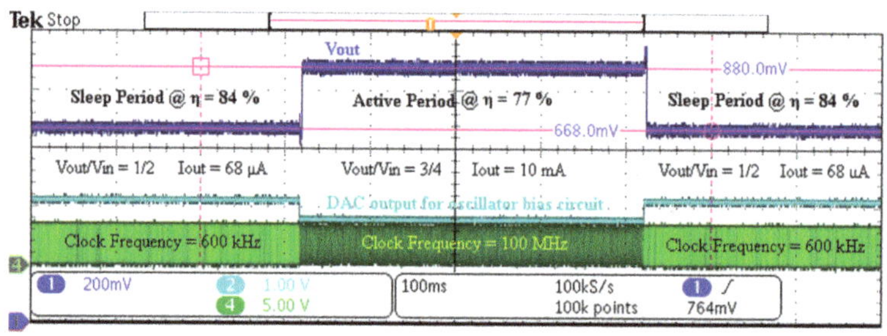

Figure 12. Measurement result for transitions between sleep and the active period of the load.

Figure 13 reflects the power consumption of the main blocks of the test chip, over the supported frequency range. It can be observed that the power consumption of the oscillator and the controller decreases with the decrease in frequency, allowing for higher conversion efficiencies at low currents. The power consumption of the SC converter does not include the power supplied to the test controller. Since the test controller is designed to provide many additional test features, its power consumption does not truly reflect an optimized main controller. Moreover, the on-chip analog buffers (used for output voltage observation) consume around 20 mW.

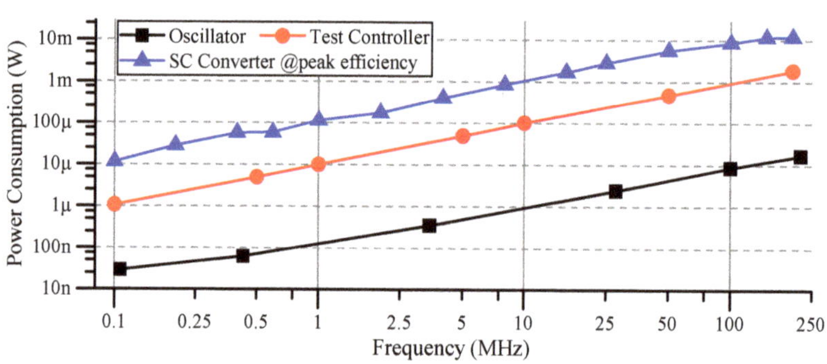

Figure 13. Test-chip power consumption at various operating frequencies.

5. Discussion

The proposed SC converter's advantage—the ability to transition smoothly between light and heavy loads, while maintaining high conversion efficiency—makes the converter well suited to loads that frequently change operating states. We have proposed a figure of merit (FOM) to evaluate a converter for wide load range operation, using:

$$\mathrm{FOM} = (\eta_{Max} + \eta_{I,Max} + \eta_{I,Min}) \times \log\left(\frac{I_{Max}}{I_{Min}}\right), \tag{11}$$

Here, η_{Max} is the peak conversion efficiency, while $\eta_{I,Max}$ and $\eta_{I,Min}$ are the conversion efficiencies at maximum and minimum load, respectively. Table 1 compares the design parameters and performance of our work with previously reported solutions. Table 1 shows that the proposed converter maintains higher efficiency compared to other works [6,10,16]. This is accomplished by the ability of the proposed converter to operate at a much wider switching frequency. Due to the limitation of existing feedback-based PFM, only limited f_{max}/f_{min} of 132 is used in [6], while the proposed converter allows for an extremely large f_{max}/f_{min} up to 1000. The capacitor banks used in this implementation offer five buck voltage conversion ratios, higher than the works in [5,6,9–11,16]. To further reduce the output ripple while having the wide load range, our architecture allows the number of phases to be increased to match [5,6]. The proposed FOM reveals the superior performance of 7.44, compared to existing works. Though the converter proposed in [9] reaches an FOM of 6.74, it uses off-chip flying capacitors and two discrete converters for sleep and active operation.

Table 1. Comparison of performance with state of the art.

Characteristics	[6]	[9]	[10]	[16]	This Work
Tech (nm)	65	800	45	130	130
No. of VCRs	3	1	1	1	5
Vin (V)	1.6–2.2	3.6–4.2	1.8	3.3	1.5
Vout (V)	0.6–1.2	1.7–2.1	0.8–1	1.2–1.5	0.4–1.1
$I_{L,Min}$ (mA)	5 [b]	0.005	0.1	1 [b]	0.01
$I_{L,Max}$ (mA)	125 [b]	2	10	53	10
C_{fly} (pF)	4797	440,000 [a]	534	2176	960
C_L (pF)	2000	1,000,000 [a]	700	1000	640
Ripple (mV)	2.2–30	100	<50	8–55	56
η_{Max} (%)	80	94	69	73	85
$\eta_{I,Max}$ (%)	73 [b]	85	48	73 [b]	80
$\eta_{I,Min}$ (%)	71 [b]	80	65	44 [b]	83
FOM	3.13	6.74	3.64	3.28	7.44
Area (mm^2)	0.842	N/A	0.16	2.04	0.59
Freq. (MHz)	0.25–33	0.05	30	40	0.1–200

[a] Off-chip Component. [b] Approximated values from graphs.

While this paper is focused on the proactive PFM, it is not limited only to proactive PFM with prior schedules. We can combine the proactive PFM with existing feedback-based PFM methods [6,11] to take advantage of both methods. Then, the feedback can allow finer adjustments in frequency, whereas the proactive PFM provides a wider frequency range, which is expected to provide further improvement.

6. Conclusions

This paper has proposed and evaluated a wide load range SC converter aiming on sensor nodes, where the load power requirements varies based on a schedule. The proposed voltage converter is proactively reconfigured to ensure high conversion efficiency, for various load current and voltage requirements. The converter utilizes multiphase operation to mitigate output ripple voltage and phase frequency modulation using an integrated oscillator to maintain high conversion efficiency across the load range. In addition, the converter employs the proposed distributed non-overlapping generation to reduce area and power consumption. The measurements demonstrate above 80% efficiency for output current ranging from 10 μA to 10 mA. Due to its wide load range, the converter is a strong candidate for driving sensor nodes, where the power requirements often change by orders of magnitude between sleep and active state. The converter architecture allows for upscaling the number of phases/capacitor banks to further reduce output ripple. Moreover, the converter architecture can be evolved to incorporate existing feedback-based PFM to fine-tune its oscillator frequency against minor load variations.

Author Contributions: Conceptualization, S.A.; methodology, S.A., S.A.A.S., H.K.; software, S.A. and S.A.A.S.; validation, S.A. and S.A.A.S.; formal analysis, S.A.; investigation, S.A. and S.A.A.S.; Writing—original draft preparation, S.A.; Writing—review and editing, S.A.A.S. and H.K.; supervision, H.K.; project administration, H.K.; funding acquisition, H.K. All authors have read and agreed to the published version of the manuscript.

Funding: This work was supported by the Grand Information Technology Research Center support program (IITP-2020-0-01462) supervised by the IITP and funded by the MSIT (Ministry of Science and ICT) of Korean government, and it was also supported by Industry coupled IoT Semiconductor System Convergence Nurturing Center under System Semiconductor Convergence Specialist Nurturing Project funded by the National Research Foundation (NRF) of Korea (2020M3H2A107678611). This work was also supported by Institute of Information & communications Technology Planning & Evaluation (IITP) grant funded by the Korea government (MSIT) (No.2020-0-01304, Development of Self-learnable Mobile Recursive Neural Network Processor Technology).

Conflicts of Interest: The authors declare no conflict of interest.

References

1. Feng, J.; Shi, X. Efficient Node and Sensed Module Management for Multisensory Wireless Sensor Networks. *Sensors* **2018**, *18*, 2328. [CrossRef] [PubMed]
2. Wan, R.; Xiong, N.; Loc, N.T. An energy-efficient sleep scheduling mechanism with similarity measure for wireless sensor networks. *Hum. Cent. Comput. Inf. Sci.* **2018**, *8*, 18. [CrossRef]
3. Chew, Z.J.; Ruan, T.; Zhu, M. Power Management Circuit for Wireless Sensor Nodes Powered by Energy Harvesting: On the Synergy of Harvester and Load. *IEEE Trans. Power Electron.* **2019**, *34*, 8671–8681. [CrossRef]
4. Lee, W.; Wang, Y.; Shin, D.; Chang, N.; Pedram, M. Power conversion efficiency characterization and optimization for smartphones. In Proceedings of the 2012 ACM/IEEE International Symposium on Low Power Electronics and Design—ISLPED '12, Redondo Beach, CA, USA, 30 July–1 August 2012; pp. 103–108.
5. Bang, S.; Seo, J.; Chang, L.; Blaauw, D.; Sylvester, D. A Low Ripple Switched-Capacitor Voltage Regulator Using Flying Capacitance Dithering. *IEEE J. Solid State Circuits* **2016**, *51*, 919–929. [CrossRef]
6. Lu, Y.; Jiang, J.; Ki, W.-H. A Multiphase Switched-Capacitor DC–DC Converter Ring with Fast Transient Response and Small Ripple. *IEEE J. Solid State Circuits* **2017**, *52*, 579–591. [CrossRef]
7. Le, H.-P.; Sanders, S.R.; Alon, E. Design Techniques for Fully Integrated Switched-Capacitor DC-DC Converters. *IEEE J. Solid State Circuits* **2011**, *46*, 2120–2131. [CrossRef]
8. Shah, S.A.A.; Arslan, S.; Kim, H. A Reconfigurable Voltage Converter with Split-Capacitor Charging and Energy Recycling for Ultra-Low-Power Applications. *IEEE Access* **2018**, *6*, 68311–68323. [CrossRef]
9. Grasso, A.D.; Palumbo, G.; Pennisi, S. Switched-Capacitor Power Management Integrated Circuit for Autonomous Internet of Things Node. *IEEE Trans. Circuits Syst. II* **2018**, *65*, 1455–1459. [CrossRef]
10. Ramadass, Y.K.; Fayed, A.A.; Chandrakasan, A.P. A Fully-Integrated Switched-Capacitor Step-Down DC-DC Converter with Digital Capacitance Modulation in 45 nm CMOS. *IEEE J. Solid State Circuits* **2010**, *45*, 2557–2565. [CrossRef]
11. Jeon, H.; Kim, K.; Kim, Y.-B. Fully Integrated on-Chip Switched DC–DC Converter for Battery-Powered Mixed-Signal SoCs. *Symmetry* **2017**, *9*, 18. [CrossRef]
12. Jain, R.; Geuskens, B.M.; Kim, S.T.; Khellah, M.M.; Kulkarni, J.; Tschanz, J.W.; De, V. A 0.45–1 V Fully-Integrated Distributed Switched Capacitor DC-DC Converter with High Density MIM Capacitor in 22 nm Tri-Gate CMOS. *IEEE J. Solid State Circuits* **2014**, *49*, 917–927. [CrossRef]

13. Current Ranger|LowPowerLab. Available online: https://lowpowerlab.com/guide/currentranger/ (accessed on 2 October 2019).
14. Zimmer, B.; Lee, Y.; Puggelli, A.; Kwak, J.; Jevtic, R.; Keller, B.; Bailey, S.; Blagojevic, M.; Chiu, P.-F.; Le, H.-P.; et al. A RISC-V vector processor with tightly-integrated switched-capacitor DC-DC converters in 28nm FDSOI. In *2015 Symposium on VLSI Circuits (VLSI Circuits)*; IEEE: Kyoto, Japan, 2015; pp. C316–C317.
15. Craig, K.; Shakhsheer, Y.; Arrabi, S.; Khanna, S.; Lach, J.; Calhoun, B.H. A 32 b 90 nm Processor Implementing Panoptic DVS Achieving Energy Efficient Operation from Sub-Threshold to High Performance. *IEEE J. Solid State Circuits* **2014**, *49*, 545–552. [CrossRef]
16. Lee, J.-Y.; Kim, G.-S.; Oh, K.-I.; Baek, D. Fully Integrated Low-Ripple Switched-Capacitor DC–DC Converter with Parallel Low-Dropout Regulator. *Electronics* **2019**, *8*, 98. [CrossRef]

sensors

MDPI

Communication

A Digital Improvement—Trimming a Digital Temperature Sensor with EEPROM Reprogrammable Fuses

Anca Mihaela Vasile (Dragan) [1,2,*], Alina Negut [1], Adrian Tache [1] and Gheorghe Brezeanu [2]

[1] ON Semiconductor Romania, 020983 Bucharest, Romania; Alina.Negut@onsemi.com (A.N.); AdrianMacarie.Tache@onsemi.com (A.T.)
[2] Faculty of Electronics, Politehnica University of Bucharest, Telecommunications and Information Technology, 999032 Bucharest, Romania; gheorghe.brezeanu@dce.pub.ro
* Correspondence: AncaMihaela.Vasile@onsemi.com

Abstract: An EEPROM (electrically erasable programmable read-only memory) reprogrammable fuse for trimming a digital temperature sensor is designed in a 0.18-μm CMOS EEPROM. The fuse uses EEPROM memory cells, which allow multiple programming cycles by modifying the stored data on the digital trim codes applied to the thermal sensor. By reprogramming the fuse, the temperature sensor can be adjusted with an increased trim variation in order to achieve higher accuracy. Experimental results for the trimmed digital sensor showed a +1.5/−1.0 °C inaccuracy in the temperature range of −20 to 125 °C for 25 trimmed DTS samples at 1.8 V by one-point calibration. Furthermore, an average mean of 0.40 °C and a standard deviation of 0.70 °C temperature error were obtained in the same temperature range for power supply voltages from 1.7 to 1.9 V. Thus, the digital sensor exhibits similar performances for the entire power supply range of 1.7 to 3.6 V.

Keywords: integrated circuits; EEPROM reprogrammable fuses; memory cells; trimming techniques with fuses; digital temperature sensor; temperature sensor with digital serial interface

Citation: Vasile (Dragan), A.M.; Negut, A.; Tache, A.; Brezeanu, G. A Digital Improvement—Trimming a Digital Temperature Sensor with EEPROM Reprogrammable Fuses. *Sensors* **2021**, *21*, 1700. https://doi.org/10.3390/s21051700

Academic Editor: Francesc Serra-Graells

Received: 18 January 2021
Accepted: 25 February 2021
Published: 2 March 2021

1. Introduction

Digital temperature sensors are suitable for thermal and power management systems of PCs, laptops, and smartphones. A low-cost and high-accuracy temperature sensor with a digital interface is desired nowadays due to the high demand for electronic smart gadgets [1–4]. Furthermore, the use of temperature detectors with digital output has been recently reported in smart sensor networks, image sensors, Internet of Things devices, and biomedical applications [5–9].

Modern CMOS smart temperature sensors are categorized according to the sensing device (BJT, MOS in subthreshold region and resistor) or the physical principle on which the temperature is detected (bandgap voltage and thermal diffusion (TD)) [7,8]. Increased accuracy, high precision, and output linearity with low power consumption are some of the most important targets to achieve when designing such integrated sensors [10–20]. The aim to meet these requirements with low production costs becomes more and more challenging these days. The limitation for achieving higher accuracy for a thermal sensor is based on a trade-off between the production costs for calibration and the required precision [21].

In deep submicron processes, nonideal factors like the temperature coefficient of the devices, components mismatch, or absolute deviation of resistance affect the performance of a smart sensor. Furthermore, process spread and packaging stress play important roles in limiting the accuracy of a thermal sensor [22–24]. Thus, calibration methods and trimming techniques are required to achieve the imposed performances of a digital temperature sensor.

A calibration procedure provides information about the accuracy of a thermal sensor [22–26]. Most circuits are calibrated to two well-defined temperature points, after which the circuit is adjusted to minimize the temperature error by trimming techniques [23]. A one-point calibration has the advantage of lower production costs [25].

Smart sensors are usually calibrated by comparing them with a reference thermometer of known accuracy. The calibration can be done at the wafer level or after packaging. Regarding wafer-level calibration, the temperature of a complete wafer is stabilized and measured using a number of reference thermometers mounted on the wafer chuck [26]. Several electrical tests and temperature readings from the chip are performed, followed by adjustments in order to meet its performance requirements. Calibration after packaging implies achieving the same temperature for every individually packaged IC (integrated circuit) as for the reference thermometer in a thermally conducting medium, such as a liquid bath or a metal block [26].

After calibration at wafer level or after packaging, smart temperature sensors usually require an adjustment for the targeted parameter by applying a digital trim code [27]. Conventional methods consist of a permanent modification of the IC by laser trimming or by altering metal fuse links [28]. Nowadays, a one-time programmable fuse (OTP fuse) is often used for trimming a thermal sensor due to its ability to store the values of the trim codes in a data latch and, for instance, to allow two states for the digital trim code [23,27,29].

For any of these trimming techniques, once the fuse is trimmed, it cannot return to its original state [1]. Thus, a more complex trim involves several programming cycles. Furthermore, trimming an integrated sensor with OTP fuses requires a lot of extra pads, which are not accessible to the user, in order to store the digital trim codes required for calibration [27]. Using EEPROM memory cells (EEcells), the fuse can be reprogrammed, allowing multiple programming cycles for trimming the digital temperature sensor. The endurance of an EEPROM memory cell, without altering its precision in time, covers around 1,000,000 programming cycles, while its data retention exceeds 100 years [2]. With this technique, the thermal sensor can be trimmed in an increased trim variation with multiple digital codes, offering an efficient way to achieve its performance requirements. For instance, the benefit of using the proposed EEPROM technique is that it allows an increased number of digital trim codes for calibrating the circuit, with a low production cost.

In comparison with OTP fuses, an EEPROM fuse offers multiple advantages, such as an increased number of programming cycles and no extra pads required for trimming, which reduce the area consumption of the IC. Furthermore, the reprogrammable fuse allows retrimming when the IC's specifications are changed by the beneficiary. Additionally, testing/trimming time is drastically reduced, resulting in a lower production cost. Moreover, an important advantage of using EEPROM fuses includes the possibility of choosing the number of fuses used to achieve the desired accuracy of the thermal sensor.

In this paper, a digital trimming technique with reprogrammable fuses for a digital temperature sensor (DTS) is proposed. The fuse uses EEPROM memory cells, which allow multiple programming cycles by altering the stored data of the digital trim codes. Thus, the digital sensor can be adjusted with an increased trim variation in order to achieve higher accuracy. The thermal sensor with the digital trim is designed and implemented in a 0.18-μm CMOS EEPROM process.

2. EEPROM Reprogrammable Fuse

An EEPROM reprogrammable fuse is proposed in Figure 1 [1]. The circuit includes a fuse-sensing part formed by two controlled current paths, *LEFT* and *RIGHT*, with four switching transistors, an output S-R latch, and two EEPROM memory cells, *EEcell_L* and *EEcell_R*, and an EEcell Control Logic & Programming Block [1]. The state of the fuse is controlled by the command signal *CMD*, while a programming signal is provided by the EEcell Control Logic & Programming Block.

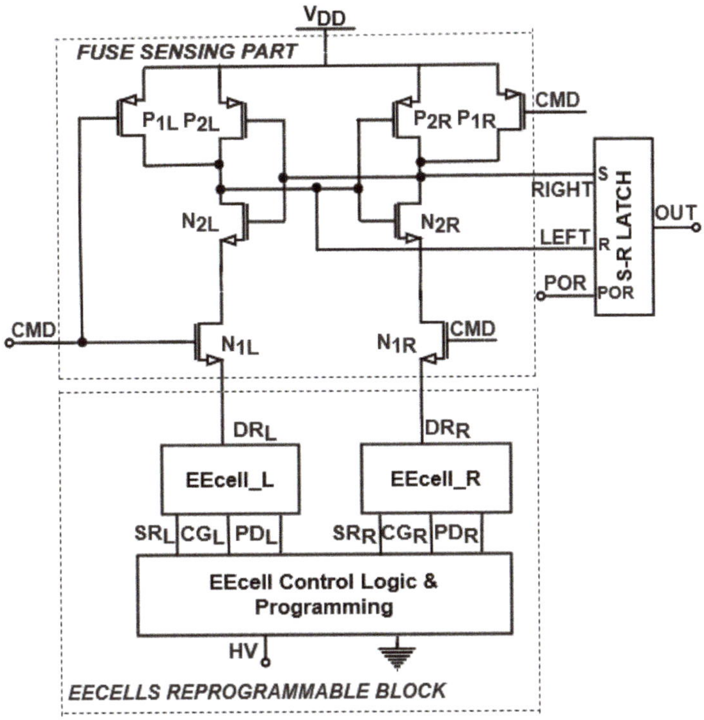

Figure 1. Schematic of the EEPROM reprogrammable fuse [1].

EEcells have electrically isolated gates, storing data in the form of a charge on the floating gate (FG). The charge is transported to the FGs in the programming operation. The EEcells have four terminals: drain read (*DR*), control gate (*CG*), source read (*SR*), and programmable drain (*PD*).

By appropriate programming controlled by the EEcell Control Logic & Programming Block, a high voltage (HV) is applied to the control gate or the programmable drain terminal. When programming the EEcells with "1", the FG potential has a positive value, determining *DR* to provide a path to ground. In the scenario of "0", a negative value is stored on the floating gate, switching the *DR* signal to high, while the path to ground is disconnected. In order to maintain proper operation of the fuse-sensing part, *EEcell_L* and *EEcell_R* are programmed with complementary data.

The novel fuse presented in Figure 1 was designed using a 0.18-µm CMOS EEPROM process with low voltage (2V) transistors, while its operation was tested through Synopsys HSPICE® (HSPICE is a trademark of Synopsys Inc. in the US and/or other countries.) simulations, with the resulting waveforms shown in Figure 2. The circuit is supplied at power supply voltage V_{DD} = 2 V, while for programming the EEcells, a high voltage (HV) is applied to the EEPROM memory cells.

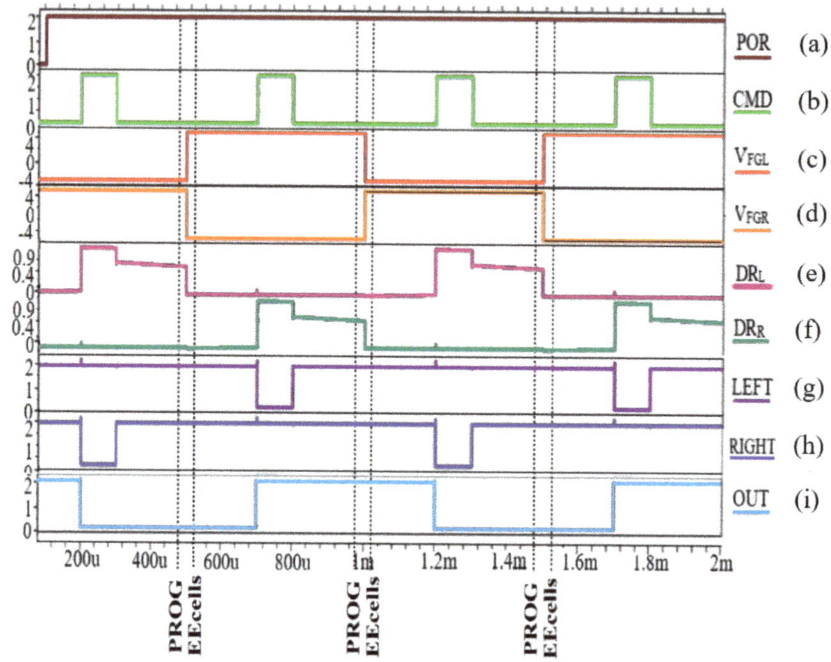

Figure 2. Detailed EEPROM reprogrammable fuse waveforms: (**a**) *POR*, (**b**) *CMD*, (**c**) V_{FGL}, (**d**) V_{FGR}, (**e**) DR_L, (**f**) DR_R, (**g**) *LEFT*, (**h**) *RIGHT*, (**i**) *OUT* [1].

After power-up and initialization of the output S-R latch, *POR* switches from "0" to logic "1" (Figure 2a). When *CMD* is asserted (*CMD* = 1; Figure 2b), the fuse enters evaluation mode. In the first scenario, *EEcell_L* is programmed with "0" and a negative value is stored on the FG (V_{FGL} = −4 V; Figure 2c), while *EEcell_R* is programmed with "1" (V_{FGR} = 4 V; Figure 2d). In this case, the current path to ground is provided by the right branch (DR_R = 0; Figure 2f), pulling *RIGHT* to "0" (Figure 2h), while *LEFT* is tied to V_{DD} (Figure 2g). Thus, the output latch stores "0", while the output of the fuse will be in "0" logic (Figure 2i) [1].

The second EEcell programming is done with complementary data. When the fuse is evaluated again (Figure 2b), a positive value is stored on *FG_L* (V_{FGL} = 4 V; Figure 2c), and the floating gate potential of *EEcell_R* has a negative value (V_{FGR} = −4 V; Figure 2d). At this time, DR_L is pulled to ground (Figure 2e), while DR_R switches to 0.9V (Figure 2f). Thus, *LEFT* is pulled to ground (Figure 2g), while the *RIGHT* signal will be in logic "1" (Figure 2h). For this scenario, the S-R latch will memorize "1" logic (OUT = 1; Figure 2i) [1].

Two similar scenarios for reprogramming the fuse are represented in Figure 2, showing the capability to change the fuse multiple times by reprogramming the EEcells. This provides the advantage of being able to trim a smart sensor with various digital trim codes until the desired parameters are obtained. Thus, the proposed fuse is used for trimming a digital temperature sensor in order for it to exhibit increased accuracy.

In addition to the benefit of being able to reprogram the fuse, the proposed trimming technique achieves lower area consumption compared to a metal fuse implementation. The digital trim of the temperature sensor presented in Figure 3, which includes 16 proposed EEPROM fuses, occupies only 0.030067 mm^2 of the total chip area of 2.07 mm^2. A metal fuse trimming technique for another temperature sensor implementation, which contains 7 trim pads, occupies 0.05077 mm^2 of the 0.3195 mm^2 chip area, representing almost $\frac{1}{4}$ of total area consumption. Thus, the proposed fuse technique has twice-lower area consumption than the metal fuse implementation and allows more digital trim codes that can be reprogrammed.

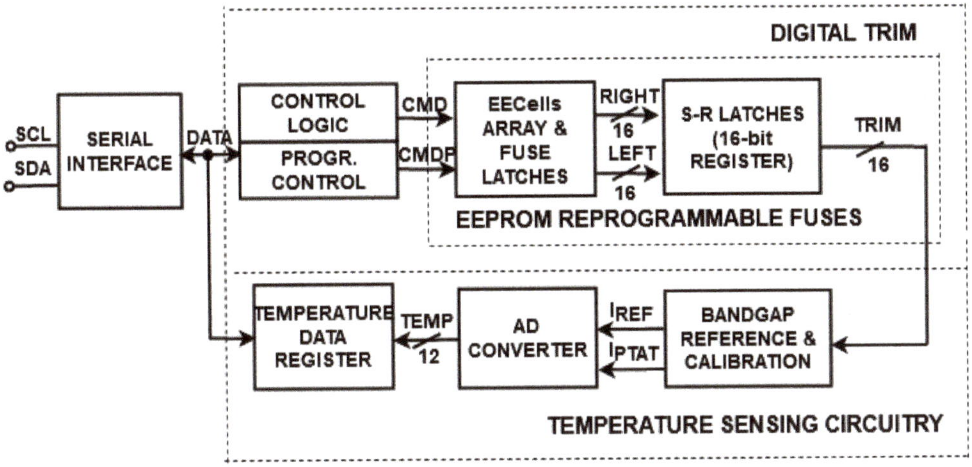

Figure 3. Block schematic of digital temperature sensor (DTS) [1].

3. Trimming a Digital Temperature Sensor with EEPROM Fuses

A digital temperature sensor (DTS) with the proposed EEPROM reprogrammable fuse is shown in Figure 3. The smart sensor includes a serial interface, a digital trim block, and temperature sensing circuitry. The interface communicates with the trimming circuitry in order to provide the digital trim codes required for adjusting temperature sensing. Furthermore, after detecting the temperature in a digital representation, the sensing circuitry communicates with the interface by sending the data stored in a temperature data register [30].

The interface has two serial communication lines: a serial clock line, *SCL*, which is an input pin, and a serial data line, *SDA*, a bidirectional pin. The serial interface sends the data for controlling 16 EEPROM reprogrammable fuses (Figure 1) with a *CMD* signal (for evaluating the fuse) and a *CMDP* signal (for programming the EEcells) [1]. Accordingly, with the programmed EEcells (Figure 2), the trimming circuitry offers digital trim codes from 0 × 0000 (minTRIM) to 0 × FFFF (maxTRIM), allowing the digital sensor to be adjusted in order to achieve increased accuracy [1].

The core of the temperature-sensing circuitry is a bandgap reference with calibration, which provides a PTAT current (I_{PTAT}), and a reference current (I_{REF}). The AD converter (analog to digital converter) compares the analog values and offers the digital representation of the temperature (*TEMP*). The digital value is stored in the temperature data register, which is sent to the serial interface.

A detailed view of the temperature-sensing circuitry of the DTS is illustrated in Figure 4. The bandgap reference and calibration include two identical BJT transistors, Q_1 and Q_2, which are biased with I_1 and pI_1 currents [26]. The positive to absolute temperature values (V_{PTAT}, I_{PTAT}) are obtained by subtracting the base-emitter voltages of the sensing devices (V_{BE1}, V_{BE2}) [26]. The reference voltage (V_{REF}) is expressed as the sum between the resulting positive voltage V_{PTAT} and the base-emitter voltage V_{BE2}. In order to acquire the reference current (I_{REF}), a buffer (*OPAMP*) and the resistor (R_{TRIM}) are used [31]. The bandgap reference and calibration are trimmed by adjusting I_{REF} in order to improve the accuracy of the digital temperature sensor.

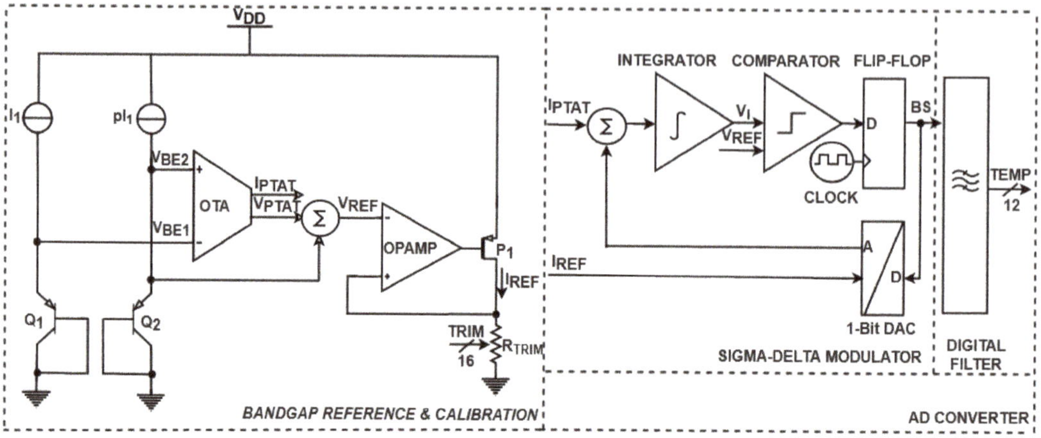

Figure 4. Detailed temperature sensing circuitry of DTS.

The AD converter of the DTS described in Figure 4 comprises a sigma-delta modulator and a digital filter. The sigma-delta modulator processes the analog currents I_{PTAT} and I_{REF} and generates a bit stream signal (BS) [32]. The integrator of the sigma-delta modulator stores the difference between the analog currents I_{PTAT} and I_{REF}, while the rising voltage V_I is compared with V_{REF}. The output of the comparator is then sampled by a clocked flip flop, which synchronizes the received data with the clock. The resulting signal (*BS*) is fed back into the system by a 1-bit DAC converter, which acts as a switch for the loop. When BS is in logic "0", only I_{PTAT} will be processed by the integrator, while in logic "1", the difference between the analog values will be taken into account. The output of the sigma-delta modulator is then processed through the digital filter, which generates a filtered multibit digital signal by oversampling and decimation techniques [33,34]. Thus, the temperature detected by the BJT transistors is represented in a digital format (*TEMP* [0:11]).

The digital temperature sensor depicted in Figure 3 was designed using a 0.18-μm CMOS EEPROM process. The system operates at supply voltages from 1.7 to 3.6 V. The operation of the trimmed DTS was observed through Synopsys HSPICE simulations and wafer-level and encapsulated IC measurements. A DC sweep analysis for the trimmed current I_{REF} in the temperature range of −20 to 125 °C, with minTRIM and maxTRIM digital codes, is depicted in Figure 5a. The reference current can be adjusted between 7.5 and 8.5 μA, depending on the digital trim code.

The temperature error of the DTS applied to I_{REF} is shown in Figure 5b. A negative slope of the error vs. temperature variation results from trimming I_{REF} with maxTRIM, while a positive slope is observed by applying minTRIM. By modifying the slope of the temperature error through the trim applied to I_{REF}, an increased accuracy can be achieved. Thus, an optimal digital trim code is available for the reference current, which provides a minimum temperature error.

The DTS's measurements were performed at the wafer level and on ceramic encapsulated ICs. For testing the untrimmed sensor's performances, wafer-level measurements were carried out at room and hot temperatures (25 and 85 °C). The temperature error for 5 measured untrimmed DST samples is represented in Figure 6. For half of the samples, a variation of −2 °C inaccuracy can be observed for both measured temperatures, while a maximum −9.5 °C variation of temperature error is obtained. Furthermore, the investigated DTS samples show a +13.75/−5 °C inaccuracy for the untrimmed DTS.

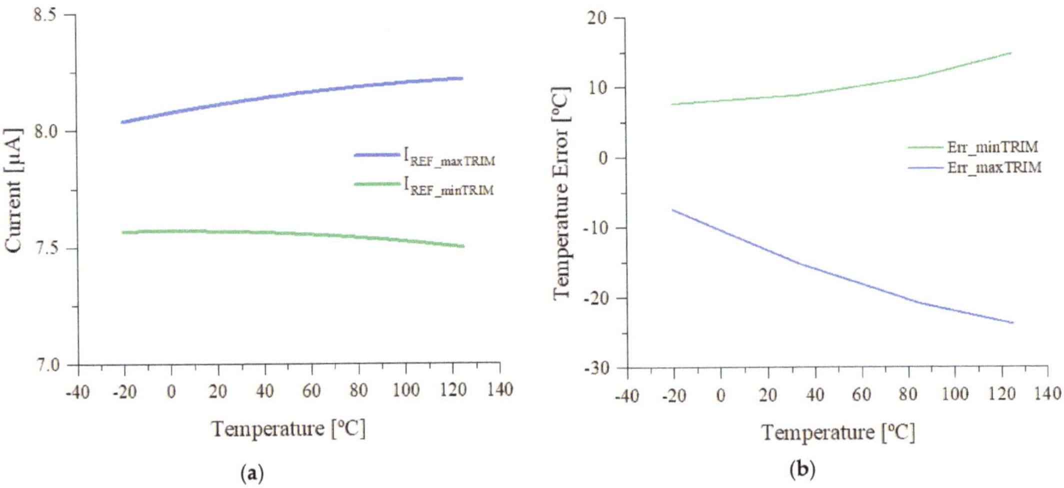

Figure 5. (**a**) Simulated dependence of the I_{REF} with trim variation and temperature for DTS (**b**) simulated temperature error of DTS with trim variation.

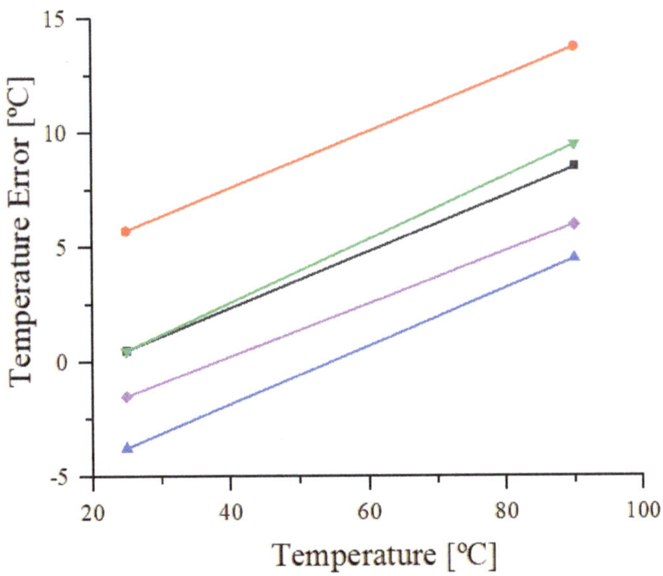

Figure 6. Measured temperature error of 5 untrimmed DTS samples at 1.8 V.

In order to evaluate the trimmed sensor's inaccuracy for the entire temperature range, measurements on encapsulated ICs were investigated. The test setup involves the test system Maverick-II and the micro-bath calibrator Fluke7103 [35], which control the environmental temperature precisely. The one-point calibration method follows the steps described in Figure 7. The circuit is calibrated at 85 °C. If accuracy is achieved, the process is finished and the optimum digital trim code is found. Otherwise, the investigations continue until accuracy is obtained or the trim variation reaches all its digital trim codes.

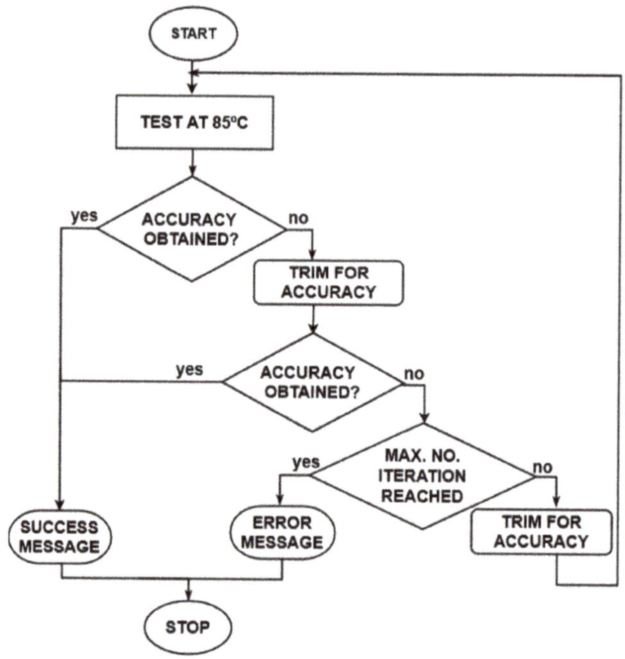

Figure 7. One-point calibration diagram.

I_{REF} is adjusted with the optimal digital trim code during the test process for each fabricated sample. DTS samples (supplied at 1.8 V) with the optimal trim code were measured in the temperature domain of −20 to 125 °C. The temperature error of the DTS for 5 encapsulated ICs is illustrated in Figure 8. A fairly low inaccuracy of +1/−0.75 °C was obtained. Furthermore, for most of the samples, the temperature error varied by ±0.5 °C for the entire temperature range. Thus, a 0.24 °C average mean with a 0.44 °C standard deviation of the measured error was obtained for the full temperature range. The advantage of using this proposed digital trim is the possibility of adjusting each sample with an optimal digital trim code in order to minimize the effect of device spread for a given technology. Additionally, in comparison with the measurements of the untrimmed sensor depicted in Figure 6, the inaccuracy illustrated in Figure 8 is more than 5 times lower for the entire investigated temperature range.

Further investigations of the digital temperature sensor's inaccuracy were performed by measuring 20 DTS samples. The measured error dependence as a function of temperature for three power supply values is presented in Figure 9a–c. The full DTS temperature range was tested. The means and standard deviations of the temperature errors for the measured samples at each supply voltage are shown in Table 1.

Table 1. Mean and standard deviation of the temperature error for 20 measured DTS samples.

Parameter	Power Supply [V]	Temperature Range [°C] −20~125
Mean [°C]	1.7	0.30
	1.8	0.44
	1.9	0.35
Standard Deviation [°C]	1.7	0.71
	1.8	0.71
	1.9	0.69

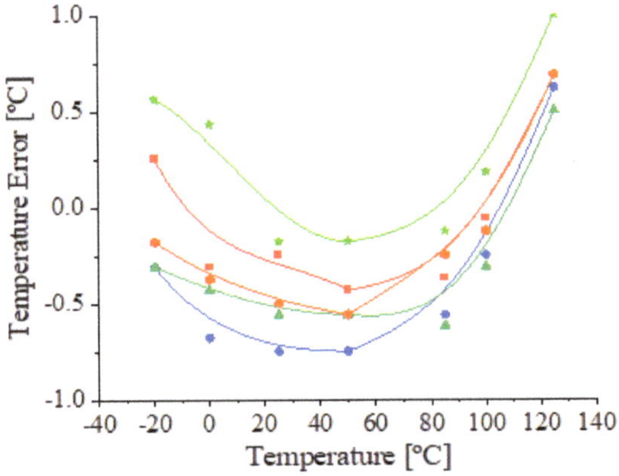

Figure 8. Measured temperature error of 5 trimmed DTS samples at 1.8 V.

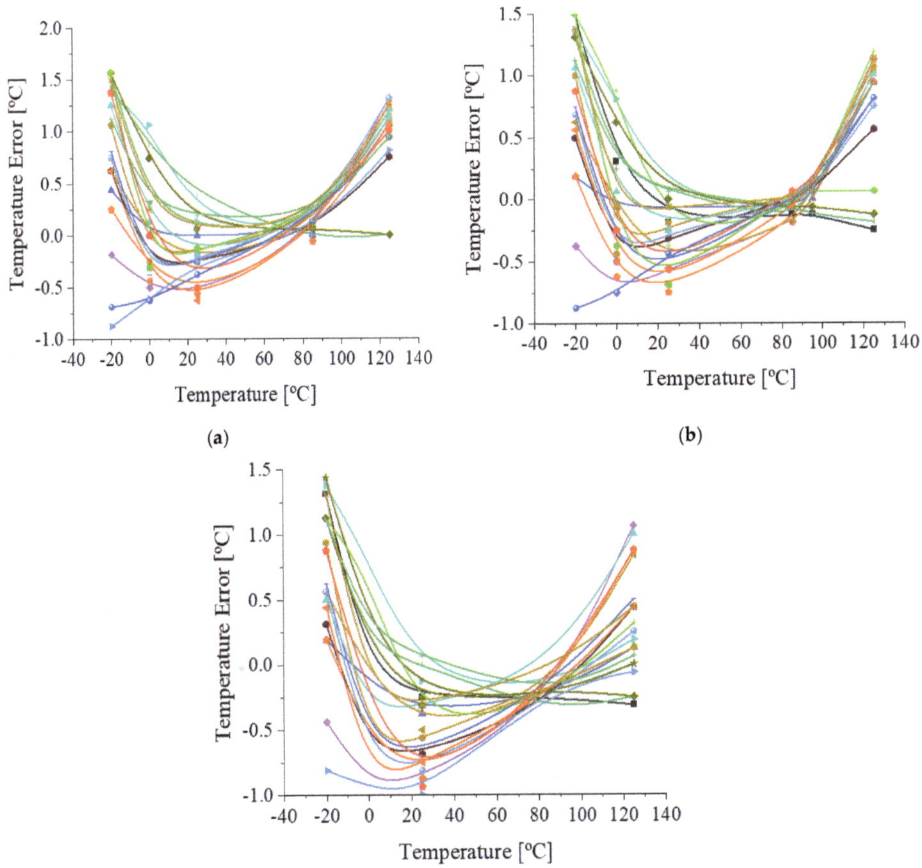

Figure 9. (**a**) Measured temperature error of 20 trimmed DTS samples at 1.7 V; (**b**) measured temperature error of 20 trimmed DTS samples at 1.8 V; (**c**) measured temperature error of 20 trimmed DTS samples at 1.9 V.

The measured error data of samples supplied at 1.8 V have similar dependence at high temperatures (Figures 8 and 9b). At low temperatures, the inaccuracy increases up to +1.5 °C (Figure 9b) for less than half the samples. Thus, a mean value of 0.44 °C, with a standard deviation of 0.71 °C, for the 20 measured temperature errors is observed in Table 1 in the temperature range of −20 to 125 °C at 1.8 V.

At a lower power supply voltage (Figure 9a), the inaccuracy at −20 °C increases to 1.56 °C, while for high temperatures, it varies between +0.75 to +1.3 °C in almost all cases. At 1.9 V (Figure 9c), the error reported by most of the measured samples varied by ±1 °C in the full temperature range. With regards to the mean and standard deviation of the temperature error reported at 1.7 and 1.9 V, they have similar values to the ones obtained at 1.8 V (Table 1).

The performances of the trimmed thermal sensor (Figure 3) are compared with recently reported digital temperature sensors [4–25] in Table 2. The experimental results of 25 measured DTS samples are in good agreement with the majority of referred data [4–6,11–15,18,19,21,24], providing low inaccuracy in a wide temperature range. Furthermore, the proposed trimmed DTS can be supplied with an extended domain of supply voltages. The presented digital sensor achieves its performance by calibration to just one point, while the inaccuracies obtained in [7–9,11,13,18,20–22,24] are performed using the two-point calibration method. With regards to power consumption, the investigated sensor reaches an acceptable value in comparison with [11,12,18,22].

Table 2. Comparison with recently reported digital temperature sensors.

	Type	Process [nm]	Power Supply [V]	Temperature Range [°C]	Inaccuracy [°C]	Calibration	No. Samples	Power [μW]
This Work	BJT	180	1.7~3.6	−20~125	+1.56/−1.0	One point	25	850
[4]	TD	55	0.8~1.3	−40~125 −10~110	±0.70 (3σ)/±0.94 (3σ) ±1.38 (3σ)/±1.64 (3σ)	Two points One point	4 4	9.3 9.8
[5]	CIS pixels	180	3.3	−20~80	±1.30 (3σ)	Two points	3	36
[6]	CIS pixels	180	3.3	0~100	±1.40 (3σ)	Two points	3	144
[7]	MOS	130	1.3	−20~85	±0.60(3σ)	Two points	10	6
[8]	TD	130	1.3	−20~85	±0.60 (3σ)	Two points	10	6
[9]	MOS	180	1.0	0~100	+0.29/−0.98	Two points	10	22.3
[10]	BJT	160	1.8	−40~180	±0.25 (3σ)	One point	24	9.75
[11]	MOS	350	3.3	0~90	+0.7/−1.35	Two points	3	3000
[12]	BJT	180	-	16~87	+0.68/−0.8	One point	3	586
[13]	RES	65	0.6~1.0	−20~120	±1.5 (3σ)/0.80 (3σ)	Two points	16	0.1
[14]	RES	65	0.6~1.2	−45~85	+1.6/−1.0 ±4	Two points One point	8 8	47.2
[15]	RES	65	1.0	0~100	+1.5/−1.1	One point	12	36
[16]	BJT&MOS	22	1.0	−30~120	±1.07 (3σ)	One point	38	50
[17]	TD	180	1.2	−40~120	+1.0/−1.0	One point	4	3
[18]	BJT	14	1.35	0~100	+1.0/−1.5	Two points	52	1600
[19]	MOS	60	1.0	20~80	±3	One point	4	14
[20]	BJT	-	3.0~5.5	−20 ~55	±0.15	Two points	14	-
[21]	MOS	180	0.8	−20 ~80	+1.2/−0.9	Two points	9	11
[22]	BJT	180	1.8	−55 ~125	±0.3 (3σ)	Two points	20	8280
[23]	BJT	180	1.8–5.5	20 ~50	±0.1 (3σ)	One point	15	16
[24]	MOS	180	1.2	0~100	+1.5/−1.4	Two points	4	0.071
[25]	RES	65	0.83–1.35	−50~105	±1.2 (3σ)	One point	20	32.5

4. Conclusions

A 0.18-μm CMOS reprogrammable fuse using EEcells is proposed for trimming a digital temperature sensor. The fuse uses EEPROM memory cells, which allow multiple programming cycles by altering the stored data for digital trim codes applied to the smart sensor. By reprogramming the fuse, the digital sensor can be adjusted with an increased trim variation in order to achieve higher accuracy. The operation of the trimmed DTS was validated by Synopsis HSPICE simulations and wafer-level and encapsulated IC measurements. A +1.5/−1.0 °C inaccuracy in the −20 to 125 °C range was obtained for 25 DTS measured samples at 1.8 V by one-point calibration, while the mean was centered at 0.44 °C, with a standard deviation of 0.71 °C. The digital sensor exhibits similar results for a power supply range of 1.7 to 3.6 V. Thus, the DTS's performance is in fairly good agreement with recently reported temperature sensors, and the proposed trimming technique can be used in multiple presented applications.

Author Contributions: Conceptualization, A.M.V., A.N., and A.T.; methodology, A.M.V. and A.N.; software, A.M.V. and A.N.; validation, A.N., A.T., and G.B.; formal analysis, A.M.V.; investigation, A.M.V. and A.N.; resources, A.M.V. and A.T.; data curation, A.M.V.; writing—original draft preparation, A.M.V.; writing—review and editing, A.N.; visualization, A.T. and G.B.; supervision, A.T. and G.B. All authors have read and agreed to the published version of the manuscript.

Funding: This research received no external funding.

Institutional Review Board Statement: Not applicable.

Informed Consent Statement: Not applicable.

Data Availability Statement: Data sharing not applicable.

Conflicts of Interest: The authors declare no conflict of interest.

References

1. Dragan, A.; Negut, A.; Tache, A.M.; Brezeanu, G. A Reprogrammable Fuse with EEcells for trimming a Temperature Sensor. In Proceedings of the 2020 International Semiconductor Conference (CAS), Sinaia, Romania, 7–9 October 2020; pp. 111–114.
2. Datasheet of N34TS04 Circuit. Available online: https://www.onsemi.com/pub/Collateral/N34TS04-D.PDF (accessed on 25 February 2021).
3. Datasheet of N34TS108 Circuit. Available online: https://www.onsemi.com/pub/Collateral/N34TS108-D.PDF (accessed on 25 February 2021).
4. Tang, Z.; Fang, Y.; Shi, Z.; Yu, X.-P.; Tan, N.N.; Pan, W. A 1770-μm^2 Leakage-Based Digital Temperature Sensor with Supply Sensitivity Suppression in 55-nm CMOS. *IEEE J. Solid-State Circuits* **2019**, *55*, 781–793. [CrossRef]
5. Xie, S.; Prouza, A.A.; Theuwissen, A. A CMOS-Imager-Pixel-Based Temperature Sensor for Dark Current Compensation. *IEEE Trans. Circuits Syst. II Express Briefs* **2019**, *67*, 255–259. [CrossRef]
6. Xie, S.; Theuwissen, A.J.P. A CMOS Image Sensor With Thermal Sensing Capability and Column Zoom ADCs. *IEEE Sens. J.* **2019**, *20*, 2398–2404. [CrossRef]
7. Ballo, A.; Bruno, G.; Grasso, A.D.; Vaiana, M. A Compact CMOS Temperature Sensor for On-Chip Thermal Monitoring. In Proceedings of the 2020 International Conference on Electrical, Communication, and Computer Engineering (ICECCE), Istanbul, Turkey, 12–13 June 2020; pp. 1–5.
8. Ballo, A.; Bruno, G.; Grasso, A.; Vaiana, M. A Compact Temperature Sensor with a Resolution FoM of 1.82 pJ·K2. *IEEE Trans. Instrum. Meas.* **2020**, *69*, 1. [CrossRef]
9. Setiabudi, A.; Tamura, H.; Tanno, K. CMOS Temperature Sensor with Programmable Temperature Range for Biomedical Applications. *Int. J. Electr. Comput. Eng.* **2018**, *8*, 946–953. [CrossRef]
10. Yousefzadeh, B.; Makinwa, K.A.A. A BJT-Based Temperature-to-Digital Converter With a ±0.25 °C 3σ -Inaccuracy From −40 °C to +180 °C Using Heater-Assisted Voltage Calibration. *IEEE J. Solid-State Circuits* **2019**, *55*, 369–377. [CrossRef]
11. Chen, C.-C.; Chen, C.-L.; Fang, W.; Chu, Y.-C. All-Digital CMOS Time-to-Digital Converter with Temperature-Measuring Capability. *IEEE Trans. Very Large Scale Integr. (VLSI) Syst.* **2020**, *28*, 2079–2083. [CrossRef]
12. Deotti, D.; Ramirez, J.L.; Fruett, F. Design and Characterization of a Smart Temperature Sensor. In Proceedings of the 2020 IEEE 11th Latin American Symposium on Circuits & Systems (LASCAS), San Jose, Costa Rica, 25–28 February 2020; pp. 1–4.
13. Pelzers, K.; Xin, H.; Cantatore, E.; Harpe, P. A 2.18-pJ/conversion, 1656-μm^2 Temperature Sensor With a 0.61-pJ·K^2 FoM and 52-pW Stand-By Power. *IEEE Solid-State Circuits Lett.* **2020**, *3*, 82–85. [CrossRef]
14. Park, H.; Kim, J. A 0.8-V Resistor-Based Temperature Sensor in 65-nm CMOS With Supply Sensitivity of 0.28 °C/V. *IEEE J. Solid-State Circuits* **2018**, *53*, 906–912. [CrossRef]
15. Xin, H.; Andraud, M.; Baltus, P.; Cantatore, E.; Harpe, P. A 174 pW–488.3 nW 1 S/s–100 kS/s All-Dynamic Resistive Temperature Sensor With Speed/Resolution/Resistance Adaptability. *IEEE Solid-State Circuits Lett.* **2018**, *1*, 70–73. [CrossRef]

16. Lu, C.-Y.; Ravikumar, S.; Sali, A.D.; Eberlein, M.; Lee, H.-J. An 8b subthreshold hybrid thermal sensor with ±1.07 °C inaccuracy and single-element remote-sensing technique in 22nm FinFET. In Proceedings of the 2018 IEEE International Solid—State Circuits Conference (ISSCC), San Francisco, CA, USA, 11–15 February 2018; pp. 318–320.
17. Azcona, C.; Calvo, B.; Medrano, N.; Celma, S. 1.2 V–0.18-μm CMOS Temperature Sensors With Quasi-Digital Output for Portable Systems. *IEEE Trans. Instrum. Meas.* **2015**, *64*, 2565–2573. [CrossRef]
18. Oshita, T.; Shor, J.; Duarte, D.E.; Kornfeld, A.; Zilberman, D. Compact BJT-Based Thermal Sensor for Processor Applications in a 14 nm tri-Gate CMOS Process. *IEEE J. Solid-State Circuits* **2015**, *50*, 799–807. [CrossRef]
19. Xie, S.; Ng, W.T. Digital integrated temperature sensors for VLSI thermal management. In Proceedings of the 2014 12th IEEE International Conference on Solid-State and Integrated Circuit Technology (ICSICT), Guilin, China, 28–31 October 2014; pp. 1–4.
20. Lewis, G.D.; Merken, P.; Vandewal, M. Enhanced Accuracy of CMOS Smart Temperature Sensors by Nonlinear Curvature Correction. *Sensors* **2018**, *18*, 4087. [CrossRef] [PubMed]
21. Someya, T.; Islam, A.K.M.M.; Sakurai, T.; Takamiya, M. An 11-nW CMOS Temperature-to-Digital Converter Utilizing Sub-Threshold Current at Sub-Thermal Drain Voltage. *IEEE J. Solid-State Circuits* **2019**, *54*, 613–622. [CrossRef]
22. Yousefzadeh, B.; Makinwa, K.A.A. 9.3 A BJT-based temperature sensor with a packaging-robust inaccuracy of ±0.3 °C (3σ) from −55 °C to +125 °C after heater-assisted voltage calibration. In Proceedings of the 2017 IEEE International Solid-State Circuits Conference (ISSCC), San Francisco, CA, USA, 5–9 February 2017; pp. 162–163.
23. Deng, C.; Sheng, Y.; Wang, S.; Hu, W.; Diao, S.; Qian, D. A CMOS Smart Temperature Sensor With Single-Point Calibration Method for Clinical Use. *IEEE Trans. Circuits Syst. II Express Briefs* **2015**, *63*, 136–140. [CrossRef]
24. Jeong, S.; Foo, Z.; Lee, Y.; Sim, J.-Y.; Blaauw, D.; Sylvester, D. A Fully-Integrated 71 nW CMOS Temperature Sensor for Low Power Wireless Sensor Nodes. *IEEE J. Solid-State Circuits* **2014**, *49*, 1682–1693. [CrossRef]
25. Lee, Y.; Choi, W.; Kim, T.; Song, S.; Makinwa, K.A.A.; Chae, Y. A 5800-μm2 Resistor-Based Temperature Sensor with a One-Point Trimmed Inaccuracy of ±1.2 °C (3σ) From −50 °C to 105 °C in 65-nm CMOS. In Proceedings of the ESSCIRC 2019—IEEE 45th European Solid State Circuits Conference (ESSCIRC), Cracow, Poland, 23–26 September 2019; pp. 68–71.
26. Meijer, G.; Pertijs, M.; Makinwa, K. 2—Calibration and Self-Calibration of Smart Sensors. In *Smart Sensor Systems: Emerging Technologies and Applications*; John Wiley & Sons, Ltd.: Hoboken, NJ, USA, 2014. [CrossRef]
27. Meijer, G.C.; Wang, G.; Heidary, A. *Smart Temperature Sensors and Temperature Sensor Systems*; Elsevier BV: Amsterdam, The Netherlands, 2018; pp. 57–85.
28. Sabate, A.C.; Nordin, N.; Jimenez, B. Fuse trim links physical analysis methodology. In Proceedings of the 2016 IEEE 23rd International Symposium on the Physical and Failure Analysis of Integrated Circuits (IPFA), Singapore, 18–21 July 2016; pp. 100–104. [CrossRef]
29. Wong, C.-C.; Chang, S.-P.; Tseng, C.-H.; Chen, W.-S.; Chang, S.-J. Communication—Diffusion Break-Assisted Programming Mode for Active Electrically Programmable Fuse. *ECS J. Solid State Sci. Technol.* **2018**, *7*, Q109–Q111. [CrossRef]
30. Dragan, A.M.; Enache, A.; Negut, A.; Tache, A.M.; Brezeanu, G. An improved digital output buffer for a digital temperature sensor with an I2C high speed interface. *Solid State Electron. Lett.* **2019**, *1*, 147–151. [CrossRef]
31. Chen, Y.; Tan, X.; Yu, B.; Li, C.; Guo, Y. A new all-in-one bandgap reference and robust zero temperature coefficient (TC) point current reference circuit. In Proceedings of the 2017 IEEE 12th International Conference on ASIC (ASICON), Guiyang, China, 25–28 October 2017; pp. 541–544.
32. Ramirez, J.L.; Tiol, J.P.; Deotti, D.; Fruett, F. Delta-Sigma modulated output temperature sensor for 1V voltage supply. In Proceedings of the 2019 IEEE 10th Latin American Symposium on Circuits & Systems (LASCAS), Armenia, Colombia, 24–27 February 2019; pp. 249–252. [CrossRef]
33. Mankani, S.K.; Sajjanar, S.; Aradhya, H.R. Power and area optimization of decimation filter for application in Sigma Delta ADC. In Proceedings of the 2016 International Conference on Circuits, Controls, Communications and Computing (I4C), Bangalore, India, 4–6 October 2016; pp. 1–5.
34. Caldwell, T.; Shibata, H. High-speed oversampled continuous-time analog-to-digital converters. In Proceedings of the 2017 IEEE 60th International Midwest Symposium on Circuits and Systems (MWSCAS), Boston, MA, USA, 6–9 August 2017; pp. 1001–1004.
35. Datasheet of Fluke7103 Calibrator. Available online: https://instrumentation.com/PDFs/fluke%207103%20user%20guide.pdf (accessed on 25 February 2021).

sensors

MDPI

Communication

A Low-Power 12-Bit 20 MS/s Asynchronously Controlled SAR ADC for WAVE ITS Sensor Based Applications

Khuram Shehzad, Deeksha Verma, Danial Khan, Qurat Ul Ain, Muhammad Basim, Sung Jin Kim, Behnam Samadpoor Rikan, Young Gun Pu, Keum Cheol Hwang, Youngoo Yang and Kang-Yoon Lee *

Department of Electrical and Computer Engineering, Sungkyunkwan University, Suwon 16419, Korea; khuram1698@skku.edu (K.S.); deeksha27@skku.edu (D.V.); danialkhan@skku.edu (D.K.); quratulain@skku.edu (Q.U.A.); basim@skku.edu (M.B.); sun107ksj@skku.edu (S.J.K.); behnam@skku.edu (B.S.R.); hara1015@skku.edu (Y.G.P.); khwang@skku.edu (K.C.H.); yang09@skku.edu (Y.Y.)

* Correspondence: klee@skku.edu; Tel.: +82-31-299-4954

Abstract: A low power 12-bit, 20 MS/s asynchronously controlled successive approximation register (SAR) analog-to-digital converter (ADC) to be used in wireless access for vehicular environment (WAVE) intelligent transportation system (ITS) sensor based application is presented in this paper. To optimize the architecture with respect to power consumption and performance, several techniques are proposed. A switching method which employs the common mode charge recovery (CMCR) switching process is presented for capacitive digital-to-analog converter (CDAC) part to lower the switching energy. The switching technique proposed in our work consumes 56.3% less energy in comparison with conventional CMCR switching method. For high speed operation with low power consumption and to overcome the kick back issue in the comparator part, a mutated dynamic-latch comparator with cascode is implemented. In addition, to optimize the flexibility relating to the performance of logic part, an asynchronous topology is employed. The structure is fabricated in 65 nm CMOS process technology with an active area of 0.14 mm^2. With a sampling frequency of 20 MS/s, the proposed architecture attains signal-to-noise distortion ratio (SNDR) of 65.44 dB at Nyquist frequency while consuming only 472.2 μW with 1 V power supply.

Keywords: asynchronous control logic; successive approximation register (SAR); wireless access in vehicular environments (WAVE); low power consumption; capacitive digital to analog converter (CDAC)

Citation: Shehzad, K.; Verma, D.; Khan, D.; Ain, Q.U.; Basim, M.; Kim, S.J.; Rikan, B.S.; Pu, Y.G.; Hwang, K.C.; Yang, Y.; et al. A Low-Power 12-Bit 20 MS/s Asynchronously Controlled SAR ADC for WAVE ITS Sensor Based Applications. *Sensors* **2021**, *21*, 2260. https://doi.org/10.3390/s21072260

Academic Editor: Francesc Serra-Graells

Received: 21 February 2021
Accepted: 21 March 2021
Published: 24 March 2021

1. Introduction

Internet of Things (IoT) is considered as a challenging technology and next growth engine that will have an everlasting effect in the semiconductor field. IoT devices have capability to connect a plenty of different end systems. IoT based techniques applied on traffic management systems result in an intelligent and advanced transportation system. A wireless access in vehicular environments (WAVE) is a protocol related to vehicle communications and provides an efficient, and reliable radio communications in an intelligent transportation system (ITS). In most of the ITS application, the WAVE protocol system has been designed in such a way to allow one vehicle to communicate with other vehicles (V2V), to other device (V2R), or to infrastructures (V2I) via dedicated short-range communications (DSRC) [1,2]. WAVE protocol has the potential to carry out an authentic and competent V2V, V2I and V2R communications to facilitate the mobility, safety and environmental applications. It consists of an on-board equipment (OBE) and a roadside equipment (RSE), wirelessly connected to provide an intelligent system. An on-board equipment (OBE) generally should offer low power, low cost, low design complexity, good reliability, and high energy efficiency. A fully-integrated RF-SoC is a most suitable option to meet the above mentioned requirements.

Figure 1 plots the top block diagram of highly integrated 5.8 GHz DSRC transceiver system. It satisfies the aforementioned requirements without any externally connected block like low-noise amplifier (LNA) and external power amplifier (PA) [3]. The main building blocks consists of a matching network (M.N), single pole double throw (SPDT) switch, an inductively generated low-noise amplifier (LNA) to amplify the input signal, a mixer (MIXER), a 12-bit ADC, a 12-bit DAC, a received signal strength indicator (RSSI), a variable gain amplifier (VGA) with a low pass filter (LPF) and power amplifier (PA) [4]. An integrated SAR ADC allows transceiver to communicate with the digital baseband [5].

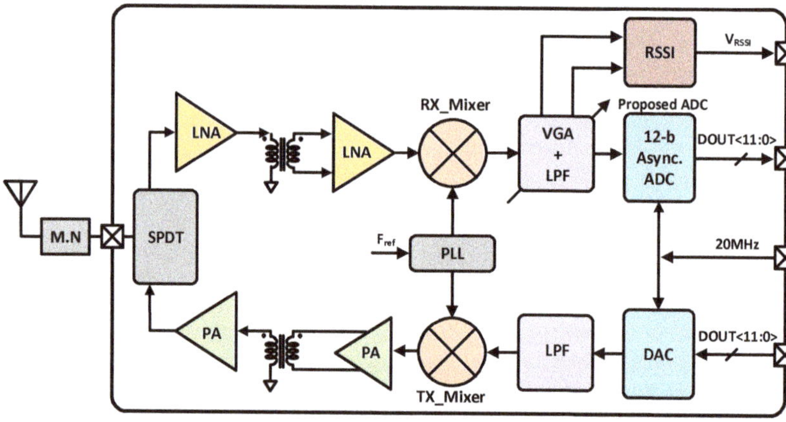

Figure 1. Top architecture of the DSRC transceiver.

As a result of the breakneck advancement in wireless technologies, a number of communication standard applications including ITS transceivers require on-chip ADCs with a sampling speed of few tens of MS/s and a resolution of more than 10 bits. The converters for finest communication systems, which include wireless local area networks (WLANs) based on IEEE standards protocol IEEE 802.11 require comparatively higher resolution of more than 10-bit and sampling rate of about few tens of MS/s [6].

Successive approximation register (SAR) analog-to-digital converters (ADCs) have been proven to be energy efficient in achieving moderate resolution and speed range [7,8] having a single comparator structure with no static power consumption and comparatively a simple structure. Recently, SAR algorithm based ADCs have also been used for higher speed and medium resolution applications by time interleaving multiple sub-SAR channels replacing traditionally implemented flash or pipeline structures [9]. However, with the increased number of bits, limitations due to comparator noise become severe which make SAR ADC as a difficult approach to implement for high resolution [10,11]. An energy-efficient prototype for high resolution is implemented front-end sampling switch, which results in eliminating the timing skew [12]. For high resolution ADCs, capacitive DAC consumes very high switching energy [13], for noise filtering integrator-based amplifier is used in [14]. Re-configurability and bandwidth scalability is achieved in [15] SAR ADC at a cost of comparatively high power consumption. A top-plate sampling increases the precision for 12-bit due to the implemented bootstrap switch. For a smaller overall capacitance, a DAC configurable binary window switching technique is implemented in [16]. However, it is lagging behind in terms of energy efficiency. For fully differential architecture, several techniques have been implemented to decrease the capacitor array size without digital calibration [17]. To reduce the switching energy and improve the linearity, floating DAC switching technique is presented in [18]. In [19,20], a binary-window DAC switching technique is presented to decrease switching error and DAC non-linearity at the cost of excessive power consumption. To decrease the distortion introduced by threshold voltage and parasitic capacitance, a linearity enhancement switch is implemented in [21].

A top-plate sampling technique is used to reduce the capacitor array size by half in [22], but it can cause non-linearity and common mode dependency upon input. A bottom plate sampling method is presented in [23] to reduce the overall size of the capacitor array.

This paper presents a 12-bit, 20 MS/s asynchronously controlled fully differential SAR ADC for wide-band WAVE based DSRC transceiver systems. To improve the static and dynamic performance of ADC, various techniques have been implemented. For 12-bit ADC, the implemented switching technique with CMCR switching conversion reduces the switching energy of DAC by 56.3% as compared to the conventional CMCR switching technique. The top-plate sampling results in increased settling because the influence of charge injection is reduced due to the aligned switching (AS) and detect-and-skip (DAS) switching technique. The implemented bootstrap switching technique improves the static performance of ADC. A constant DC shift and gain error can be introduced by the sustainable charge injection error, and sampling linearity will not deteriorate by implemented bootstrap switches. To decrease the power consumption and kickback noise of ADC, the proposed dynamic latch comparator with cascode is used.

The top configuration of the proposed asynchronous SAR ADC architecture is depicted in Section 2. The sub-blocks of the proposed ADC, such as the proposed capacitive DAC with a modified CMCR switching method, bootstrap switching, and dynamic latch comparator with cascode are explained in Section 3. The measured results and the performance summary of the presented ADC architecture is discussed in Section 4, and finally, we conclude our brief in Section 5.

2. The Top-Block Diagram of Proposed ADC Architecture

The presented configuration of SAR ADC is depicted in Figure 2. The presented architecture contains a comparator, SAR logic, clock generator, binary-weighted capacitive DAC and bootstrap switch. To improve the common-mode noise rejection and to reduce the noise of supply voltage we have implemented a fully differential architecture of SAR ADC. By bootstrap switches, the differential input signal is sampled at the bottom plate of capacitive DAC. According to the comparator decision and output digital code stored by the modified asynchronously controlled SAR logic, which controls the capacitive DAC switches.

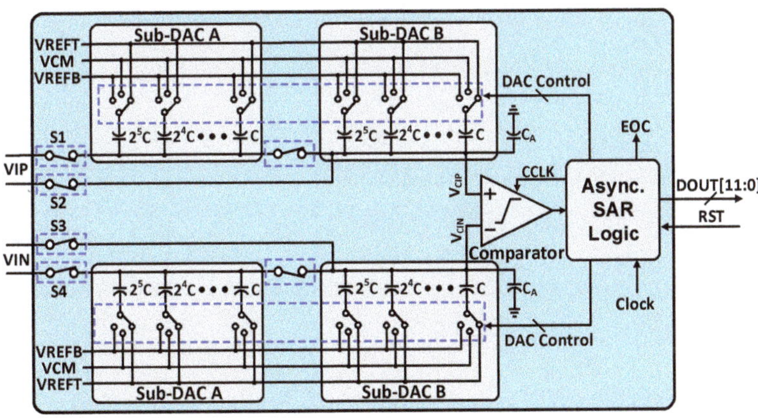

Figure 2. Top block diagram of proposed 12-bit asynchronous SAR ADC.

3. Circuit Implementation

3.1. Capacitive DAC with Modified CMCR Switching Technique

In a conventional switching scheme, for N-bit resolution, SAR ADC usually requires 2^N number of unit capacitors. The number of the unit capacitor can be reduced by optimizing the capacitive DAC's switching sequence, which is broadly explored, such as common mode based switching, set-and-down [24,25], and so on. The Area and power consumption of the capacitive DAC are significantly large for the high-resolution ADC such as over 10-bit resolution. To lower the capacitance from the DAC part, we adopt the common-mode charge recovery (CMCR) switching method [26]. With this switching, we use the possible minimum size of the unit capacitor in a capacitive DAC layout. An example of a 3-bit CMCR switching sequence is represented in Figure 3.

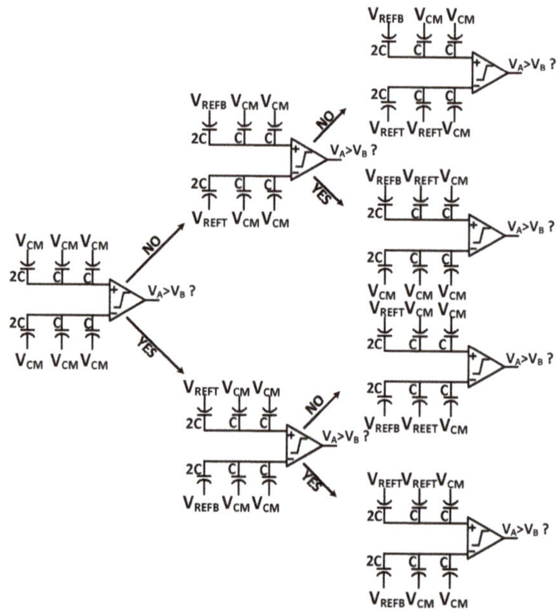

Figure 3. A 3-bit example of CMCR switching scheme.

The CMCR switching technique operation is based on the common-mode voltage scheme anticipated by the last comparison cycle. An additional bit is monotonously converted to the lowest capacitor of capacitive DAC to V_{REFT} from V_{CM}. Due to differential implementation, noise can be eliminated by the CMCR switching technique, furthermore, this switching method introduced the ripple in the LSB conversion by V_{CM}. Although, VCM can easily implement 1-bit accuracy, and it is effective for the reduction of the cost of sampling switches and DAC. We propose a switching technique based on the CMCR switching method for 12-bit SAR ADC as shown in Figure 4. Switches S_1~S_4 are input sampling switches which sample the input signal to the sub-DACs. The driving requirement of the SAR logic and comparator must be satisfied by the DAC control switches. For the 12-bit ADC, large switches are needed to achieve the charge sharing within the restricted time, because the peak-to-peak value of voltage discrepancy is V_{REF} on the top plate of the capacitive DAC, which causes the large power dissipation. Besides, the loss of switching energy in the proposed switching technique for the first ten comparison cycle is 2/4 times in comparison with the 10-bit CMCR switching method for the DAC capacitance increase.

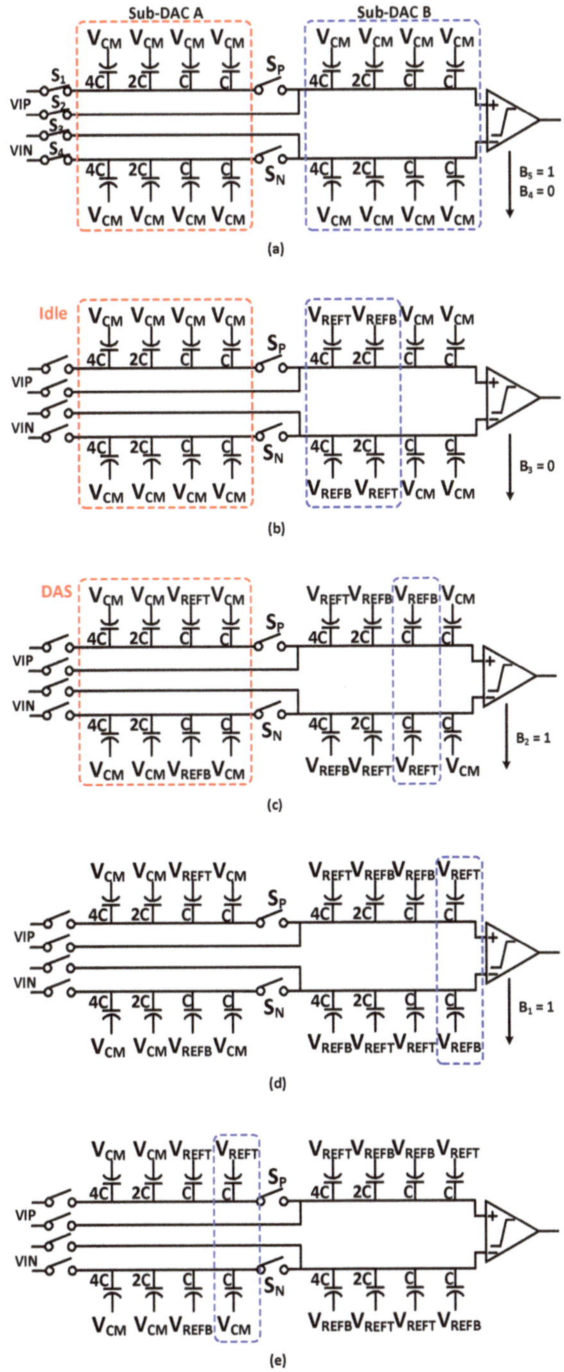

Figure 4. Example of proposed switching technique for 6-bit SAR architecture (**a**) Sampling phase (**b**) First two conversion cycles (**c**) 3rd conversion cycles (**d**) 4th conversion cycle (**e**) 5th conversion cycle.

The proposed DAC switching sequence use the detect-and-skip (DAS) technique, and aligned switching (AS) technique to minimize the switching energy loss and the size of DAC controlling switches S_1~S_4, in the comparison phase [14]. The proposed switching procedure prosecute with two steps:

(1) LSB conversion by the whole sub-DACs.
(2) MSB conversion by one sub-DAC.

The proposed switching technique for 6-bit SAR ADC which contains two sub-DACs is represented in Figure 4. In the implemented DAC, with the CMCR switching technique ADC start the conversion of sampling signal after the sampling phase, while sub-DAC A does not consume any switching energy because it is idle. By using the DAS and AS technique, the data transferred to sub-DAC A after the generation of the first three bits (B3~B5). Simultaneously, B2 determines by the comparator. When the sub-DAC B's LSB capacitors are switched by B2, the switches S_P and S_N are switched on in the 4th comparison cycle. During the 3rd comparison cycle, AS sets up and sub-DACs switching does not require any additional settling time. In the end, by the CMCR switching technique, the bits (B1~B0) are converted. By the proposed switching method, we are able to reduce the loss of switching energy from the capacitive DAC part, because the generation of the first nine bits is done by only one sub–DAC, and others are idle. The comparison between the conventional switching and the proposed switching energy versus output code is shown in Figure 5. The proposed switching method consumes 56.3% less energy when compared to the conventional CMCR switching method. The voltage variation is very small at the top plate, and DAC controlling switches S_1~S_4 requirements are reduced because they turn on after the achievement of nine bits. The switching energy is very efficient for the DAS and AS techniques. The dynamic performance with behavioral simulation is done in MATLAB® of proposed switching technique with 1% unit capacitor mismatch is shown in Figure 6. The static performance differential non-linearity (DNL), and integral non-linearity (INL) behavioral model of proposed switching technique with 1% unit capacitor mismatch is shown in Figure 7.

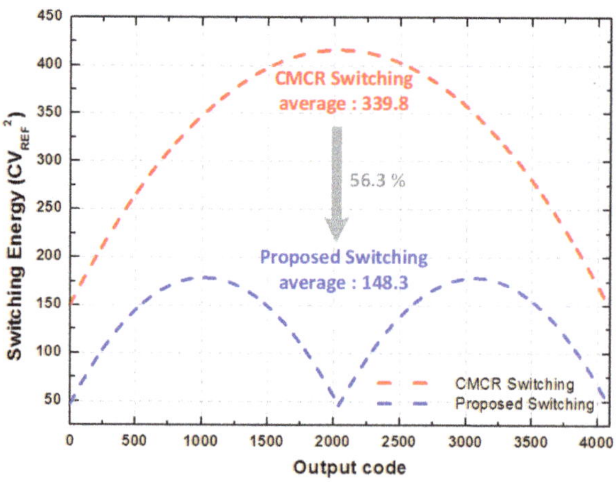

Figure 5. Comparison between switching energies of the proposed and conventional switching technique.

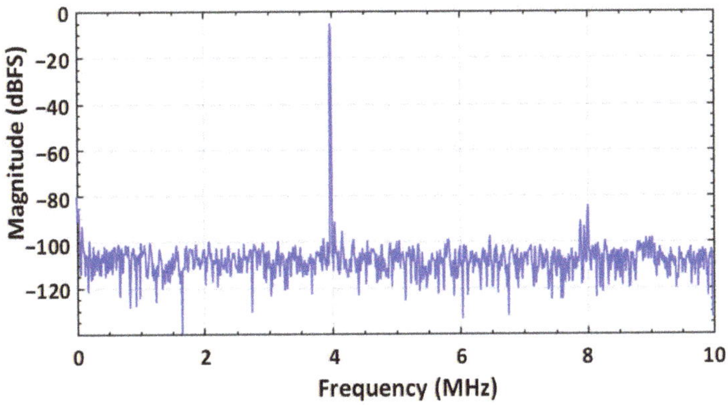

Figure 6. Dynamic performance with behavioral model of proposed switching technique with 1% unit capacitor mismatch.

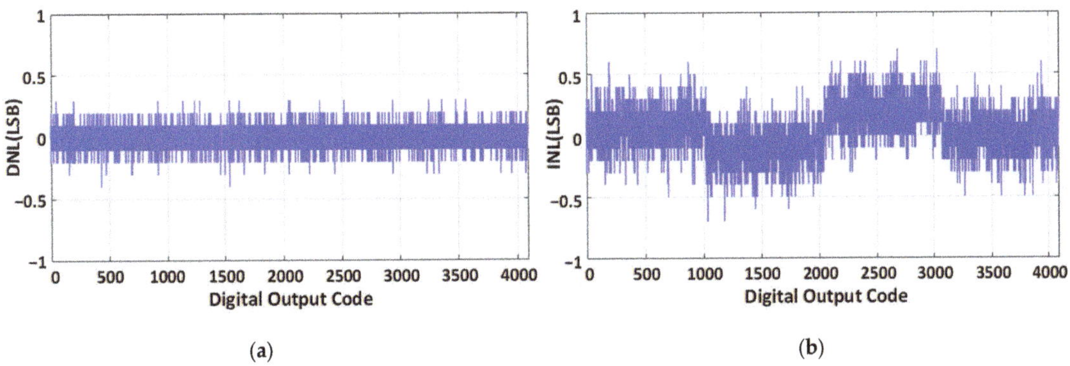

Figure 7. Static performance with behavioral model of proposed switching technique with 1% unit capacitor mismatch (**a**) differential non-linearity (DNL), and (**b**) integral non-linearity (INL).

For 12-bit, unit capacitor size is calculated with 1 V power supply by considering the capacitor mismatch and thermal noise power from DAC and due to sampling. The effective noise power due to sampling and from DAC is calculated by following the Equations (1) and (2) respectively,

$$v_{ns}^2 = \frac{2KT}{C_{SAM}} \tag{1}$$

$$v_{nd}^2 = \frac{2KT(C_{DAC})}{(C_{SAM})(C_A)} \tag{2}$$

where, v_{ns}^2 is noise because of sampling,v_{nd}^2 is effective noise from DAC, K is Boltzman's constant, T is the temperature, C_{SAM} is the total sampling capacitance, C_A is the intentional grounding capacitor for attenuation, and C_{DAC} is the overall DAC capacitance. The total sampling capacitance is 286 C, where C is the unit capacitance with a value of 15 fF.

3.2. Bootstrap Switch

Figure 8 shows the employed schematic of bootstrap switching, which is improved as proposed in [27]. The implemented bootstrap switch operates at the supply voltage. The gate body voltage (V_{GB}) of transistor M11 will be twice the supply voltage. Deep N-well (DNW) transistors M10–M13 are used to reduce the risk of enhancing reliability and failure. Transistors M10–M12 are turned on during the sampling phase. During this phase, the gate source voltage (V_{GS}) and V_{GB} of transistor M11 abide to supply voltage. During the sampling period, the implemented procedure is also competent to increase the sampling linearity and abolish the body effect because the body source voltage (V_{BS}) of transistor M11 stays at zero. When the substrate of transistor M11 goes to zero, while M13 is turned on then transistors M10–M12 are turned off during the conversion phase. In this way, we can ensure the separation of input V_{IP}/V_{IN} from the output V_{OUT}, because both drain substrate pn junction and source substrate pn junction are inversely-biased.

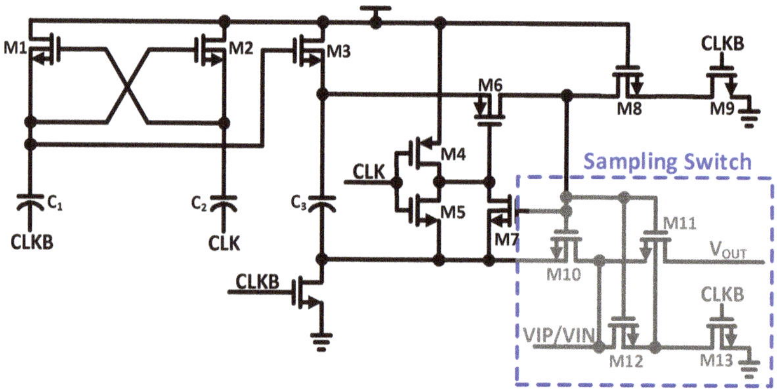

Figure 8. Bootstrap Switching Schematic.

For the differential architecture, we assume that the bootstrap switches matching accuracy is sufficient, since small common mode variation is caused by this, and the clock feed-through effect can be ignored. By embracing the implemented bootstrap switch architecture, we can alleviate the body-effect impact. On the differential inputs, a constant DC shift and gain error can be introduced by the sustainable charge injection error, and sampling linearity will not deteriorate. Hence, the clock feed-through and the charge injection's negative effect is attenuated by the differential architecture.

3.3. Dynamic Latched Comparator

To decrease the power consumption of ADCs, dynamic latched comparators are frequently used [28]. Several issues have been considered during the comparator designing such as; due to the comparator's clock operation, kickback noise affects the CDAC top plate. During the monotonic switching, offset voltage V_{offset} dependent on the V_{CM} and V_{offset} generated by device mismatch. The clock transition of the comparator distributes the comparator differential input V_{CIP} and V_{CIN}. In the proposed dynamic latched comparator, when the comparator clock signal CCLK goes to high then the input difference is settled. By the clock feed-through, at the comparator input, kickback noise is created during the CCLK transition. When the comparator input V_{CIP} and V_{CIN} sort out to a stable voltage then there is a recovery period. The comparator begins to sort out the variance between inputs, during this recovery period. Decision error can cause by a small asymmetry in the recovery period.

The implemented architecture of the comparator is depicted in Figure 9a. Due to the process variation, mismatch and hysteresis can exist in the comparator because of the use of transistors M13 and M16. Therefore, to minimize the hysteresis common centroid layout is used. To reduce the kickback noise and common mode dependent offset calibration the proposed comparator is designed. Cascode transistors M2, M5, and M6 shield the input transistors M3 and M4 to reduce the kickback noise of the comparator. The aspect ratio of cascode transistors M2, M5, and M6 is small so that these transistors operate in the saturation region and increase the output resistance of these transistors. Increased output resistance attenuates the large voltage step produced in result of the transition from

CCLK. By the circuit simulation, we choose the bias voltages V_{B1} and V_{B2} and size of the cascode transistors. We control the current of transistors M5 and M6 through bias voltage V_{B2}. During the CCLK transition, the peak current reduced through transistors M5 and M6 when bias voltage V_{B2} is reduced. Furthermore, we optimize the size of input transistors M3 and M4. The kickback peak value is reduced by using the size optimization and cascode transistors. Figure 9b represents the DAC output voltages V_{CIP} and V_{CIN}. When CCLK goes to high from low, the input difference is minimized in the dynamic latched comparator. At the comparators input, kickback noise is generated by clock feed-through of CCLK transition. When the comparators input V_{CIP} and V_{CIN} settles to a stable voltage, then there is a recovery period and in this time period input difference of comparator start to resolve, decision error can cause by a small asymmetry in the recovery period.

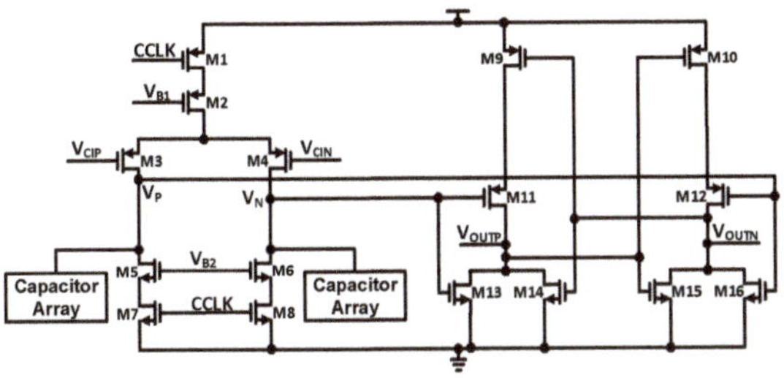

(a)

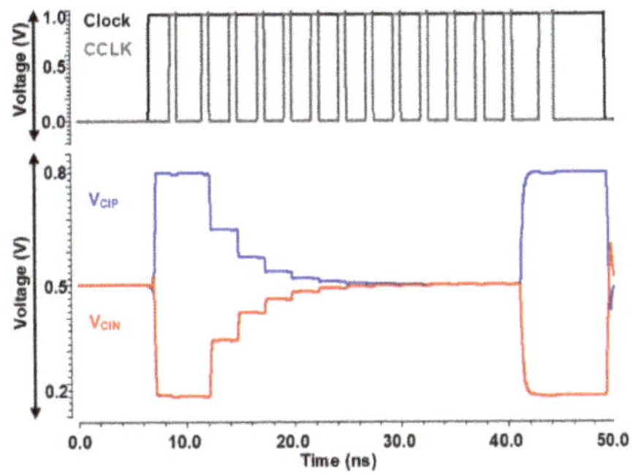

(b)

Figure 9. (**a**) Schematic of Dynamic Latched Comparator using Cascode to reduce kick-back (DNL), and (**b**) Waveform represented the CDAC settling with comparator.

Comparator offset calibration is performed by using binary-weighted capacitor array to control voltage offset V_{offset}. We select digital approach instead of analog offset calibration, because it requires additional DAC [29]. The voltage offset V_{offset} consists the dynamic and static offset of the comparator. V_{offset} of comparator can be derived as:

$$V_{offset} = \Delta V_{TH3,4} + \frac{V_{SG} - |V_{TH3,4}|}{2}\left(\frac{\Delta(W/L)_{3,4}}{(W/L)_{3,4}} + \frac{\Delta R_{load}}{R_{load}}\right) \tag{3}$$

where $V_{TH3,4}$ is the threshold voltage, $\Delta V_{TH3,4}$ is the threshold mismatch, ΔR_{load} is the load resistance mismatch, and $\Delta(W/L)_{3,4}$ is the physical dimension mismatch between transistors M3 and M4.

3.4. Asynchronous SAR Logic Processing

Asynchronous SAR control logic avoids the need of high frequency external clock signal as all conversions are carried out in a single clock cycle. To optimize the flexibility relating to the performance of logic part, an asynchronous topology is employed as shown in Figure 10.

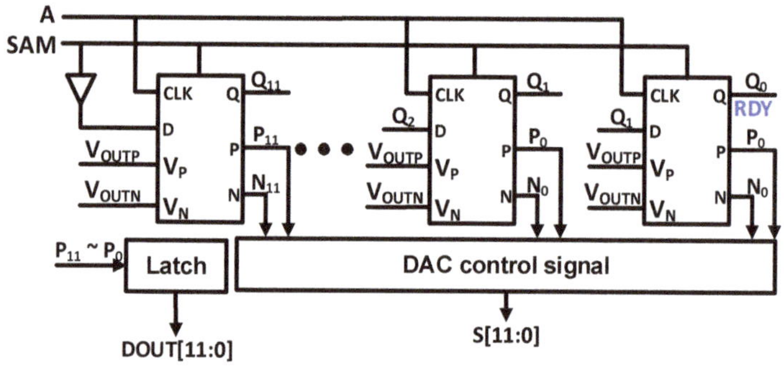

Figure 10. SAR logic block diagram.

To optimize the DAC switching and conversion time, we have added the digitally controllable delay cells for each conversion. Schematic and logic explanation of modified asynchronous clock generator CCLK is presented in Figure 11. The comparator output reset to VDD, when CCLK signal is high which is controlled by sampling signal SAM. SAM is the modified Clock signal with changed duty cycle. After sampling phase SAM signal goes to low and CCLK makes the comparator starts working after T_1 time. The comparator's outputs generate a high signal A through NAND1. After T_2 time, CCLK goes to high which results in resetting the comparator for further comparison. After T_3 time, the comparator's outputs goes to high which generates a low signal A and comparator is triggered. Digitally controllable delay cell is added in time T_3, to optimize the conversion speed, depending upon the settling time of the DAC.

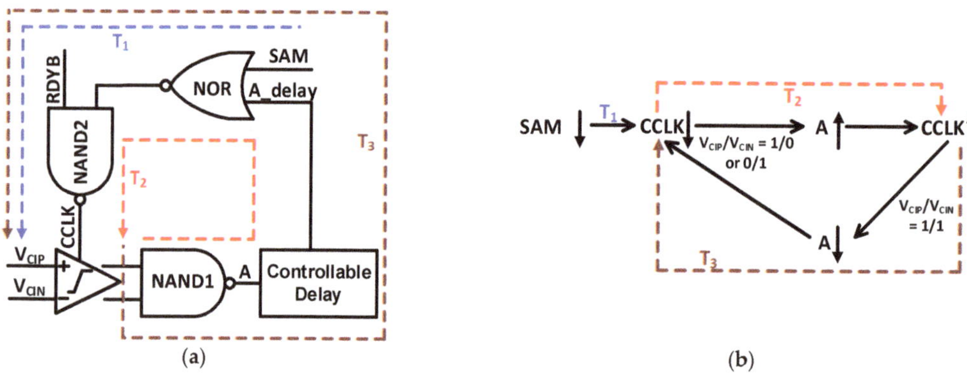

Figure 11. Asynchronous CCLK generator (**a**) Schematic of asynchronous clock generator with controllable delay (**b**) Logic description of CCLK generation.

4. Measurement Results

In a one poly six metal (1P6M) 65 nm CMOS technology, the fabricated prototype occupies an active chip area of 0.14 mm^2 as shown if Figure 12. The marked contents of the die micrograph correspond to each sub block of the proposed SAR ADC. The measured dynamic performance of the proposed ADC at two different frequencies is shown in Figure 13. The FFT spectrum shows that it achieves an ENOB of 10.98 bit at 4 MHz input frequency and 10.58 bit at around Nyquist input frequency with a sampling rate of 20 MS/s and an input single with a peak-to-peak voltage range of 600 mV.

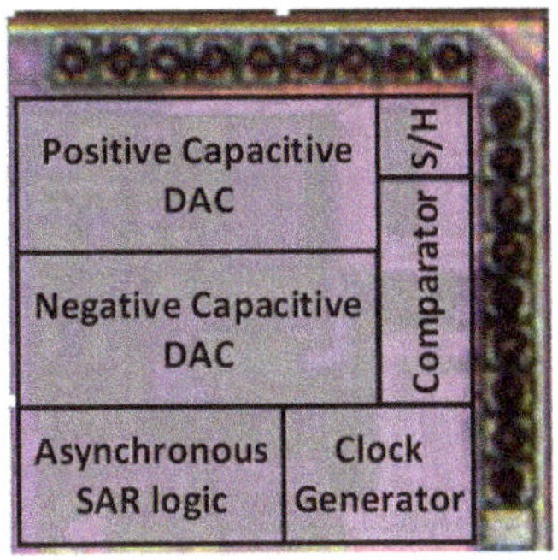

Figure 12. Die photograph of the implemented ADC.

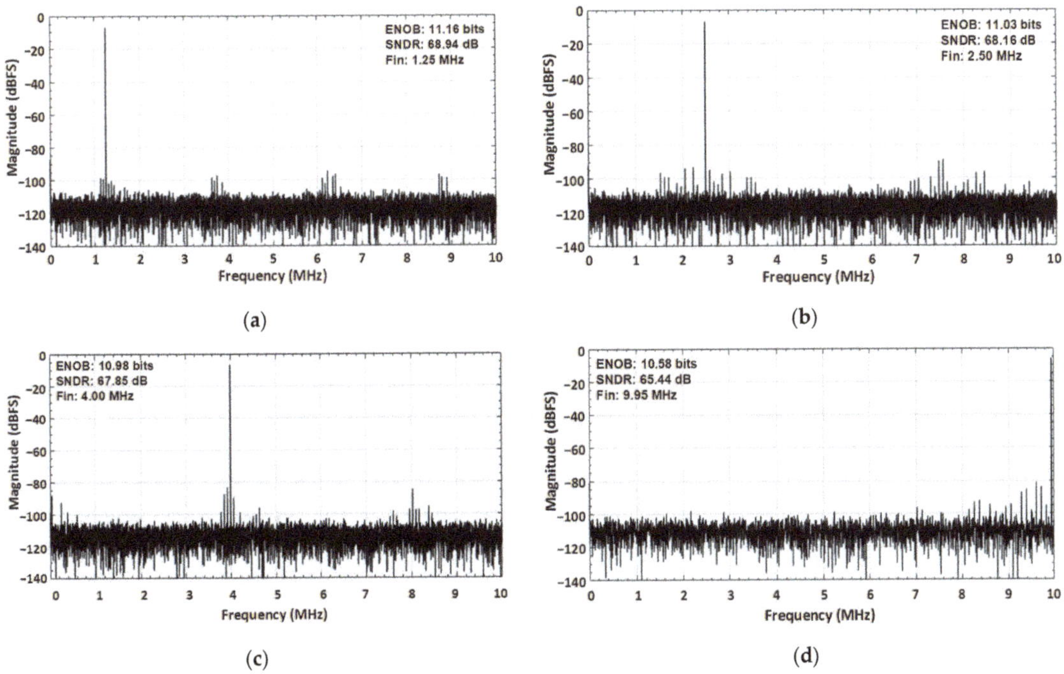

Figure 13. Measured dynamic performance for two different input frequencies at the sampling rate of 20 MS/s (**a**) for 1.25 MHz input frequency (**b**) for 2.50 MHz input frequency (**c**) for 4.00 MHz input frequency, and (**d**) for nyquist input frequency of 9.95 MHz.

Figure 14 presents the measured static performance. The peak differential non-linearity (DNL) and integral non-linearity (INL) values are +0.6/−0.6 LSB and +0.9/−0.9 LSB, respectively.

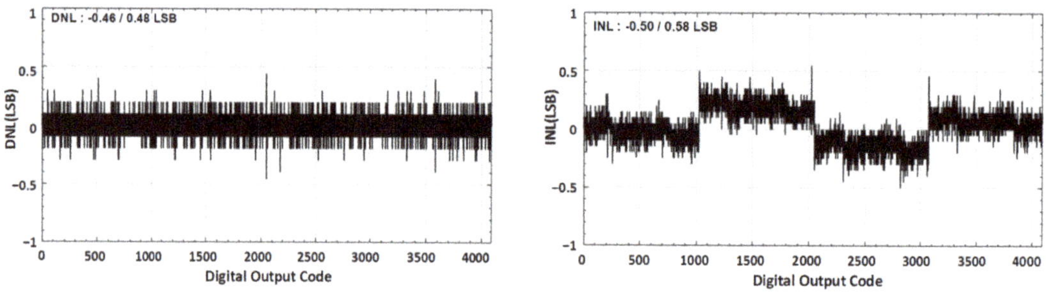

(**a**) (**b**)

Figure 14. Static performance parameter (**a**) DNL, and (**b**) INL.

Figure 15a presents the trend of SFDR and SNDR versus the applied input signal frequencies at 20 MS/s with 1 V of power supply. ENOB variation with respect to input signal frequency is shown in Figure 15b. The breakdown of power consumption with respect of sub blocks is presented in Figure 16.

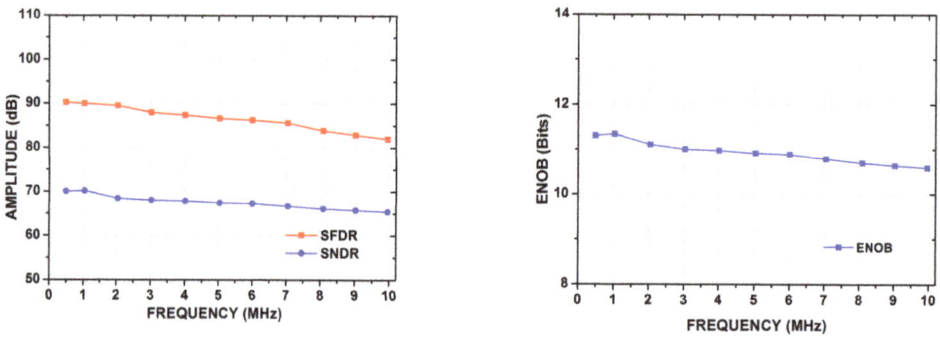

(a) (b)

Figure 15. (**a**) Input frequency versus SNDR and SFDR. (**b**) Input frequency versus ENOB.

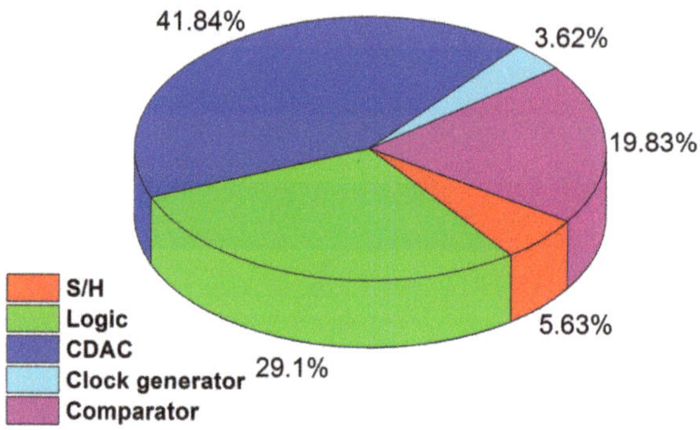

Figure 16. Breakdown of power consumption of ADC.

Table 1 presents the performance summary of the proposed architecture and its comparison with the other state of the art architectures [15,16,19–21]. It is evident that the proposed architecture exhibits a competitive performance in terms of energy efficiency and linearity. To evaluate the overall performance of the proposed ADC, the commonly used parameter, Figure of Merit (FOM), is used as

$$FOM = \frac{P_{ADC}}{\min\{F_S, 2 \times ERBW\} 2^{ENOB}} \tag{4}$$

where, F_S denotes the sampling rate and P_{ADC} is the power consumed by the structure. The proposed structure achieves a FOM of 15.42 fJ/conv. step.

Table 1. Performance summary and comparison.

Parameter	[15]	[16]	[19]	[20]	[21]	This Work
Process (nm)	180	180	180	180	180	65
Supply Voltage (V)	1.8	1.5	1.5	1.2	1.8	1
Resolution (bit)	12	12	12	12	12	12
Sampling Rate (MS/s)	20	20	10	40	10	20
SNDR (dB)	64.6	59.1	63.8	62.5	66.9	65.44
ENOB (bits)	10.44	9.52	10.31	10.09	10.82	10.58
DNL (LSB)	−0.51/0.445	−0.65/0.58	1.05	2.33	0.69	−0.46/0.48
INL (LSB)	−1.01/0.98	−1.06/1.04	1.38	3.1	1.15	−0.50/0.58
Power Consumption (μW)	1770	1220	600	1320	820	472.2
FOM (fJ/conv. step)	63.7	83	47.2	30.4	44.2	15.42

5. Conclusions

A low power 12-bit, 20 MS/s asynchronously controlled SAR ADC was fabricated with one poly six metal (1P6M) 65 nm CMOS technology to be used in WAVE protocol based intelligent transportation system. Several techniques have been proposed to optimize the architecture with respect to power consumption and performance. To alleviate the switching energy problem of the DAC part, the proposed switching method which employs CMCR switching technique is implemented in CDAC part. A mutated dynamic latch comparator with cascode is implemented to make certain a high speed operation with low power consumption and to overcome the kick back issue. Moreover, the presented modified asynchronous topology in control logic part optimizes the flexibility relating to the performance of logic part. The structure have an active area of 0.14 mm^2. The presented SAR ADC was operated at a sampling rate of 20 MS/s, attaining a peak SNDR level of 65.44 dB with a peak ENOB of 10.58 bits at Nyquist frequency. While consuming only 472.2 μW of power with 1 V power supply, the proposed architecture achieved a FOM of 15.42 fJ/conv. step.

Author Contributions: Conceptualization, K.S. and D.V.; methodology, K.S.; software, K.S., D.V. and S.J.K.; validation, K.S., D.V. and Q.U.A.; formal analysis, K.S., D.V. and M.B.; investigation, K.S., D.V. and D.K.; resources, K.S.; data curation, K.S. and D.V.; writing—original draft preparation, K.S. and K.-Y.L.; writing—review and editing, K.S., D.K., Y.G.P. and K.-Y.L.; visualization, K.S., D.V., B.S.R. and K.-Y.L.; supervision, K.C.H., Y.Y. and K.-Y.L.; project administration, K.-Y.L. All authors have read and agreed to the published version of the manuscript.

Funding: This research received no external funding.

Institutional Review Board Statement: Not applicable.

Informed Consent Statement: Not applicable.

Data Availability Statement: Not applicable.

Acknowledgments: This research was supported by the MOTIE (Ministry of Trade, Industry & Energy) (10080622) and KSRC (Korea Semiconductor Research Consortium) support program for the development of the future semiconductor device.

Conflicts of Interest: The authors declare no conflict of interest.

References

1. Electronic Toll Collection-Dedicated Short Range Communication—Part 1: Physical Layer. China Nat. Standard GB/T 20851.1–2007, 19 March 2007.
2. Kwon, K.; Choi, J.; Choi, J.; Lee, K.; Ko, J. A 5.8 GHz integrated cmos dedicated short range communication transceiver for the Korea/Japan electronic toll collection system. *IEEE Trans. Microw. Theory* **2010**, *58*, 2751–2763. [CrossRef]
3. Ali, I.; Asif, M.; Rehman, M.R.U.; Khan, D.; Yingge, H.; Kim, S.J.; Pu, Y.; Yoo, S.-S.; Lee, K.-Y. A Highly Reliable, 5.8 GHz DSRC Wake-Up Receiver with an Intelligent Digital Controller for an ETC System. *Sensors* **2020**, *20*, 4012. [CrossRef] [PubMed]
4. Bai, Z.; Kim, S.; Rad, R.E.; Lee, K. A 5.8 GHz Adaptive CMOS Image Rejection Mixer for DSRC Transceiver. In Proceedings of the 2020 International Conference on Electronics, Information, and Communication (ICEIC), Barcelona, Spain, 19–22 January 2020; pp. 1–3.
5. Kim, S.J.; Kim, D.G.; Oh, S.J.; Lee, D.S.; Pu, Y.G.; Hwang, K.C.; Yang, Y.; Lee, K.Y. A Fully Integrated Bluetooth Low-Energy Transceiver with Integrated Single Pole Double Throw and Power Management Unit for IoT Sensors. *Sensors* **2019**, *19*, 2420. [CrossRef] [PubMed]
6. Xia, B.; Valdes-Garcia, A.; Sanchez-Sinencio, E. A 10-bit 44-MS/s20-mW configurable time-interleaved pipeline ADC for a dual-mode802.11b/Bluetooth receiver. *IEEE J. Solid-State Circuits* **2006**, *41*, 530–539. [CrossRef]
7. Harpe, P.; Zhang, Y.; Dolmans, G.; Philips, K.; de Groot, H. A 7-to-10 b 0-to-4 MS/s flexible SAR ADC with 6.5-to-16 fJ/conversion-step. In Proceedings of the IEEE International Solid-State Circuits Conference Digest of Technical Papers (ISSCC), San Francisco, CA, USA, 19–23 February 2012; pp. 472–473.
8. Tai, H.-Y.; Hu, Y.-S.; Chen, H.-W.; Chen, H.-S. 11.2 A 0.85 fJ/conversion-step 10 b 200 kS/s subranging SAR ADC in 40 nm CMOS. In Proceedings of the 2014 IEEE International Solid-State Circuits Conference Digest of Technical Papers (ISSCC), San Francisco, CA, USA, 9–13 February 2014; pp. 196–197.
9. Kapusta, R.; Shen, J.; Decker, S.; Li, H.; Ibaragi, E.; Zhu, H. A 14 b 80 MS/s SAR ADC With 73.6 dB SNDR in 65 nm CMOS. *IEEE J. Solid-State Circuits* **2013**, *48*, 3059–3066. [CrossRef]
10. Kramer, M.J.; Janssen, E.; Doris, K.; Murmann, B. A 14 b 35 MS/s SAR ADC Achieving 75 dB SNDR and 99 dB SFDR With Loop-Embedded Input Buffer in 40 nm CMOS. *IEEE J. Solid-State Circuits* **2015**, *50*, 2891–2900. [CrossRef]
11. Miki, T.; Morie, T.; Matsukawa, K.; Bando, Y.; Okumoto, T.; Obata, K.; Sakiyama, S.; Dosho, S. A 4.2 mW 50 MS/s 13 bit CMOS SAR ADC With SNR and SFDR Enhancement Techniques. *IEEE J. Solid-State Circuits* **2015**, *50*, 1372–1381. [CrossRef]
12. Verbruggen, B.; Deguchi, K.; Malki, B.; Craninckx, J. A 70 dB SNDR 200 MS/s 2.3 mW dynamic pipelined SAR ADC in 28nm digital CMOS. In *2014 Symposium on VLSI Circuits Digest of Technical Papers*; Institute of Electrical and Electronics Engineers (IEEE): Piscataway, NJ, USA, 2014; pp. 268–269.
13. Van der Goes, F.; Ward, C.M.; Astgimath, S.; Yan, H.; Riley, J.; Zeng, Z.; Mulder, J.; Wang, S.; Bult, K. A 1.5 mW 68 dB SNDR 80 MS/s 2×interleavedpipelined SAR ADC in 28 nm CMOS. *IEEE J. Solid-State Circuits* **2014**, *49*, 2835–2845. [CrossRef]
14. Lim, Y.; Flynn, M.P. A 1 mW 71.5 dB SNDR 50 MS/s 13 bit Fully Differential Ring Amplifier Based SAR-Assisted Pipeline ADC. *IEEE J. Solid-State Circuits* **2015**, *50*, 2901–2911. [CrossRef]
15. Shen, Y.; Zhu, Z.; Liu, S.; Yang, Y. A Reconfigurable 10-to-12-b 80-to-20-MS/s Bandwidth Scalable SAR ADC. *IEEE Trans. Circuits Syst. I Regul. Pap.* **2017**, *65*, 51–60. [CrossRef]
16. Chung, Y.-H.; Zeng, Q.-F.; Lin, Y.-S. A 12-bit SAR ADC with a DAC-Configurable Window Switching Scheme. *IEEE Trans. Circuits Syst. I Regul. Pap.* **2019**, *67*, 358–368. [CrossRef]
17. Zhang, M.; Noh, K.; Fan, X.; Sanchez-Sinencio, E. A 0.8–1.2 V 10–50 MS/s 13-bit Subranging Pipelined-SAR ADC Using a Temperature-Insensitive Time-Based Amplifier. *IEEE J. Solid-State Circuits* **2017**, *52*, 2991–3005. [CrossRef]
18. Martens, E.; Hershberg, B.; Craninckx, J. A 69-dB SNDR 300-MS/s Two-Time Interleaved Pipelined SAR ADC in 16-nm CMOS FinFET With Capacitive Reference Stabilization. *IEEE J. Solid-State Circuits* **2018**, *53*, 1161–1171. [CrossRef]
19. Chung, Y.-H.; Yen, C.-W.; Tsai, P.-K. A 12-bit 10-MS/s SAR ADC with a binary-window DAC switching scheme in 180-nm CMOS. *Int. J. Circuit Theory Appl.* **2018**, *46*, 748–763. [CrossRef]
20. Chung, Y.-H.; Yen, C.-W.; Tsai, P.-K.; Chen, B.-W. A 12-bit 40-MS/s SAR ADC with a Fast-Binary-Window DAC Switching Scheme. *IEEE Trans. Very Large Scale Integr. (VLSI) Syst.* **2018**, *26*, 1989–1998. [CrossRef]
21. Liu, S.; Shen, Y.; Zhu, Z. A 12-Bit 10 MS/s SAR ADC with High Linearity and Energy-Efficient Switching. *IEEE Trans. Circuits Syst. I Regul. Pap.* **2016**, *63*, 1616–1627. [CrossRef]
22. Liu, C.-C.; Chang, S.-J.; Huang, G.-Y.; Lin, Y.-Z. A 10-bit 50-MS/s SAR ADC with a Monotonic Capacitor Switching Procedure. *IEEE J. Solid-state Circuits* **2010**, *45*, 731–740. [CrossRef]
23. Zhu, Y.; Chan, C.-H.; Chio, U.-F.; Sin, S.-W.; Martins, R.P.; Maloberti, F. A 10-bit 100-MS/s Reference-Free SAR ADC in 90 nm CMOS. *IEEE J. Solid-State Circuits* **2010**, *45*, 1111–1121. [CrossRef]
24. Verma, D.; Shehzad, K.; Khan, D.; Ain, Q.U.; Kim, S.J.; Lee, D.; Pu, Y.G.; Lee, M.; Hwang, K.C.; Yang, Y.; et al. A Design of 8 fJ/Conversion-Step 10-bit 8 MS/s Low Power Asynchronous SAR ADC for IEEE 802.15.1 IoT Sensor Based Applications. *IEEE Access* **2020**, *8*, 85869–85879. [CrossRef]
25. Shehzad, K.; Verma, D.; Khan, D.; Ain, Q.U.; Basim, M.; Kim, S.J.; Pu, Y.; Hwang, K.C.; Yang, Y.; Lee, K.-Y. Design of a Low Power 10-b 8-MS/s Asynchronous SAR ADC with On-Chip Reference Voltage Generator. *Electronics* **2020**, *9*, 872. [CrossRef]
26. Verma, D.; Shehzad, K.; Khan, D.; Kim, S.J.; Pu, Y.G.; Yoo, S.-S.; Hwang, K.C.; Yang, Y.; Lee, K.-Y. A Design of Low-Power 10-bit 1-MS/s Asynchronous SAR ADC for DSRC Application. *Electronics* **2020**, *9*, 1100. [CrossRef]

27. Abo, A.; Gray, P. A 1.5 V, 10-bit, 14 MS/s CMOS pipeline analog-to-digital converter. In *1998 Symposium on VLSI Circuits. Digest of Technical Papers (Cat. No.98CH36215)*; Institute of Electrical and Electronics Engineers (IEEE): Piscataway, NJ, USA, 2002; Volume 34, pp. 599–606. [CrossRef]
28. Van Elzakker, M.; van Tuijl, E.; Geraedts, P.; Schinkel, D.; Klumperink, E.; Nauta, B. A 10-bit charge-redistribution ADC consuming 1.9 W at 1 Ms/s. *IEEE J. Solid-State Circuits* **2010**, *45*, 1007–1015. [CrossRef]
29. Yoshioka, M.; Ishikawa, K.; Takayama, T.; Tsukamoto, S. A 10-b 50-Ms/s 820-_WSAR ADC with on-chip digital calibration. *IEEE Trans. Biomed. Circuits Syst.* **2010**, *4*, 410–416. [CrossRef] [PubMed]

sensors

MDPI

Article

Analysis and Reduction of Nonlinear Distortion in AC-Coupled CMOS Neural Amplifiers with Tunable Cutoff Frequencies

Beata Trzpil-Jurgielewicz, Władysław Dąbrowski and Paweł Hottowy *

Faculty of Physics and Applied Computer Science, AGH University of Science and Technology, 30-059 Kraków, Poland; btrzpil@agh.edu.pl (B.T.-J.); W.Dabrowski@ftj.agh.edu.pl (W.D.)
* Correspondence: hottowy@agh.edu.pl

Abstract: Integrated CMOS neural amplifiers are key elements of modern large-scale neuroelectronic interfaces. The neural amplifiers are routinely AC-coupled to electrodes to remove the DC voltage. The large resistances required for the AC coupling circuit are usually realized using MOSFETs that are nonlinear. Specifically, designs with tunable cutoff frequency of the input high-pass filter may suffer from excessive nonlinearity, since the gate-source voltages of the transistors forming the pseudoresistors vary following the signal being amplified. Consequently, the nonlinear distortion in such circuits may be high for signal frequencies close to the cutoff frequency of the input filter. Here we propose a simple modification of the architecture of a tunable AC-coupled amplifier, in which the bias voltages V_{gs} of the transistors forming the pseudoresistor are kept constant independently of the signal levels, what results in significantly improved linearity. Based on numerical simulations of the proposed circuit designed in 180 nm technology we analyze the Total Harmonic Distortion levels as a function of signal frequency and amplitude. We also investigate the impact of basic amplifier parameters—gain, cutoff frequency of the AC coupling circuit, and silicon area—on the distortion and noise performance. The post-layout simulations of the complete test ASIC show that the distortion is very significantly reduced at frequencies near the cutoff frequency, when compared to the commonly used circuits. The THD values are below 1.17% for signal frequencies 1 Hz–10 kHz and signal amplitudes up to 10 mV peak-to-peak. The preamplifier area is only 0.0046 mm^2 and the noise is 8.3 μV_{rms} in the 1 Hz–10 kHz range. To our knowledge this is the first report on a CMOS neural amplifier with systematic characterization of THD across complete range of frequencies and amplitudes of neuronal signals recorded by extracellular electrodes.

Keywords: CMOS neural amplifier; AC coupling; pseudoresistor; nonlinear distortion; area-efficient design

Citation: Trzpil-Jurgielewicz, B.; Dąbrowski, W.; Hottowy, P. Analysis and Reduction of Nonlinear Distortion in AC-Coupled CMOS Neural Amplifiers with Tunable Cutoff Frequencies. *Sensors* **2021**, *21*, 3116. https://doi.org/10.3390/s21093116

Academic Editors: Francesc Serra-Graells, Michele Dei, Kyoungrok Cho and Federico Alimenti

Received: 28 February 2021
Accepted: 21 April 2021
Published: 30 April 2021

1. Introduction

Multielectrode neural interfaces are widely used in basic neuroscience research [1–4] and for development of advanced brain-computer interfaces and neuroelectronic prostheses [5–7]. Taking advantage of large numbers of closely spaced microelectrodes, such systems make it possible to record activity of large neuronal populations with resolution of individual neurons, providing new insights into processing and coding of information in the brain circuits. Systems with several hundred to a few thousand of channels are now routinely used for recording the brain activity in live animals [8,9] and a prototype device with tens of thousands of recording channels was reported recently [10]. Systems dedicated to large-scale recording of brain activity in human are also being developed [11].

The neural signals acquired by extracellular electrodes are of two types. First, the action potentials (APs) can be recorded from individual neurons located close to the sensing electrode. An action potential is generated by a neuron when the total input signal received by this cell—either from sensory circuits of the central nervous system like the eyes or ears, or from other neurons—exceeds a specific threshold [12]. The APs recorded by extracellular electrodes have forms of short pulses with frequency spectrum from 300 Hz to 5 kHz and amplitudes range 50 μV_{pp}–2 mV_{pp} (peak-to-peak). Second, the electrodes can record local field potentials (LFPs) that are primarily generated by ionic currents that occur at the synapses—the physiological connections between neurons—when the information is transferred between cells; however, other processes also contribute to the LFPs [13]. The LFPs are low-frequency oscillations (1–300 Hz) with amplitudes up to 10 mV_{pp}.

The readout electronics for modern multielectrode systems is routinely designed as application specific integrated circuits (ASIC) that can comprise hundreds or thousands of recoding amplifiers on a single chip. High-fidelity recording of neuronal signals requires that the noise within the AP range (300 Hz–10 kHz) and the LFP range (1–300 Hz) is not much higher than 5 μV. Also, signals up to ~10 mV_{pp} should be recorded with the total harmonic distortion on the order of 1% or lower [14]. Minimization of the dissipated power and silicon area is also critical for the design of neuronal interfaces with very large number of recording channels.

One technical difficulty in electrical recording of neural signals is related to the large DC voltage at the input of the amplifier that results from electrochemical interactions between the electrode and the tissue [15]. The recording circuit must cut off this DC electrode voltage with high-pass filter with lower cutoff frequency typically on the order of 1 Hz and amplify the remaining AC signals with a gain on the order of 40 dB. Most of the multichannel integrated neural amplifiers are based on the architecture proposed in [15] (Figure 1a); in some designs, the circuit is followed by another amplification stage. The gain of the circuit shown in Figure 1a is given by the C_{ina}/C_{fa} ratio and the cutoff frequency of the AC-coupling circuit is defined by the $R_{fa} \times C_{fa}$ product [15].

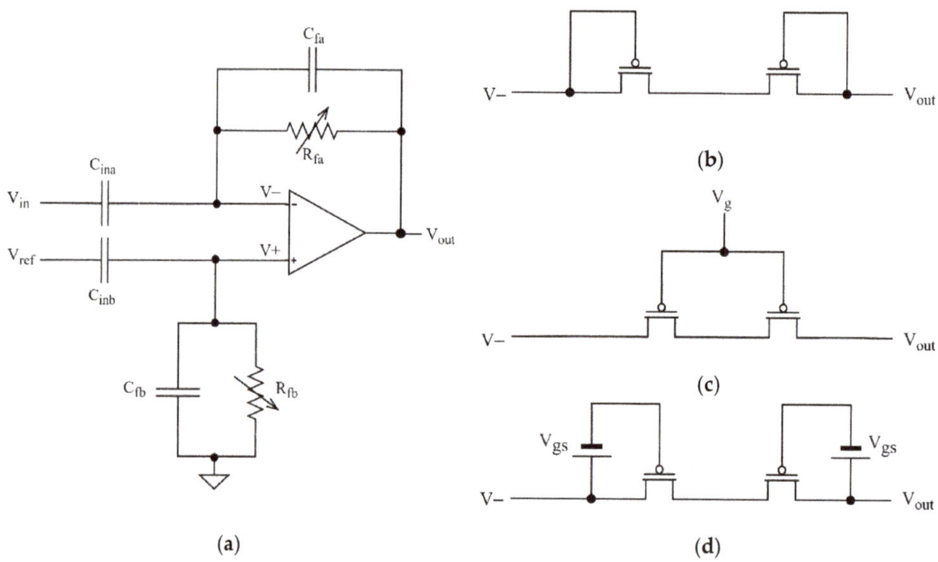

Figure 1. Architecture of an AC-coupled neural amplifier: (**a**) Schematic (**b**–**d**) Implementations of the pseudoresistor: (**b**) diode-connected; (**c**) subthreshold with variable V_{gs}; (**d**) subthreshold with fixed V_{gs}. For the following sections we impose $C_{ina} = C_{inb}$, $C_{fa} = C_{fb}$, $R_{fa} = R_{fb}$. The distinction between the names of elements of identical values is important for some of the following analyses, where impact of individual elements on the system performance must be analyzed separately.

Due to silicon area restrictions, the C_{in} is typically in the range 5–20 pF and the C_f capacitance is typically in the range of tens to hundreds of fF. The feedback resistance in the TΩ range is necessary to achieve sufficiently large time constant. Such a resistance is realized by transistors connected in diode configuration [15] (Figure 1b) or in subthreshold mode that allows one to tune the channel resistance value by changing the gate voltage [16–18] (Figure 1c). Such tuning allows the user to control the cutoff frequency of the high-pass filtering and to find—for specific experimental conditions—the optimal compromise between the two requirements:

- efficacy of filtering out of the ultra-slow oscillations present in the brain [19] and electrode drifts, which improves with increasing the cutoff frequency,
- the quality of recording of low-frequency signals with minimized amplitude and phase distortions related to analog high-pass filtering, which improves with lowering of the cutoff frequency.

As the two requirements mentioned above are difficult to define quantitatively in a general way, the ability to control the cutoff frequency of the high-pass filter is a desirable—even if not mandatory—feature of a neural amplifier.

The circuitry shown in Figure 1a makes it possible to achieve very low input-referred noise values. Since the resistor R_{fa} is in the feedback loop, its thermal noise is divided by the amplifier gain when referred to the input. Although this does not apply to resistor R_{fb}, its integrated thermal noise is minimized thanks to the parallel connection with the large capacitor C_{inb} (detailed analysis of the thermal noise introduced by the AC-coupling circuit is presented in Section 4).

Three alternative approaches to remove the electrode offset have been proposed. First, a high-pass RC filter can be used to remove the DC voltage at the input of the amplifier that works without feedback loop [14,20]. This solution has worse noise performance than the circuit shown in Figure 1a, since the thermal noise source from the large resistance is located directly at the amplifier input. The rms (root mean square) value of noise can be minimized by setting the cutoff frequency to very low values—much below the frequency range of interest [20]—but this would compromise on filtering out very slow signals, as discussed previously. As a result, open-loop neural amplifiers use larger input capacitors (20 pF and more) to improve the signal-to-noise ratio (SNR). This comes at the cost of increased area, which is not optimal for systems with high channel counts.

Chopper stabilization is another technique to remove electrode offset in neural amplifiers [21,22]. This method uses modulation to shift the spectrum of the input signal to higher frequencies and to minimize the problem of $1/f$ noise in the amplification circuits [23]. However, high complexity of the design (further increased by a dedicated feedback loop to boost the input impedance, which is low in such circuits) results in increased circuit area.

The third alternative approach to remove the electrode DC voltage is to use a low-pass filter in the feedback loop of the amplifier, which allows for subtraction of the low-frequency signal components (including the offset) from the input signal by a differential amplifier [24,25]. The low-pass filter requires using additional operational amplifier in the feedback loop that increases the circuit area and power. Using a sigma-delta modulation for low-pass filtering was proposed in [25] with a promise of reduced circuit area in case of using higher-density CMOS process. However, the presented amplifier design in 130 nm technology has a total area (amplifier + ADC) per channel at ~0.05 mm^2, which is 3–4 times more than the most area-efficient existing designs. At the same time, using very high-density technologies for a large-scale design (with thousands of recording channels) that is not expected to be produced in high volume is impractical because of very high cost of fabrication.

In overall, the architecture shown in Figure 1 remains the gold standard for modern neural amplifiers [26–28]. Many designs based on this circuit idea, with excellent noise performance and low power consumption, have been reported; for a review see [29,30]. However, a weak point of the pseudoresistor is its poor linearity. The pseudoresistor is placed in the feedback loop and the voltage drop across this resistance is identical to the amplifier output voltage. Within the range of the output voltage swing (several hundreds of mV to 1 V) the effective resistance of a pseudoresistor may differ by several orders of magnitude [15]. For the circuit with tunable cutoff frequency (Figure 1c) the main reason for the nonlinearity is the modulation of V_{gs} bias voltage by the continuously changing voltage across the resistor. This nonlinear behavior affects the THD of the circuit, in particular for input signals of large amplitudes and frequencies close to the cutoff frequency of the AC-coupling circuit.

Very few of the published articles on neural amplifiers discuss this problem. Kassiri et al. [31] analyzed several subthreshold two-transistor configurations with fixed-V_{gs} voltage for the pseudoresistors and achieved significant reduction of nonlinear distortion compared with the classic diode-based architecture. However, the distortion level reported in that paper is still high for 0.5 V_{PP} voltage swing across the resistors. Another paper by the same group includes measurements of the THD vs. signal frequency for the AC-coupled amplifier. The reported value was 3% for the signal frequency equal to cutoff frequency, and input signal amplitude of 1.4 mV_{PP} [32]. The distortion for larger amplitudes was not shown. To our knowledge, this is the only published measurement of THD as a function of signal frequency for amplifiers of this class.

It is important to note that the THD value of neural amplifiers is typically reported in the literature for the frequency of 1 kHz. Since this frequency is about three orders of magnitude higher than the cutoff frequency of the AC coupling circuit, the impedance of the feedback loop at this point is entirely defined by the capacitor C_{fa}. The THD value defined this way describes the performance of the operational amplifier used for the circuit but—as we show in this paper—it is not related to the distortion produced by the pseudoresistors at low frequencies. To our best knowledge, there has been no multichannel AC-coupled neural amplifier described in the literature with low distortion (~1% THD or less) reported consistently for the complete range of frequencies of extracellular neuronal signals.

In this paper we discuss an improved tunable AC-coupling architecture for CMOS neural amplifiers, based on pseudoresistors built from transistors working with fixed gate-source voltage, which yields low-distortion (~1%) for input signals ranging from 1 Hz to 10 kHz and with amplitudes up to 10 mV_{PP}. Based on numerical simulations we describe in detail the mechanism of nonlinear distortion generation and scaling of the THD with the amplifier gain, the cutoff frequency setting, and sizing of the feedback transistors and input capacitors. We also discuss the impact of the AC coupling circuit parameters on the noise performance of the recording system and the noise-distortion design trade-off. Finally, we present results of post-layout simulations of an area-efficient neural preamplifier in an 180 nm PD-SOI technology for verification of the proposed AC coupling architecture at the level of complete integrated circuit.

2. Materials and Methods

The work reported here is based on numerical simulations performed in Cadence Virtuoso 6.1.6 (Cadence Design Systems, San Jose, CA, USA). The design has been implemented in an 180 nm PD-SOI technology from XFAB (X-FAB Semiconductor Foundries GmbH, Erfurt, Germany) and all simulations have been performed for the selected technology. Also, for comparison of the key circuit parameters simulations have been performed for other processes (65 and 350 nm). The data analysis and visualization were done using MATLAB (MathWorks, Natick, MA, USA).

2.1. Circuits Used for Simulations

To identify the dominant sources of noise and distortion in the circuit and to optimize the design of the proposed amplifier, the analysis of the circuit has been performed in four stages: (a) using a schematic based on ideal passive components (resistor and capacitors) and an ideal differential amplifier, (b) using a transistor level models for the pseudoresistor and an ideal differential amplifier, (c) adding a transistor level models for the amplifier, and (d) using the post layout extracted circuit including parasitic components.

For the analysis of noise contribution of the AC-coupling circuit (Section 4) we used a simplified version of the schematic presented in Figure 1a, with the pseudoresistors replaced by ideal resistors. This allowed us to analyze the thermal noise of resistors R_{fa} and R_{fb} separately from other noise sources of the pseudoresistors—like the flicker noise—

which could be dependent on technology and sizing of pseudoresistors. For all the other simulations of noise and nonlinear distortion discussed in the paper, the pseudoresistors built from PMOS transistors were used.

For the analysis of nonlinear distortion and noise introduced by the pseudoresistors (Sections 3 and 4) we used the schematic shown in Figure 1a, with ideal operational amplifier. This allowed us to analyze signal distortions introduced by the pseudoresistor independently of distortions caused by the operational amplifier. The open-loop gain of the operational amplifier was set to 100 dB; increasing the gain beyond this value did not affect the simulation results.

For simulations discussed in Section 6 we used the operational amplifier designed in 180 nm PD-SOI technology. The analysis of nonlinear distortion was based on circuits extracted from the chip layout. For the Monte Carlo simulations, we combined post-layout extracted circuits of a single channel with schematics of all the off-channel circuits to reduce the computational cost of simulations. We confirmed that results of such simulations for nominal circuits (without Monte Carlo sampling) were undiscernible from simulations based on chip-scale extracted circuits.

2.2. Methodology of I-V Characteristics Analysis

To characterize linearity of the I-V characteristics of pseudoresistors (Section 3.1) we performed DC simulations. One terminal (named *V-* in Figure 1c,d) of the pseudoresistor was grounded, and the other terminal (named V_{out} in Figure 1c,d) was connected to varying voltage. Both polarities of the changing voltage were analyzed. For the variable-V_{gs} configuration the gates of both transistors were shorted and connected to voltage fixed at a constant value with respect to ground. For the fixed-V_{gs} configuration the gate-source voltages were applied using DC voltage sources as shown in Figure 1d.

2.3. Methodology of THD Analysis

All the analyses of the nonlinear distortion of the AC-coupled amplifier were based on transient simulations. We routinely analyzed the THD as a function of the input signal frequency. We ran the simulation separately for every frequency of the input signal within the frequency range of interest. For each simulation the input signal was a stationary sinewave. The THD numbers were calculated offline by taking rms value of the first five harmonics.

We used 15 frequencies points per decade for majority of the performed analyses of THD. For the Monte Carlo analyses we used 5 frequencies points per decade. Initial analyses of distortion (Section 3.2) and final verification of the preamplifier (Section 6) were performed in a wide frequency range (0.1 Hz–10 kHz). For detailed discussion of the distortion introduced by the pseudoresistors we present the results limited to frequency range 0.1–10 Hz where these distortions are most prominent. For plots combining the THD and noise characteristics we use range 0.1–100 Hz to show critical features of all these curves. For plots of the THD we used a piecewise cubic hermite interpolating polynomial (PCHIP) interpolation [33].

With exception of the initial analysis presented in Section 3.2, we used signal amplitude of 10 mV_{pp} for all the simulations of nonlinear distortion. This is the worst case scenario from the distortion point of view, as this is the largest realistic amplitude of the input signal, and the THD values for the analyzed circuit consistently increase with the signal amplitude (Section 3). We note that in experimental practice the recorded signals may have lower amplitudes, in particular at high frequencies. However, the goal of this study is to propose the circuit architecture that is relevant for high-fidelity recording of a complete range of neuronal signals, including large-amplitude voltage oscillations.

2.4. Methododology of Noise Analysis

The analysis of the equivalent input noise of the AC-coupled amplifier were based on noise simulations. We routinely analyzed the output-referred noise spectrum and amplifier's closed loop gain as a function of the input signal frequency. The input referred noise was integrated within three ranges: from 1 to 300 Hz (LFP range), from 300 Hz to 10 kHz (AP range) and from 1 Hz to 10 kHz (full range).

3. Amplifier Architecture and Sources of Nonlinear Distortion

3.1. DC Analysis of Linearity of Pseudoresistors

As discussed in Section 1, one critical problem with tunable AC-coupled neural amplifier designs in CMOS technology (shown in Figure 1c) is the poor linearity of the pseudoresistors. The main reason for this poor linearity is the variability of the gate-source voltage of the transistors forming pseudoresistors when AC signal is amplified by the circuit. We compared linearity of the I-V characteristics for two configurations of pseudoresistor: a standard configuration based on two PMOS transistors in symmetric configuration (Figure 1c) and alternative configuration with transistors working with fixed-V_{gs} voltage (Figure 1d) using DC simulations. All the transistors had identical dimensions (W/L = 1 µm/40 µm). The small-signal resistances for both pseudoresistors were identical.

Figure 2a shows the I-V characteristics and Figure 2b shows the incremental resistance values for both solutions. The linearity of the fixed-V_{gs} configuration is much better and particularly good within ±100 mV range of voltage across the resistor. We therefore propose to use the fixed-V_{gs} configuration for the low-distortion amplifier design. We note that such a circuit can be practically realized by including only a single resistor to the amplifier circuit (and additional off-channel circuitry that can be shared by multiple channels), and the required value of this resistor (several hundred of kΩ) means it can be realized in typical CMOS process with small and perfectly linear polysilicon resistor. The complete design of such circuit is presented in Section 6.

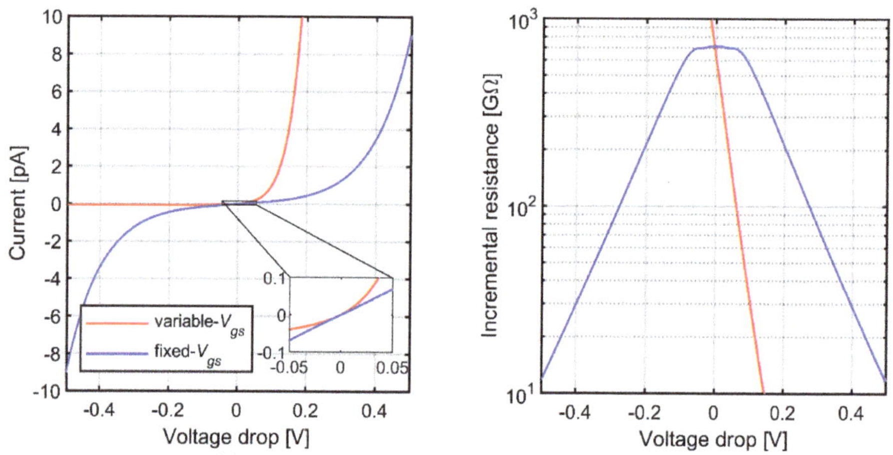

(**a**) (**b**)

Figure 2. (**a**) Simulated current-voltage relationship for variable-V_{gs} and symmetric fixed-V_{gs} pseudoresistors (**b**) Incremental resistance for both implementations.

Having in mind that the designed circuit aims at low-distortion recording of signals up to 10 mV_{pp}, we propose to take advantage of very good linearity of the fixed-V_{gs} pseudoresistor in the ±100 mV range of the output voltage and to set the amplifier gain value K = 20 V/V. We use this value as the nominal gain setting for the following sections, although we analyze impact of the gain on the noise performance (in Section 4) and on distortion (in Section 5). To achieve the required gain of the complete recording circuit (~100) we plan to use a second amplification stage following the presented preamplifier. The second stage can be DC coupled to the output of the preamplifier to avoid additional nonlinearity. The contributions of the second amplifier stage to both noise and distortion of the recording circuit would be much lower than that of the preamplifier and are not analyzed in this paper.

3.2. THD-vs-Frequency Characteristics of AC-Coupled Amplifier with Tunable Cutoff Frequency

It is expected that the large difference in the linearity of the I-V characteristics for the variable-V_{gs} and fixed-V_{gs} pseudoresistor configurations is reflected in the nonlinear distortions of the complete AC-coupled amplifier processing an AC signal. In particular, one can suspect that highly nonlinear I-V characteristic for the variable-V_{gs} pseudoresistor configuration should result in very high distortion, making this configuration impractical for amplification of signals up to 10 mV_{PP}. To evaluate these effects, we used Spectre transient simulations (Cadence Design Systems, San Jose, CA, UAS) to find the THD-vs-frequency curves for the AC-coupled amplifier (Figure 1a) using the standard tunable and improved tunable pseudoresistor configurations (Figure 1c,d, respectively). The operational amplifier and capacitors C_{in} and C_f used in these simulations are ideal elements, therefore the only source of nonlinearity is the pseudoresistor in the feedback loop. Transistor models for 180 nm PD-SOI technology were used for the pseudoresistor.

The results are presented in Figure 3. We present the THD values in the range limited to 3%—we assume that higher THD values disqualify given curve from further analysis, so the corresponding fragments of the THD curves are not shown to avoid confusion. The THD values for the variable-V_{gs} pseudoresistor configuration almost reach the 3% threshold already for 1 mV_{PP} signal amplitude (Figure 3a). The shapes of the THD curves for this configuration exhibit a single maximum nearby cutoff frequency of the AC-coupling circuit. This is easily explained by the increase of the I-V curve nonlinearity for larger voltage drop across the resistor (Figure 2). On one hand, for frequencies below the cutoff frequency of the AC-coupling circuit the signal amplification is reduced which implies lower THD values. On the other hand, for signal frequencies much above the cutoff frequency the impedance in the feedback loop is dominated by the C_{fa}, therefore the THD values are also lower than for frequencies closer to the cutoff frequency. Combination of these two effects results in the single-maximum shape of the THD curve. The overall increase of distortion for larger signal amplitudes (Figure 3a) is also an expected consequence of highly nonlinear I-V characteristic for large voltages (Figure 2).

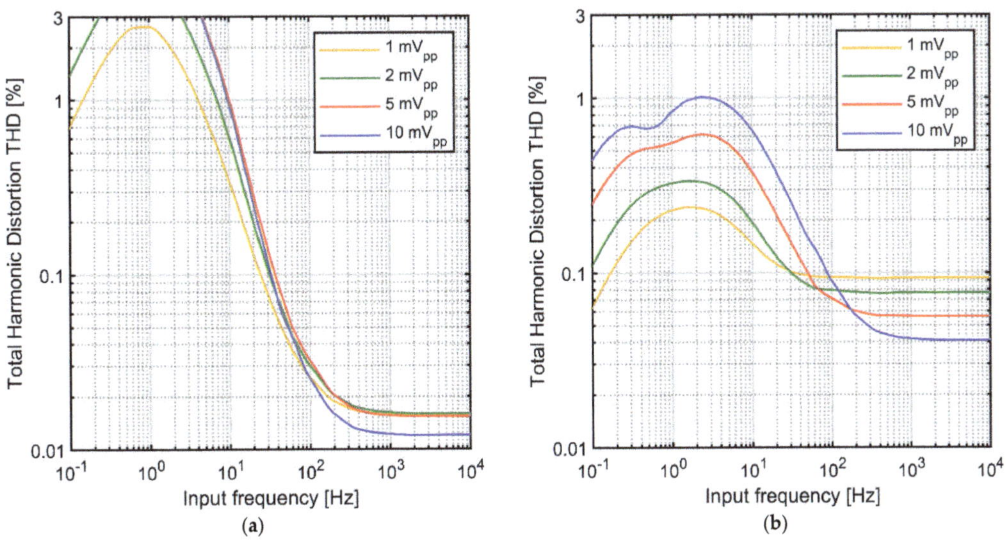

Figure 3. Simulated THD vs. signal frequency for AC-coupled neural amplifier (with 1 Hz cutoff frequency) and for various input signal amplitudes: (**a**) using the variable-V_{gs} pseudoresistor (**b**) using the symmetric fixed-V_{gs} pseudoresistor.

We note that although these results clearly show that the variable-V_{gs} pseudoresistor configuration is not compatible with design requirements for the neural amplifier considered in this paper (signal amplitudes with frequencies down to 1 Hz and amplitudes up to 10 mV_{pp}) it may be a reasonable option for different applications. For example, if the recording of low-frequency LFPs is not critical, the cutoff frequency of the AC-coupling circuit can be increased. Since the LFPs amplitude follows, in general, the $1/f$ dependence [19] the signal amplitude in the critical frequency range—that is, close to the cutoff frequency—would be lower than 10 mV and could be amplified with low distortion.

Another potential application is in experiments performed in-vitro, where signals from slices of neural tissue or dissociated neuronal cultures are acquired. As the LFPs produced by such neuronal populations have much lower amplitudes than that observed in the in-vivo measurements, the variable-V_{gs} configuration could be a sensible option for the pseudoresistor design. In fact, several designs of CMOS neural amplifiers with variable-V_{gs} pseudoresistors were reported in the literature [16,34,35] (the details of the pseudoresistor configurations differ between specific designs and the linearity performance may also vary to some degree). That said, for the design requirements considered here, the performance of this configuration with respect to nonlinear distortions is not acceptable.

The levels of THD for the fixed-V_{gs} configuration of the pseuderesistor are significantly lower than for the standard pseudoresistor, especially in the low frequency range around the cutoff frequency (Figure 3). This is expected considering improved linearity of the I-V characteristic for this configuration (compare Figure 2). However, the THD vs. frequency curves for this configuration show particular profiles with two local maxima, which are clearly separated for the highest input signal level of 10 mV_{pp}. To explain this shape one must consider additional source of nonlinearity in the circuit, associated with the gate capacitances of the transistors comprising the pseudoresistors.

3.3. *Impact of Capacitive Gate Currents of Pseudoresistors on THD*

A simplified model of the feedback loop for the fixed-V_{gs} pseudoresistor is shown in Figure 4a. Since the transistors are biased in the deep subthreshold region the gate capacitances are practically equal to the gate-bulk capacitances. These capacitances are not perfectly linear and depend on the V_{gb} voltage. The AC current I_{gb} is present in both transistor B (current flow between the output of the amplifier and the common bulk of both transistors) and transistor A (current flow between the common bulk and the external voltage source V_{gs1}—Figure 4a). We note that if the gate-bulk currents were identical for both transistors, all the AC current flowing from the output of the amplifier to the gate of transistor B would go to the external voltage source V_{gs1}. This would be equivalent to additional capacitive load of the amplifier output with no impact on the impedance of the feedback loop, therefore, no impact on THD. However, closer examination of the simulation results shows that the effect is more complex.

In order to confirm that the gate capacitances indeed contribute significantly to the impedance of the feedback loop we show in Figure 4 the results of transient simulations for a sinewave signal of 2.5 Hz for three different sizes of transistors composing the pseudoresistor. The cutoff frequency for each size of transistors was set at 1 Hz by tuning of the V_{gs} value. The plots show the output voltage, the drain-source currents and the gate-bulk currents for transistors A and B. As expected, the gate currents scale linearly with the gate area. The current is driven by the AC component of the gate-bulk voltage, which is equal to V_b–V_{ref} for transistor A and V_{out}–V_b for transistor B. Since the C_{gb} also changes with V_{gb}, the I_{gb} is not linear function of V_{gb} and it must include harmonics of the base frequency of the signal. As visible in Figure 4 the harmonics for transistors A and B are similar in amplitudes but different in phase (Figure 4e,f), and the differential current I_{gbB}–I_{gbA} includes almost entirely the harmonic frequencies (with very low component of the base frequency—Figure 4g). It means that while the base frequency component of the I_{gbB} current flows in almost 100% through C_{gbB} and C_{gbA} to external voltage source V_{gs1}, the current path for the harmonic components of I_{gbB} closes via the drain-source resistances of transistors A and B. As a result one should expect the increase of the THD values. This effect will be reduced for very low frequencies—where the gate currents become very small compared with drain-source currents—and for frequencies much above the cutoff frequency, where the feedback loop impedance becomes dominated by C_f. For the amplitude of 10 mV_{pp} the nonlinear gate currents result in second maximum in the THD curve located at ~2.5 Hz and become the dominating source of distortion. For smaller amplitudes these two maxima overlap.

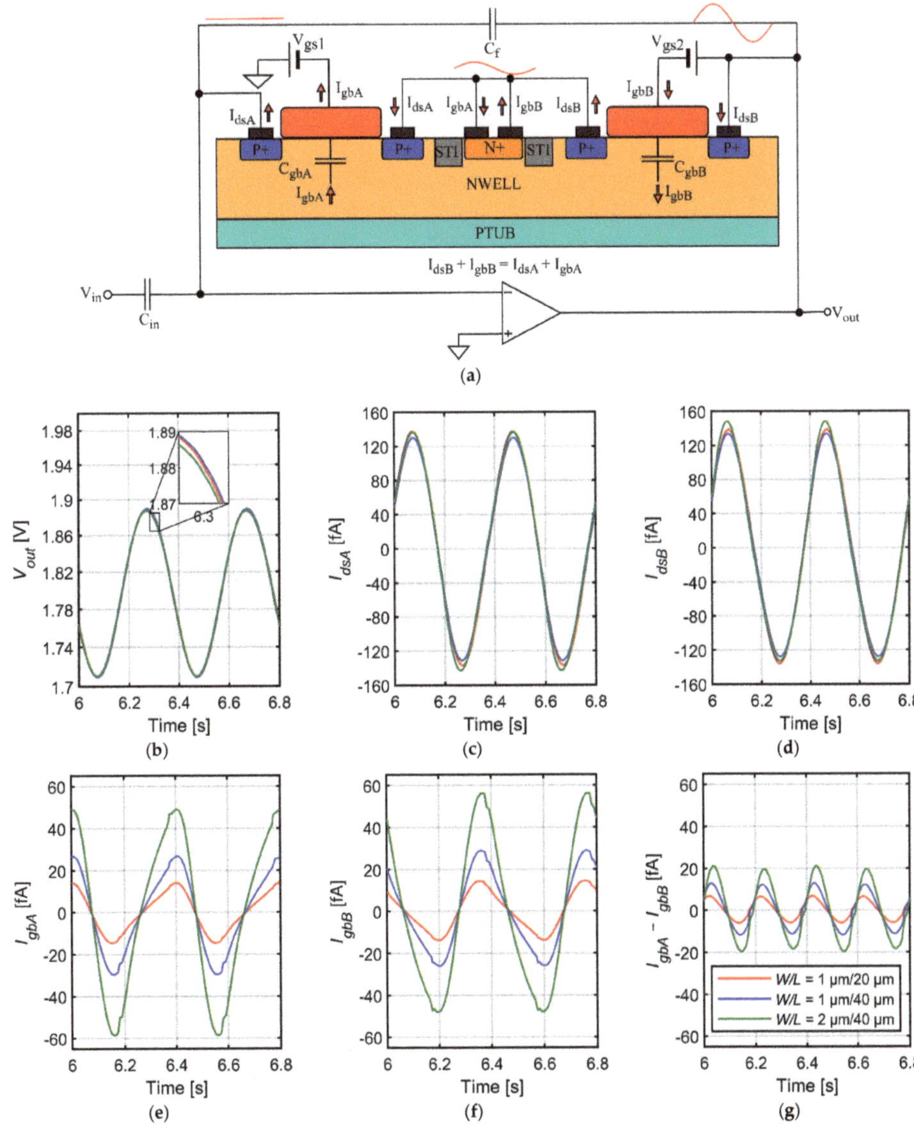

Figure 4. Results of transient simulations for a sinewave signal of 2.5 Hz for three different sizes of transistors composing the pseudoresistor: (**a**) Design of the preamplifier with a model of the fixed-V_{gs} pseudoresistor; (**b**) Output voltage; (**c–f**) Drain-source currents and the gate-bulk currents for transistors A and B; (**g**) Difference between gate-bulk currents for transistors A and B.

There are two additional issues with the results shown in Figure 3 that require separate comments. First, the nonlinear distortion generation due to AC gate currents, described above, must be also present in the standard variable-V_{gs} pseudoresistor configuration (Figure 3a). However, with large distortion values caused by the I-V characteristic, this effect is simply too subtle to be visible in the THD plots. Second, the THD curves for the fixed-V_{gs} pseudoresistor configuration show inversion of the THD vs. amplitude dependence at high frequencies—that is, the THD values decrease when signal amplitude increases. We note, however, that absolute values of the signal harmonics decrease for lower amplitudes. These values are in sub-microvolt range (when referred to the input) for signal amplitude 1 mV_{pp}, and will be even lower for smaller signals. Therefore, the effect has no practical implications and is not analyzed in more details here.

3.4. Scaling of THD with Gate Area and Oxide Thickness of Transistors Forming Pseudoresistors

The contribution of gate current to the circuit distortion is expected to scale with the gate area of transistors A and B. Figure 5 shows the simulated THD value as a function of input signal frequency for various sizing of the transistors and large signal amplitude (10 mV_{pp}). As expected, the height of the second peak (at 2–3 Hz for different curves) scales with the product $W \times L$; in contrast, the first peak is associated with the nonlinearity of the I-V curve (at ~0.3 Hz) scales with the transistor channel length. These observations may lead to a conclusion that reducing the gate area of transistors forming the pseudoresistor is a way for reduction of the THD. However, we note that excessive reduction of the gate areas may result in significant mismatch of transistors A and B and disturbed symmetry of I-V curve for positive and negative voltages. This would lead to increased THD values. For this reason, we accept the gate width $W = 1$ μm and length $L = 40$ μm for the following analyses.

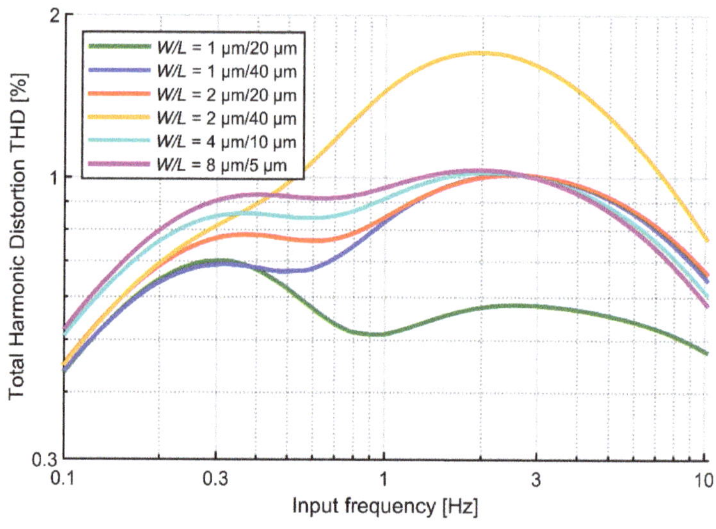

Figure 5. Simulated THD versus signal frequency for the fixed-V_{gs} circuit for different dimensions of the pseudoresistor. Signal amplitude: 10 mV_{pp}. V_{gs} values were tuned for each simulation to get the same cutoff frequency (1 Hz).

Finally, the presented analysis suggests that the thickness of the gate oxide should also affect the circuit distortion. We performed simulations of the circuit shown in Figure 1a with fixed-V_{gs} pseudoresistors designed in three different CMOS technologies (350, 180 and 65 nm), using the available thick-oxide transistors for the pseudoresistors. The sizes of transistors were identical in all simulations ($W = 1$ μm, $L = 40$ μm) and the V_{gs} values were tuned for each simulation to get the same cutoff frequency (1 Hz). The results are shown in Figure 6. Consistently with the analysis presented above, transistors with larger oxide thickness (and proportionally lower gate-bulk capacitances) yield lower distortion related to gate-bulk currents. The 5 V transistors in two technologies (350 and 180 nm) result in virtually identical THD values above the cutoff frequency. Below the cutoff frequency, where THD value is determined by the nonlinearity of the I-V characteristic, the distortion does not correlate with the oxide thickness. Although the results suggest that technologies that provide higher voltage transistors (5 V) may be preferable, we note that the distortion also critically depends on the gate area. Since more advanced technologies offer in general improved transistor matching [36] it may be possible to use smaller transistors in more advanced nodes to compensate for lower oxide thickness. These aspects require further systematic studies. For this work, the 180 nm technology with 5 V transistors for pseudoresistor design was chosen.

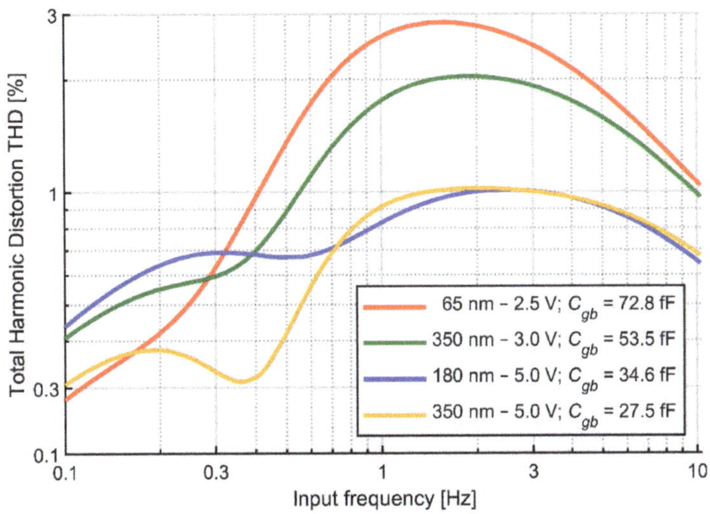

Figure 6. Simulated THD versus signal frequency for the fixed-V_{gs} circuit for different technologies. Signal amplitude: 10 mV_{PP}. V_{gs} values were tuned for each simulation to get the same cutoff frequency (1 Hz).

4. Noise Contribution of the AC-Coupling Circuit

Since a neural amplifier must be capable of recording signals with amplitudes down to tens of μV with good SNR, noise performance is a critical aspect of the design [15]. In this context, the noise contribution of the resistive elements of the AC coupling circuit must be carefully analyzed.

The results presented in this section are based on simulations of circuit shown in Figure 1a, with ideal operational amplifier and ideal resistors. This way, the thermal noise of the resistors is analyzed in separation from other sources of noise associated with the pseudoresistors (like flicker noise) or with the operational amplifier. In the Section 5 we discuss noise and nonlinear distortion based on simulations taking advantage of pseudoresistor built from PMOS transistors.

We start with analyzing the noise contribution from the feedback resistor R_{fa} (compare Figure 1a). At this point we assume the non-inverting input is at virtual ground, therefore the noise measured across the R_{fa} is equivalent to the noise contribution of this resistor measured at the output of the circuit. The total rms voltage noise of this resistor is:

$$V_{ni,rms\ R_f} = \sqrt{\frac{kT}{C_{fa}}}, \tag{1}$$

and is independent of the R_{fa} value and the cutoff frequency. However, this value comes from integration of power spectral density (PSD) curve from 0 to infinity. In order to quantify noise level in the defined frequency range, we need to look closer at the noise spectrum.

In Figure 7a we present the PSDs of the noise contributed by R_{fa}. The PSDs are shown at the amplifier output for various values of the cutoff frequency of the AC-coupling circuit. Although the tuning does not change the total rms noise, it shapes the noise spectrum. Therefore, the noise in the frequency range of interest can be reduced by shifting the cutoff frequency below this range; this is discussed in more detail later in this section.

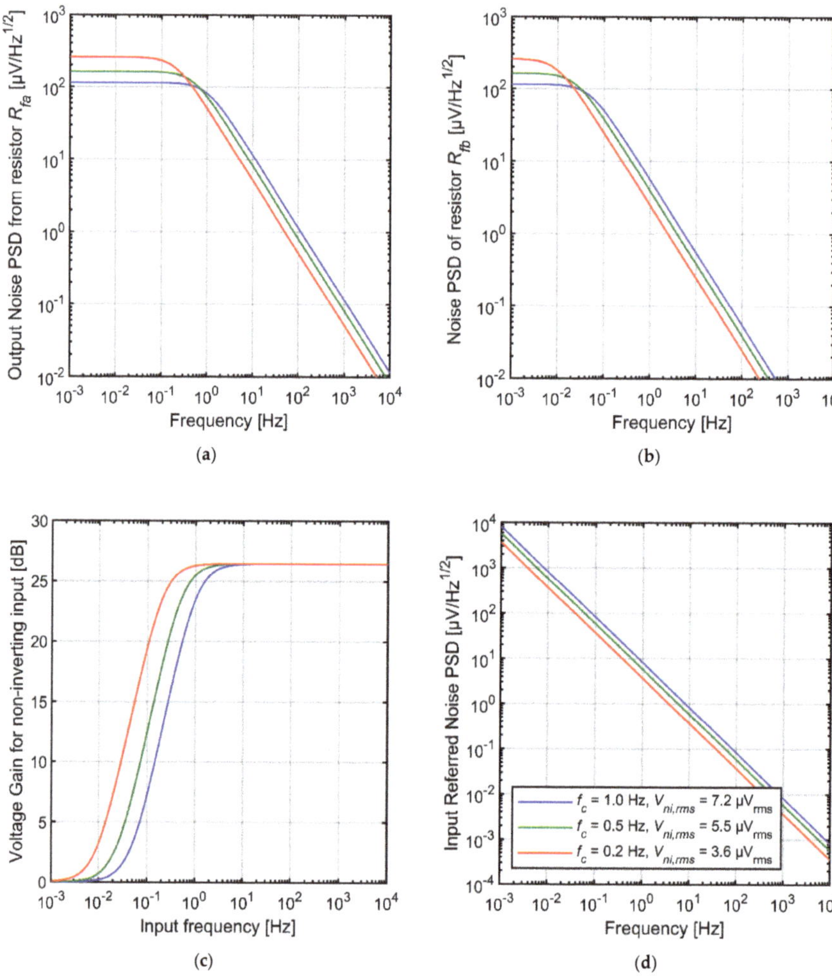

Figure 7. Characteristics of the AC-coupling circuit thermal noise for different settings of the AC-coupling circuit cutoff frequency (**a**) output noise PSD from resistor R_{fa} (**b**) noise PSD of resistor R_{fb} measured across this resistor (**c**) AC gain of the circuit with respect to non-inverting input of the operational amplifier (**d**) PSD of the combined input-referred noise from both resistors R_{fa} and R_{fb}.

In order to explain the noise contribution of the resistor R_{fb}, we analyzed the PSD of its thermal noise (measured across the resistor itself) and the transmittance of the complete circuit with respect to signals that appear at the noninverting input of the operational amplifier U1 (compare Figure 1a). The plots are shown in Figure 7b,c. Since the resistor R_{fb}

is shunted with large capacitance (C_{inb} + C_{fb}) the PSD values for most of the frequency range are much lower than those shown in Figure 7a. On the other hand, the $1/f$ dependence is extended toward lower frequencies as the time constant has a large value of $R_{fb} \times (C_{in} + C_{fb})$. For extremely low frequencies (below 0.1 Hz) the PSD values of the curves shown in Figure 7a,b become identical. At the same time, the circuit transmittance for this noise is equal to 1 at the very low frequencies and equalizes at value $(C_{ina} + C_{fa})/C_{fa}$ for higher frequencies. Multiplication of the respective curves shown in Figure 7b,c results in the output noise PSDs characteristics that are identical to that presented in Figure 7a. We conclude that the resistors R_{fa} and R_{fb} have identical impact on the noise performance of the circuit, both in terms of noise PSDs and the rms values.

The input-referred noise rms values calculated in 1 Hz–10 kHz frequency range are given in Table 1. One way to reduce the noise contribution of the AC-coupling circuit in specific frequency range is to set very low cutoff frequency. Some designs take advantage on this by shorting the gate and the source of the pseudoresistors, which leads to very large resistances of R_{fa} and R_{fb}. Such a solution will result in extremely low cutoff frequency in the range of tens of mHz or even lower and greatly reduced noise from feedback resistors above 1 Hz. Unfortunately, such filters do not remove the very slow and large-amplitude drifts of the electrode voltage from the signal, and this can lead to saturation of the recording amplifier. For this reason some users prefer circuits with tunable cutoff frequency which is set closer to the frequency range used for analyses, as discussed in the Introduction. Nevertheless, careful optimization of the filter time constant for specific experimental conditions may potentially be very useful in reducing the system noise.

Table 1. Input-referred noise rms for various parameters of the circuit. The rms values are calculated in 1 Hz–10 kHz frequency range. Assumptions: $C_{ina} = C_{inb} = C_{in}$, $C_{fa} = C_{fb} = C_f$, $R_{fa} = R_{fb} = R_f$.

Characteristic Parameters	C_{in}, C_f [F]	AC Gain—K [V/V]	Cutoff Frequency of the AC-Coupling Circuit [Hz]	Equivalent Input Wide-Band Noise [μV_{rms}]
Variable cutoff frequency (variable Rf)	4 p, 200 f	20	1.0	7.2
	4 p, 200 f	20	0.5	5.5
	4 p, 200 f	20	0.2	3.6
Variable AC gain (variable C_f)	4 p, 200 f	20	1.0	7.2
	4 p, 80 f	50	1.0	4.6
	4 p, 40 f	100	1.0	3.2
Variable design area (variable capacitors)	4 p, 200 f	20	1.0	7.2
	8 p, 400 f	20	1.0	5.1
	12 p, 600 f	20	1.0	4.2

The noise can be also reduced by decreasing the feedback capacitance C_{fa} and consequently increasing the preamplifier gain (Table 1). We speculate that for this reason virtually all the reported CMOS neural amplifiers use higher gain of the first amplification stage than the circuit proposed in this work. However, as increasing the gain results in higher voltage drop across the feedback resistor R_{fa}, it is expected that the side effect of such a solution will be higher signal distortion, particularly for large input signals. We provide more detailed discussion on distortion-vs-noise trade-off in Section 5.

Finally, an easy way to reduce the noise level is to increase the values of capacitors C_{ina}, C_{inb}, C_{fa} and C_{fb} (Table 1). Unfortunately, this comes at the price of circuit area. In reality, the total silicon area of many reported neural amplifiers is primarily defined by the capacitors—namely, both the input capacitors C_{ina} and C_{inb} (Figure 1). For design aiming for a compact amplifier footprint the capacitances must be kept as small as possible.

The analysis presented in this section and the noise values given in Table 1 lead to conclusion that that the proposed design parameters ($C_{ina} = C_{inb}$ = 4 pF, $C_{fa} = C_{fb}$ = 200 fF, K = 20) should allow for reasonably good noise performance, comparable with advanced multichannel neural amplifiers reported in the literature [10,28]. At the same time, the moderate gain value $K = 20$ V/V should help keeping the level of distortion under control. We therefore accept these parameters as the starting point for the following detailed analysis of the circuit distortion.

5. Design for Low Distortion, Low Noise and Small Silicon Area

In Section 3 we described the mechanism of nonlinear distortion generation based on simulations of the circuit with nominal settings ($K = 20$ V/V, $C_{in} = 4$ pF, cutoff frequency = 1 Hz). In this section we analyze how changing of these parameters affects the THD values and we discuss the distortion-noise trade-off. The results are based on simulations with an ideal operational amplifier so we can analyze distortion and noise introduced by the pseudoresistor decoupled from nonidealities of the operational amplifier itself. The pseudoresistors in these simulations are built from 5 V PMOS transistors ($W/L = 1$ μm/40 μm). We note that the results of the noise analyses presented here may be slightly different than those presented in Section 4, where only the thermal noise from an ideal resistor was considered; however, the conclusions given in Section 4 are sufficient to explain qualitatively the results presented here.

Figure 8 shows the noise PSD and the THD as a function of input signal frequency for various values of C_{in} and C_f but at fixed ratio $C_{in}/C_f = 20$. Both the noise and distortion performance benefit from larger capacitors. Increasing the C_f determines proportional decrease of the feedback resistance if the time constant of the high-pass filter is not expected to change. Obviously, the total current in the feedback loop increases proportionally to the increase of capacitors. The same applies to current in capacitor C_{fa} as well as to the drain-source currents of transistors forming the pseudoresistor (the ratio of drain-source current to current in C_{fa} for given frequency does not change, since we assumed that the time constant did not change). At the same time the gate currents for transistors A and B (Figure 4a) do not change (as both the gate capacitances and the gate-to-bulk voltages for these transistors do not change). In consequence the influence of the current I_{gbB}–I_{gbA} on the drain-source currents is reduced and the second peak in THD curves decreases, according to analysis presented in Section 3.3. On the other hand, increasing of C_{fa} and C_{ina} results automatically in reduction of thermal noise across both resistors R_{fa} and R_{fb}.

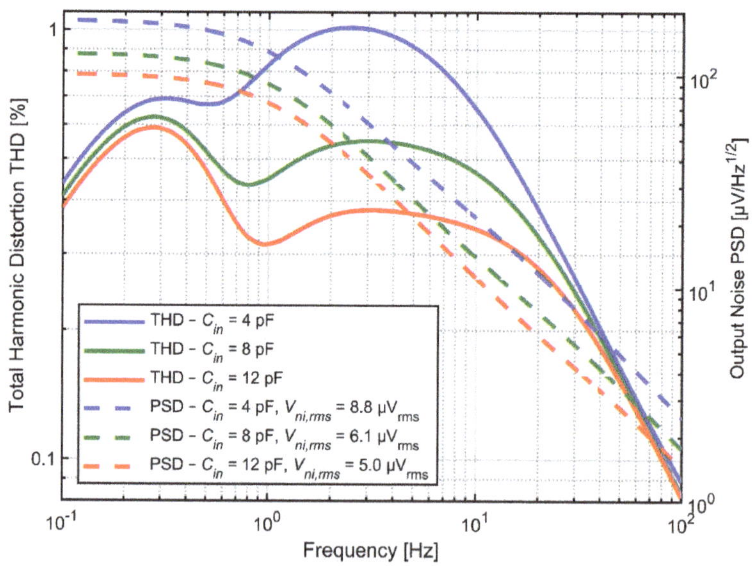

Figure 8. THD as a function of input signal frequency and output noise PSD for various values of C_{in} and C_f with fixed ratio $C_{in}/C_f = 20$. Cutoff frequency is fixed at 1 Hz.

Figure 9 presents the dependence of THD and noise spectrum on the amplifier gain. For the three gain settings the C_{in} value was fixed at 4 pF and the C_f value was set to 200, 80 or 40 fF for the gain of 20, 50 and 100 V/V, respectively. On the other hand, larger gain results in significant increase of the distortion, for two reasons. First, increased gain results in larger amplitudes of V_{out}, for which the pseudoresistor linearity becomes much worse; this effect is visible in the DC I-V curves in Figure 2. Second, the distortion related to gate-bulk nonlinear currents are also expected to increase, since these currents increase accordingly with larger gate-bulk voltages, and the total current in the feedback loop remains the same (as C_{in} does not change). The end result is that the ratio of I_{gbB}–I_{gbA} current to the drain-source currents is higher, and so are the THD values. On the other hand, since the noise rms is inversely proportional to the square-root of C_f and the gain is inversely proportional to C_f, the input-referred noise from the feedback resistor is lower if C_f is reduced. We conclude that for designs with strict limits on the silicon area, when C_{in} must be kept small, the gain of the preamplifier should be optimized for specific application in order to get the best compromise between noise and distortion values.

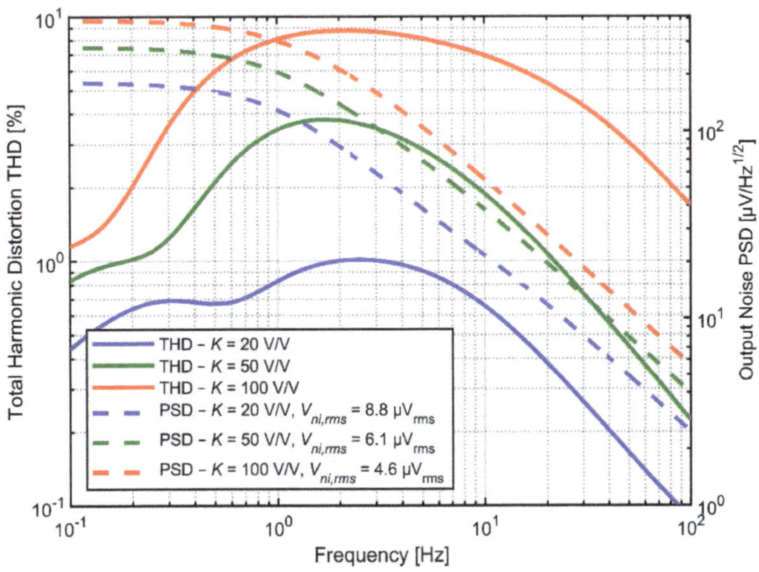

Figure 9. THD as a function of input signal frequency and output noise PSD for various values of K and C_f. C_{in} value is fixed at 4 pF and cutoff frequency is fixed at 1 Hz.

Tunable AC coupling circuit provides the opportunity to shape the spectrum of noise and distortion. One can shift the maxima of the THD curve out of the frequency range of interest by lowering the cutoff frequency (Figure 10). However, the low frequency signal components can still generate harmonics leaking into higher frequency range and can modulate higher-frequency signals due to circuit nonlinearities. It is therefore difficult to analyze the profit of the cutoff frequency decrease on the output signal distortion without knowing the spectrum of the input signal, and in particular, the power of very slow (out-of-band) oscillations of the electrode voltage. The positive effect of lowering the cutoff frequency on the noise measured above 1 Hz is straightforward, as shown in Figure 10.

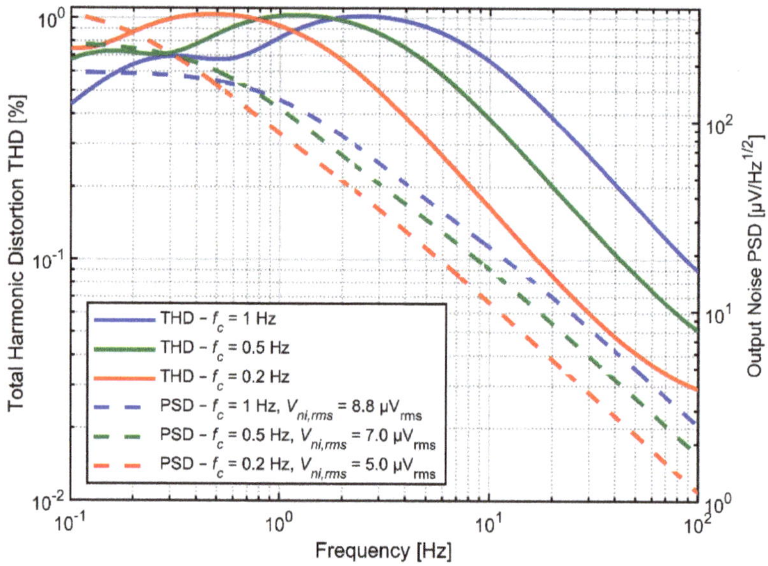

Figure 10. THD as a function of input signal frequency and output noise PSD for various values of the cutoff frequency with C_{in} = 4 pF and C_f = 200 fF (K = 20 V/V).

6. Complete Preamplifier Design

In order to verify results of our analyses in practical circuit we designed a complete neural preamplifier based on the AC coupling architecture discussed above. Figure 11a presents the block diagram of the preamplifier. We note that the V_{gs} voltage that tunes the feedback resistance is generated differently for transistors A and B. Since the source of transistor A is at virtual ground, the gate potential of transistor A can be generated off-channel and shared between all the channels of the ASIC.

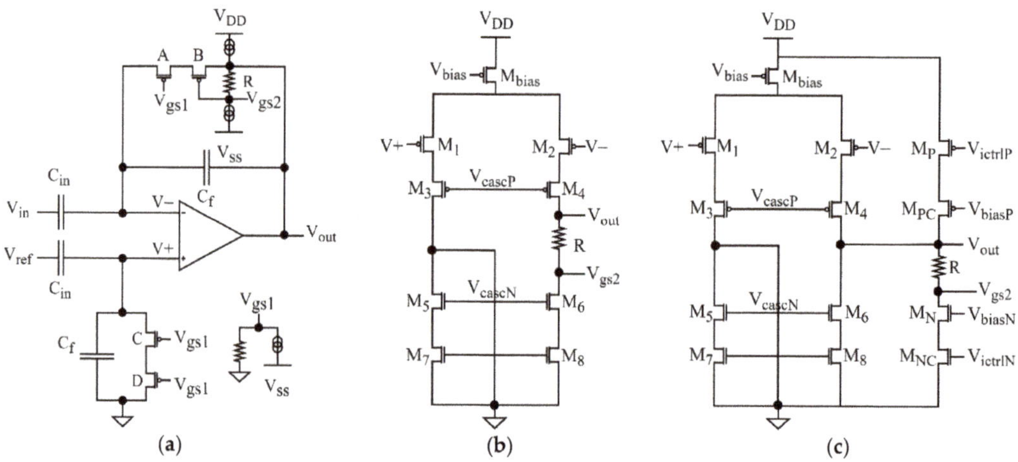

Figure 11. Design of the preamplifier: (**a**) Fixed-V_{gs} AC-coupling architecture using polysilicon resistors; (**b**) Design of the preamplifier with integrated polysilicon resistor; (**c**) Design of preamplifier for the test chip.

The gate potential of transistor B must be shifted by a constant value (V_{gs}) from the output voltage of the amplifier. This can be easily realized by integrating a single resistor into the amplifier, as shown in Figure 11b, and taking advantage of the bias current of transistor M2 to generate the V_{gs}. The required V_{gs} values are on the order of 250–400 mV (see below) so assuming the bias current of ~1 μA, the value of resistor R must be in the range of several hundreds of kΩ. Such values can be easily realized using a polysilicon resistors that typically have resistivity of a few kΩ per square (6.6 kΩ per square in case of process used in this work). Due to excellent linearity and matching properties of the polysilicon resistors, the proposed method for generation of V_{gs} value may be preferable to alternative solutions based on transistor-based voltage shifter [31]. Although tuning of the feedback resistance requires changing of the bias current which affects the thermal noise of the preamplifier (contributed mostly by transistors M_3 and M_4 in Figure 11b), this effect is negligible. The cutoff frequency scales exponentially with V_{gs} and can be shifted by an order of magnitude from its nominal value (1 Hz) with changing the bias current by ~25%. This results in a change of the thermal noise of transistors M_3 and M_4 by only ~12%.

For the design of the test integrated circuit we decoupled the controls of bias current and cutoff frequency, as shown in Figure 11c. This will allow us to measure the noise contribution of the amplifier as a function of bias current without changing the cutoff frequency. The current flowing through the polysilicon resistor is generated by cascode current sources. We used the resistor of 1 MΩ and the current of 315 nA is necessary to set the cutoff frequency to 1 Hz. For the design of the preamplifier we used the telescopic cascode architecture. Because the amplifier is designed for bidirectional neural interfaces with electrical stimulation capability, we plan to use relatively high supply voltage (3.0–3.6 V) for which the telescopic cascode architecture offers the best noise/power performance [37] and provides enough voltage headroom for the ±200 mV_{pp} ac voltage swing. The supply voltage for the simulations was set at ±1.8 V with respect to ground.

Figure 12 shows the layout of the test integrated circuit. The chip includes 14 identical channels. Each channel includes eight versions of the preamplifier differentiated by the sizing of the pseudoresistors (four versions with *W/L* respectively: 2 μm/40 μm, 1 μm/40 μm, 2 μm/20 μm, 1 μm/20 μm) and capacitors C_{in}/C_f (two versions: 4 pF/200 fF and 8 pF/400 fF). The design of the operational amplifier is identical for all 8 versions. The bias current and V_{gs} are controlled externally. The design was submitted to fabrication and the detailed measurements report will be published separately.

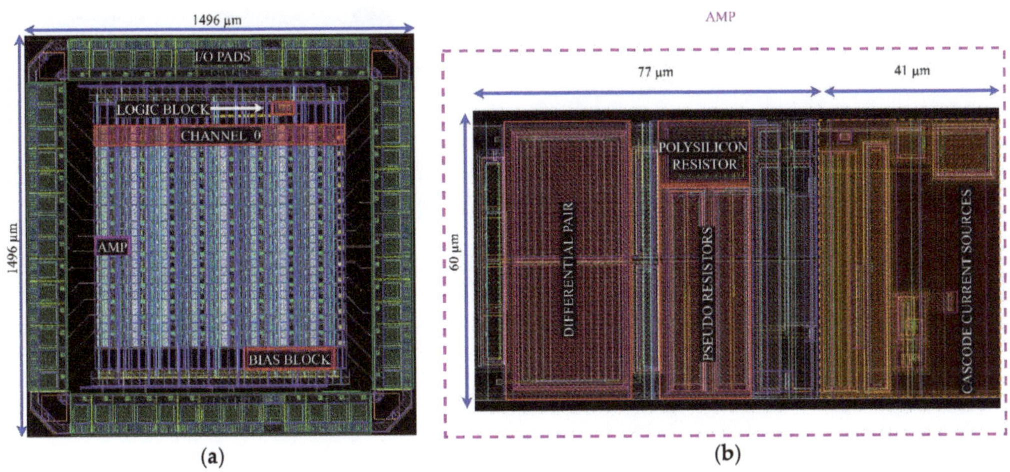

Figure 12. (**a**) Layout of the test integrated curcuit; (**b**) Layout of the preamplifier for the test chip. The elements inside the orange rectangle include the cascode transistors (M_P, M_{PC}, M_{NC}, M_N—compare with Figure 11c) and an output buffer; these blocks will not be included in the final preamplifier design shown in Figure 11b. The layout of the preamplifier is 118 μm × 60 μm (77 μm × 60 μm for the final design).

The results of noise simulations are presented in Figure 13 and Table 2. The total noise is dominated by the pseudoresistors in the LFP range (1–300 Hz) and by the preamplifier noise in the AP range (300 Hz–10 kHz). For both frequency ranges the noise on the order of 6 μV_{rms} is achievable. The results of post-layout simulations are perfectly consistent with simulations based on the schematics.

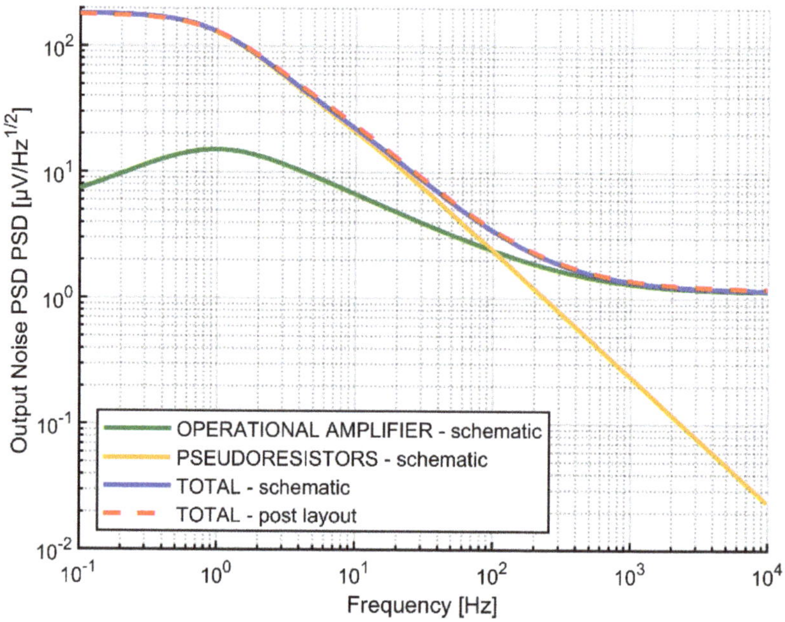

Figure 13. Output noise PSD for pseudoresistors and operational amplifier.

Table 2. Equivalent input noise [μV_{rms}]. For each bias current and cutoff frequency the noise is shown for two frequency ranges: 1–300 Hz and 300 Hz–10 kHz.

Cutoff Frequency of the AC-Coupling Circuit	Total Bias Current		
	2 μA	4 μA	6 μA
1.0 Hz	9.16 ∣ 6.18	9.03 ∣ 4.6	9.02 ∣ 3.93
0.5 Hz	7.49 ∣ 6.15	7.29 ∣ 4.57	7.26 ∣ 3.90
0.2 Hz	5.66 ∣ 6.13	5.44 ∣ 4.55	5.41 ∣ 3.87

In Figure 14 we compare the post-layout simulations of the THD-vs-frequency characteristic of the circuit presented in Figure 11c with schematic-based simulations of the circuit based on an ideal operational amplifier (discussed in the previous sections). The post-layout results show slightly higher THD peak at around 2.5 Hz (1.17% vs. 1.01%); otherwise the two graphs are very similar. Finally in Figure 15 we present the results of post-layout Monte Carlo simulations of the THD curves. The transistors mismatch leads to slight increase of THD below the cutoff frequency, which is associated with perturbed symmetry of the I-V curve for positive and negative voltages. However, the peak at ~2.5 Hz that is responsible for the global maximum of the THD characteristic, is not affected by the mismatch.

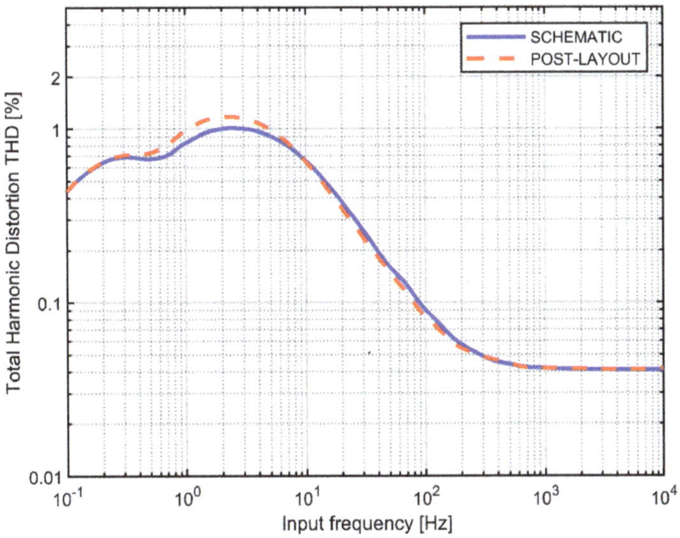

Figure 14. Simulated THD versus signal frequency for the complete AC-coupled preamplifier. Signal amplitude: 10 mV_{pp}, cutoff frequency 1 Hz.

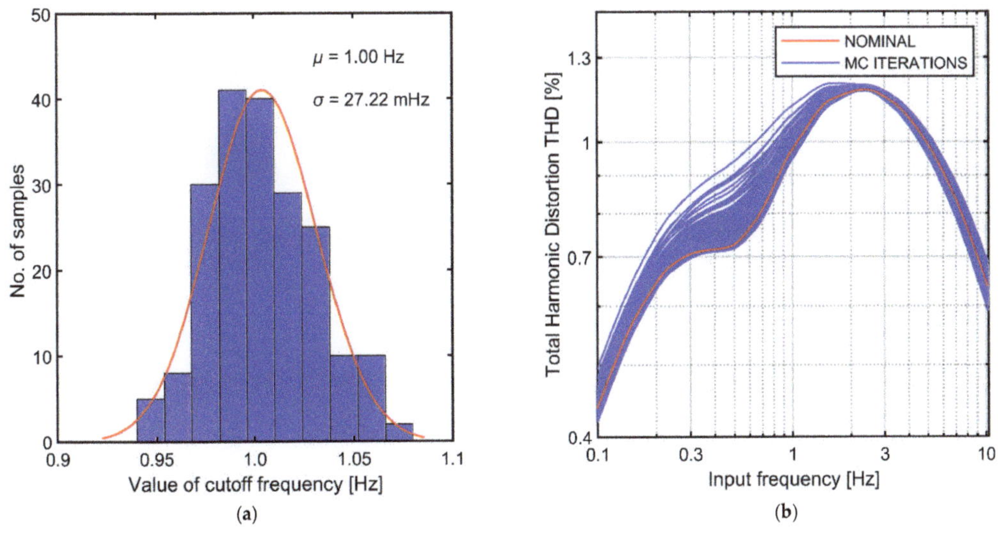

Figure 15. Results of Monte Carlo simulations of complete AC-coupled preamplifier: (**a**) Spread of the cutoff frequency; (**b**) Spread of the THD versus signal frequency characteristics.

The parameters of the test ASIC are given in Table 3. The results suggest that the proposed circuit should be capable of providing the low-distortion amplification of full range of neuronal signals, with competitive noise and power figures and very small design area. However, small corrections of the preamplifier gain and/or absolute values of capacitors C_{in} and C_f may be necessary to meet the goal of <1% THD value for large signal amplitudes (10 mV_{pp}) and across complete range of signal frequencies. These considerations will be concluded based on detailed measurements of the test chip.

Table 3. Parameter of the amplifier based on post-layout simulations. Cutoff frequency was set at 0.2 Hz.

Parameters	Values
Open loop gain	89.5 dB
AC gain	25.9 dB
Total bias current	2 μA
Power dissipation per channel	7.2 μW
Equivalent input noise in the LFP range	5.66 μV_{rms}
Equivalent input noise in the AP range	6.13 μV_{rms}
Equivalent input wide-band noise	8.34 μV_{rms}
NEF	4.55
Area single amplifier prototype	0.0071 mm^2
Area single amplifier simplified	0.0046 mm^2

7. Conclusions

In this article we present an improved AC-coupled CMOS neural amplifier that operates with low nonlinear distortion over wide range of signal frequencies (1 Hz–10 kHz) and amplitudes (up to 10 mV_{pp}). Since in the proposed circuit pseudoresistors that control the time constant of AC-coupling circuit work with fixed V_{gs} voltages, the linearity of the resistance is improved and the THD values are greatly reduced. The proposed solution requires adding of only a single compact polysilicon resistor to the preamplifier schematic. We describe the origins of nonlinear distortion and analyze the THD as a function of signal frequency and amplitude. We also analyze the impact of basic amplifier parameters (silicon area, gain and cutoff frequency) on the distortion and noise performance of the circuit. Post-layout simulations confirm that the proposed preamplifier is suitable for recording the full spectrum of electrophysiological signals with low distortion (THD < 1.17%) and competitive noise performance (~8.3 μV_{rms}). Compared with the standard solution using the AC-coupling circuit with variable V_{gs}, the circuit described here provides reduction of the THD values at low frequencies and large amplitudes by more than one order of magnitude. High-fidelity signal amplification and very compact footprint of the preamplifier (0.0046 mm^2) make the design relevant for the future CMOS-based very large scale neuroelectronic interfaces.

Author Contributions: Conceptualization, P.H.; methodology, P.H. and W.D.; simulations, B.T.-J.; data analysis, B.T.-J., P.H. and W.D.; schematic design, B.T.-J. and P.H.; layout design, B.T.-J.; writing—original draft preparation, P.H. and B.T.-J.; writing—review & editing, W.D. and P.H.; visualization, B.T.-J. and P.H.; supervision, P.H.; project administration, W.D.; funding acquisition, P.H. and W.D. All authors have read and agreed to the published version of the manuscript.

Funding: This work was supported by Polish National Science Centre grant DEC-2013/10/M/NZ4/00268 (PH). B.T.-J. has been partly supported by the EU Project POWR.03.02.00-00-I004/16.

Institutional Review Board Statement: Not applicable.

Informed Consent Statement: Not applicable.

Data Availability Statement: Not applicable.

Acknowledgments: We thank Tomasz Fiutowski for his technical support.

Conflicts of Interest: The authors declare no conflict of interest.

References

1. Beggs, J.M.; Plenz, D. Neuronal avalanches in neocortical circuits. *J. Neurosci.* **2003**, *23*, 11167–11177. [CrossRef] [PubMed]
2. Pillow, J.W.; Shlens, J.; Paninski, L.; Sher, A.; Litke, A.M.; Chichilnisky, E.J.; Simoncelli, E.P. Spatio-temporal correlations and visual signalling in a complete neuronal population. *Nature* **2008**, *454*, 995–999. [CrossRef] [PubMed]
3. Buzsáki, G.; Stark, E.; Berényi, A.; Khodagholy, D.; Kipke, D.R.; Yoon, E.; Wise, K.D. Tools for probing local circuits: High-density silicon probes combined with optogenetics. *Neuron* **2015**, *86*, 92–105. [CrossRef] [PubMed]
4. Obien, M.E.J.; Deligkaris, K.; Bullmann, T.; Bakkum, D.J.; Frey, U. Revealing neuronal function through microelectrode array recordings. *Front. Neurosci.* **2015**, *8*, 423. [CrossRef] [PubMed]

5. Lebedev, M.A.; Nicolelis, M.A.L. Brain-machine interfaces: From basic science to neuroprostheses and neurorehabilitation. *Physiol. Rev.* **2017**, *97*, 767–837. [CrossRef] [PubMed]
6. Hochberg, L.R.; Bacher, D.; Jarosiewicz, B.; Masse, N.Y.; Simeral, J.D.; Vogel, J.; Haddadin, S.; Liu, J.; Cash, S.S.; Van Der Smagt, P.; et al. Reach and grasp by people with tetraplegia using a neurally controlled robotic arm. *Nature* **2012**, *485*, 372–375. [CrossRef]
7. Gilja, V.; Nuyujukian, P.; Chestek, C.A.; Cunningham, J.P.; Yu, B.M.; Fan, J.M.; Churchland, M.M.; Kaufman, M.T.; Kao, J.C.; Ryu, S.I.; et al. A high-performance neural prosthesis enabled by control algorithm design. *Nat. Neurosci.* **2012**, *15*, 1752–1757. [CrossRef]
8. Jun, J.J.; Steinmetz, N.A.; Siegle, J.H.; Denman, D.J.; Bauza, M.; Barbarits, B.; Lee, A.K.; Anastassiou, C.A.; Andrei, A.; Aydin, Ç.; et al. Fully integrated silicon probes for high-density recording of neural activity. *Nature* **2017**, *551*, 232–236. [CrossRef]
9. Siegle, J.H.; Jia, X.; Durand, S.; Gale, S.; Bennett, C.; Graddis, N.; Heller, G.; Ramirez, T.K.; Choi, H.; Luviano, J.A.; et al. Survey of spiking in the mouse visual system reveals functional hierarchy. *Nature* **2021**, *592*, 86–92. [CrossRef]
10. Sahasrabuddhe, K.; Khan, A.A.; Singh, A.P.; Stern, T.M.; Ng, Y.; Tadić, A.; Orel, P.; LaReau, C.; Pouzzner, D.; Nishimura, K.; et al. The argo: A high channel count recording system for neural recording in vivo. *J. Neural Eng.* **2020**, *18*, 015002. [CrossRef]
11. Musk, E. An integrated brain-machine interface platform with thousands of channels. *J. Med. Internet Res.* **2019**, *21*, e16194. [CrossRef] [PubMed]
12. Stuart, G.; Schiller, J.; Sakmann, B. Action potential initiation and propagation in rat neocortical pyramidal neurons. *J. Physiol.* **1997**, *505*, 617–632. [CrossRef] [PubMed]
13. Buzsáki, G.; Anastassiou, C.A.; Koch, C. The origin of extracellular fields and currents-eeg, ecog, lfp and spikes. *Nat. Rev. Neurosci.* **2012**, *13*, 407–420. [CrossRef] [PubMed]
14. Gao, H.; Walker, R.M.; Nuyujukian, P.; Makinwa, K.A.A.; Shenoy, K.V.; Murmann, B.; Meng, T.H. HermesE: A 96-channel full data rate direct neural interface in 0.13 Mm CMOS. *IEEE J. Solid-State Circuits* **2012**, *47*, 1043–1055. [CrossRef]
15. Harrison, R.R.; Charles, C. A low-power low-noise CMOS amplifier for neural recording applications. *IEEE J. Solid-State Circuits* **2003**, *38*, 958–965. [CrossRef]
16. Shahrokhi, F.; Abdelhalim, K.; Serletis, D.; Carlen, P.L.; Genov, R. The 128-channel fully differential digital integrated neural recording and stimulation interface. *IEEE Trans. Biomed. Circuits Syst.* **2010**, *4*, 149–161. [CrossRef] [PubMed]
17. Kmon, P.; Gryboś, P. Energy efficient low-noise multichannel neural amplifier in submicron CMOS process. *IEEE Trans. Circuits Syst. I Regul. Pap.* **2013**, *60*, 1764–1775. [CrossRef]
18. Yang, T.; Holleman, J. An ultralow-power low-noise CMOS biopotential amplifier for neural recording. *IEEE Trans. Circuits Syst. II Express Briefs* **2015**, *62*, 927–931. [CrossRef]
19. Buzsáki, G.; Draguhn, A. Neuronal oscillations in cortical networks. *Science* **2004**, *304*, 1926–1929. [CrossRef]
20. Dabrowski, W.; Grybos, P.; Litke, A.M. A low noise multichannel integrated circuit for recording neuronal signals using microelectrode arrays. *Biosens. Bioelectron.* **2004**, *19*, 749–761. [CrossRef]
21. Chandrakumar, H.; Markovic, D. A high dynamic-range neural recording chopper amplifier for simultaneous neural recording and stimulation. *IEEE J. Solid-State Circuits* **2017**, *52*, 645–656. [CrossRef]
22. Sharma, K.; Sharma, R. Design considerations for effective neural signal sensing and amplification: A review. *Biomed. Phys. Eng. Express* **2019**, *5*. [CrossRef]
23. Enz, C.C.; Temes, G.C. Circuit techniques for reducing the effects of op-amp imperfections: Autozeroing, correlated double sampling, and chopper stabilization. *Proc. IEEE* **1996**, *84*, 1584–1614. [CrossRef]
24. Gosselin, B.; Ayoub, A.E.; Roy, J.F.; Sawan, M.; Lepore, F.; Chaudhuri, A.; Guitton, D. A mixed-signal multichip neural recording interface with bandwidth reduction. *IEEE Trans. Biomed. Circuits Syst.* **2009**, *3*, 129–141. [CrossRef] [PubMed]
25. Bagheri, A.; Salam, M.T.; Velazquez, J.L.P.; Genov, R. Low-frequency noise and offset rejection in dc-coupled neural amplifiers: A review and digitally-assisted design tutorial. *IEEE Trans. Biomed. Circuits Syst.* **2017**, *11*, 161–176. [CrossRef] [PubMed]
26. Uran, A.; Leblebici, Y.; Emami, A.; Cevher, V. An ac-coupled wideband neural recording front-end with Sub-1 Mm2×fj/Conv-step efficiency and 0.97 Nef. *IEEE Solid-State Circuits Lett.* **2020**, *3*, 258–261. [CrossRef]
27. Kim, S.J.; Han, S.H.; Cha, J.H.; Liu, L.; Yao, L.; Gao, Y.; Je, M. A Sub-μW/Ch analog front-end for delta -neural recording with spike-driven data compression. *IEEE Trans. Biomed. Circuits Syst.* **2019**, *13*, 1–14. [CrossRef] [PubMed]
28. Mora Lopez, C.; Putzeys, J.; Raducanu, B.C.; Ballini, M.; Wang, S.; Andrei, A.; Rochus, V.; Vandebriel, R.; Severi, S.; Van Hoof, C.; et al. A neural probe with up to 966 electrodes and up to 384 configurable channels in 0.13 μm SOI CMOS. *IEEE Trans. Biomed. Circuits Syst.* **2017**, *11*, 510–522. [CrossRef] [PubMed]
29. Zhang, M.; Tang, Z.; Liu, X.; Van der Spiegel, J. Electronic neural interfaces. *Nat. Electron.* **2020**, *3*, 191–200. [CrossRef]
30. Noshahr, F.H.; Nabavi, M.; Sawan, M. Multi-channel neural recording implants: A review. *Sensors* **2020**, *20*, 904. [CrossRef]
31. Kassiri, H.; Abdelhalim, K.; Genov, R. low-distortion super-gohm subthreshold-mos resistors for cmos neural amplifiers. In Proceedings of the 2013 IEEE Biomedical Circuits and Systems Conference (BioCAS 2013), Rotterdam, The Netherlands, 31 October–2 November 2013; pp. 270–273. [CrossRef]
32. Abdelhalim, K.; Jafari, H.M.; Kokarovtseva, L.; Velazquez, J.L.P.; Genov, R. 64-channel uwb wireless neural vector analyzer soc with a closed-loop phase synchrony-triggered neurostimulator. *IEEE J. Solid-State Circuits* **2013**, *48*, 2494–2510. [CrossRef]
33. Piecewise Cubic Hermite Interpolating Polynomial (PCHIP)—MATLAB Pchip. Available online: https://www.mathworks.com/help/matlab/ref/pchip.html (accessed on 27 February 2021).

34. Han, D.; Zheng, Y.; Rajkumar, R.; Dawe, G.S.; Je, M. A 0.45 v 100-channel neural-recording IC with Sub-MW/Channel consumption in 0.18 μ CMOS. *IEEE Trans. Biomed. Circuits Syst.* **2013**, *7*, 735–746. [CrossRef] [PubMed]
35. Tran, L.; Cha, H.K. An ultra-low-power neural signal acquisition analog front-end IC. *Microelectron. J.* **2021**, *107*, 104950. [CrossRef]
36. Lewyn, L.L.; Ytterdal, T.; Wulff, C.; Martin, K. Analog circuit design in nanoscale CMOS technologies. *Proc. IEEE* **2009**, *97*, 1687–1714. [CrossRef]
37. Ng, K.A.; Greenwald, E.; Xu, Y.P.; Thakor, N.V. Implantable neurotechnologies: A Review of integrated circuit neural amplifiers. *Med. Biol. Eng. Comput.* **2016**, *54*, 45–62. [CrossRef]

sensors

MDPI

Article

Low-Power Wireless Sensor Network Using Fine-Grain Control of Sensor Module Power Mode

Seongwon You [1], Jason K. Eshraghian [2,*], Herbert C. Iu [3] and Kyoungrok Cho [1,*]

1 Department of Information and Communication Engineering, Chungbuk National University, Chungbuk 28644, Korea; watch@cbnu.ac.kr
2 Electrical Engineering and Computer Science Department, University of Michigan, Ann Arbor, MI 48104, USA
3 School of Electrical, Electronic and Computer Engineering, University of Western Australia, Perth 6009, Australia; herbert.iu@uwa.edu.au
* Correspondence: jasonesh@umich.edu (J.K.E.); krcho@cbnu.ac.kr (K.C.)

Abstract: Wireless sensor nodes are heavily resource-constrained due to their edge form factor, which has motivated increasing battery life through low-power techniques. This paper proposes a power management method that leads to less energy consumption in an idle state than conventional power management systems used in wireless sensor nodes. We analyze and benchmark the power consumption between Sleep, Idle, and Run modes. To reduce sensor node power consumption, we develop fine-grained power modes (FGPM) with five states which modulate energy consumption according to the sensor node's communication status. We evaluate the proposed method on a test bench Mica2. As a result, the power consumed is 74.2% lower than that of conventional approaches. The proposed method targets the reduction of power consumption in IoT sensor modules with long sleep mode or short packet data in which most networks operate.

Keywords: sensor node; power mode; wireless sensor networks; power management

Citation: You, S.; Eshraghian, J.K.; Iu, H.C.; Cho, K. Low-Power Wireless Sensor Network Using Fine-Grain Control of Sensor Module Power Mode. *Sensors* **2021**, *21*, 3198. https://doi.org/10.3390/s21093198

Academic Editor: Geoff Merrett

Received: 24 March 2021
Accepted: 28 April 2021
Published: 4 May 2021

1. Introduction

In the Internet of Things (IoT), a wireless sensor network node (WSN) is a system that recognizes physical changes or signals targeting various users or environments. It is widely used in various forms for a range of purposes across home and industrial use. In general, a sensor network requires a single smart sensor node capable of detecting numerous signals such as pressure, temperature, humidity, gas flow, infrared, chemical reactions, surface, sound, steam, and others. Handling huge datasets using multiple sensory modalities is within the domain of machine learning, and interpreting the information from numerous signals is becoming increasingly important in an IoT-driven world. The common communication protocols used in WSN have been a combination of cellular and short-range wireless network technology such as Bluetooth, ZigBee, etc. [1,2]. WSN requires a long battery life, so a low-power circuit is essential with substantial resource constraints. To achieve energy efficiency, MIMO is used in the 5G environment [3]. A WSN typically has one or more connected sensors on each node and is monitoring a given physical environment with distributed multiple hops. Baniata et al. [3] used a probability sub-optimal multi-hop routing mechanism among cluster heads to increase the lifespan of the sensor network. The sensor node operates according to an event or application command, and each node communicates wirelessly. Each node has one or more sensors, a microprocessor unit, and a radio unit that receives wake-up signals. Piyare et al. [4] introduced an extension of the TSCH (Time Slotted Channel Hopping) protocol to low data rate applications using the sub-GHz frequency bands operating on TI's System-on-Chip. They employed a special schedule for the network's root nodes and their direct neighbors, as well as the option to have multiple root nodes in a single network. Most WSNs rely on small batteries, which is a serious bottleneck in the system [5,6]. Battery capacity is developing at a very slow rate compared to other technologies, such as integrated circuits or software design. Therefore, energy efficiency has been the prime goal when designing and deploying WSNs [7,8]. Bachir et al. address the challenges related to the reliability of communication and the efficient use of the node's battery in WSN [9]. The performance is improved by doubling the network packet delivery ratio. Please note that the sensor node is in an idle state most of the time. David et al. [10] provided a platform wake-up receiver (WuR) with high integrability and a low cost per node to facilitate the implementation of low-cost sensor nodes. They demonstrate the feasibility of implementing a WuR with commercially available off-chip components by demonstrating a radio frequency envelope detection (RFED) WuR on a PCB mount. The most significant power consumption savings are observed when WuRs are used in low-traffic and low-density WSNs, mainly because the main transceiver is in the sleep mode for most of the time [11]. However, this type of customized platform is not cost-effective. Therefore, low-power design of circuits, architectures, algorithms, protocols, and other elements that affect power management must be carefully considered [12–14]. The conventional radio interface or transceiver is frequently the most power-consuming element in a WSN node, dominating both the static and dynamic power consumption of the sensor [15]. The duty cycle controlling a radio receiver and transceiver is a common and well-known solution for reducing the power consumption of WSNs. It reduces the active operation time of the sensor node but increases the wake-up time when the sensor node is in a power saving state for a long time. Therefore, in real-time communication, a wireless method with a very low duty cycle may be inadequate. Similarly, Bdiri et al. [8] introduced the wake-up receiver (WuRx), which handles idle listening while keeping the main radio completely off. The main function of WuRx is to send an interrupt signal to the processing unit when receiving a radio frequency (RF) or wake-up packet (WuPt). However, the drawback is that the WuRx must always be turned on for communication with a very short waiting time. In terms of IoT MCU, ESP32 MCU [16] is conceptually similar to the FGPM proposed in this paper, but the difference is the lack of a distinction between Idle and Active modes. Our proposed FGPM distinguishes the transmit and receive stages in specific modes, which allows for fine-grain control of the duty cycle, which has been shown to be more advantageous for power consumption control. In this paper, we propose a power management method for WSNs with five states of fine-grained power modes. We evaluate the proposed method on a test bench Mica2 [17]. As a result, of increased granularity, power consumption is reduced by 74.2% when compared to conventional methods [12]. This paper is structured as follows: Section 2 will present the communication protocol in WSNs, Section 3 will propose the fine-grained power state approach, Section 4 will present simulation results and energy consumption data of our approach, with a comparison against comparable methods, before concluding the work.

2. Communication in Wireless Sensor Node

2.1. Wireless Sensor Node Architecture

Each wireless sensor node has a sensing unit which detects events in the allocated area. In a given physical environment, the WSN monitors the events distributed via multi-hops routing. It communicates with the neighboring nodes to deliver the event to the user upon detection. Each inter-node communication uses a wireless transceiver in accordance with the given network protocol [18]. Figure 1 shows a basic WSN and a block diagram of a sensor module. Each node nearby events and broadcasts a signal to the users once the event is confirmed. The node module Mica2 consists of sensing units, a processor Atemga128L with memory, and an RF (CC1000) subsystem with low supply voltage, 1.6–3.6 V [17]. The Mica2's CC1000 is a wireless data transmitter and receiver suitable for short-range communications such as gas sensors and has the advantage of easy low-power management with a simple circuit. The firmware on the microprocessor controls all sub-modules with a power management strategy.

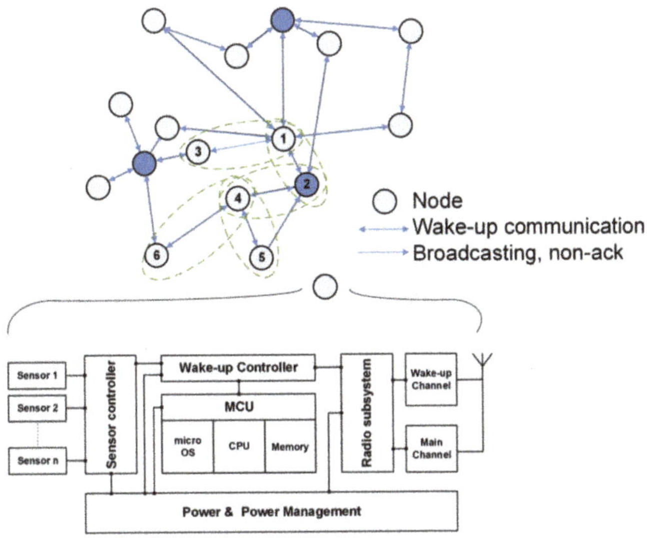

Figure 1. WSN Network connected multiple sensors and hardware block diagram of the sensor node module with a similar architecture to Mica2.

2.2. *Wireless Sensore Node Communication*

Figure 2 shows that sensor node communication is based on a handshake protocol. The first node that detects an event becomes a transmitting node (Node 1). It broadcasts a wake-up message to neighboring nodes in a fixed time slot. The node then transfers data packets to a downstream node. After packet transmission, the communication between nodes is completed by the Ack signal. When the possible communication nodes are from Node 2 to Node 4, communication is completed from the node with the highest priority (e.g., based on distance). Then, the successor node becomes a transmitting node which sends the wake-up signal to another neighboring node repeating the above procedure. When a time-out occurs due to packet loss, the scenario is restarted from the wake-up message transmission. Finally, the data on the first wake-up node is transmitted to a sink node as an alarm signal to alert the user. Additional functionalities can be integrated into the WSN as determined by the application and specifications [13]. The communication scheduling between nodes is an important issue to be considered for node power management.

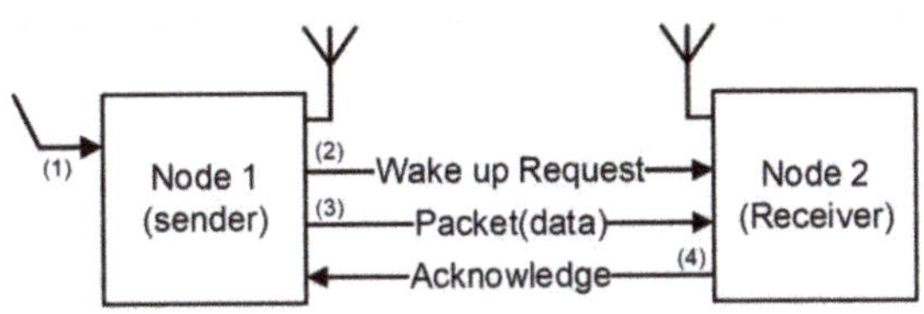

Figure 2. Simplified sensor node communication with a handshake protocol.

Figure 3 shows the node communication schedule based on the WiseMAC protocol that determines switching from sleep mode to wake-up mode. All nodes stay in the medium idle state to receive periodically wireless wake-up signals to preserve battery life. The WiseMAC (Wireless Sensor MAC) protocol is a low-power media access control protocol designed for wireless sensor networks developed based on CSMA and preamble sampling [19]. The advantage of WiseMAC is that it dynamically reduces the size of the wake-up preamble. The wake-up signal does not require high traffic and shows that it can have low-power consumption in the wake-up state. The disadvantage is that it has a slow response performance at wake-up time from sleep. However, this trade-off is tolerable under the given circumstances, where the primary aim is to have a long battery lifetime. Table 1 shows the time parameters that are defined in Figure 3 [20]. We set the medium idle checking pulse T_P such that the WSN receives a wake-up message at this time. The medium idle term T_W is the time interval of the idle time T_P. Therefore, a node broadcasts a wake-up message during T_W and the neighbor node detects an event at T_P. The receiver node (Node 2) returns an Ack signal during T_c when it receives the lossless data packet. Please note that Ack includes the wake-up schedule of the receiver node as piggyback data. When the Ack is completed without any irregularities, the Node 1 transmits a wake-up message (wRx, wake-up Rx) during T_P+T_{idle} according to the received wake-up schedule of the Node 2. This can be applied to the WiseMAC protocol. A time-out occurs since the Ack does not normally arrive during inter-node communication. Node 1 regards it as a packet loss or wake-up failure and transmits the wake-up message and packet again in the next schedule [13,21].

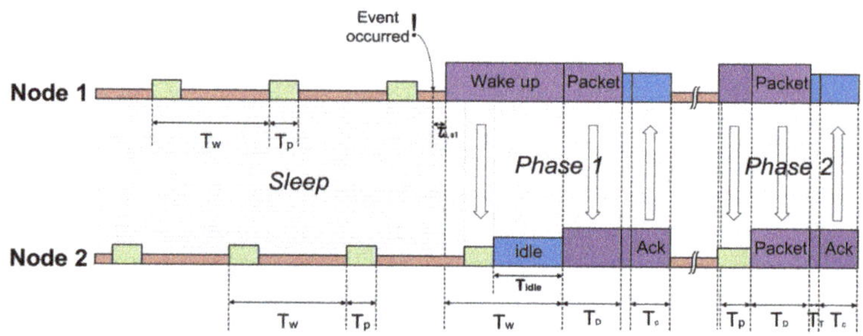

Figure 3. Node communication considering wake-up schedule.

Table 1. Node communication time parameters based on WiseMAC protocol.

Parameter	Description	Value
T_W	Medium idle term	1000 ms
T_D	Packet exchange	16 ms
T_c	Ack exchange	3.2 ms
T_T	Turnaround time between RX and TX	0.4 ms
T_P	Event sensing time	120 ms
T_{idle}	Received wake-up and waiting for Packet receive	880 ms

3. Proposed Fine Grain Five States Power Mode

3.1. Power Management

The sensor node has three operating modes: sleep, idle, and run, as shown in Figure 4. Sleep mode waits for a wake-up event, idle mode is for receiving data or standby state to receive a specific command, and run mode is for executing a specific command. In this paper, we divide the sleep mode into three modes: a deep-sleep mode (state 0), a semi-idle sensor (state 1), and a semi-idle wRx (state 2, using wake-up Rx channel). These correspond to a sleep mode which uses minimum power, a sleep mode using only the sensing unit, and a sleep mode using the wake-up Rx channel. The sleep mode transits to idle mode (s3) when a wake-up event is detected or transits to run mode (s4) according to the node schedule.

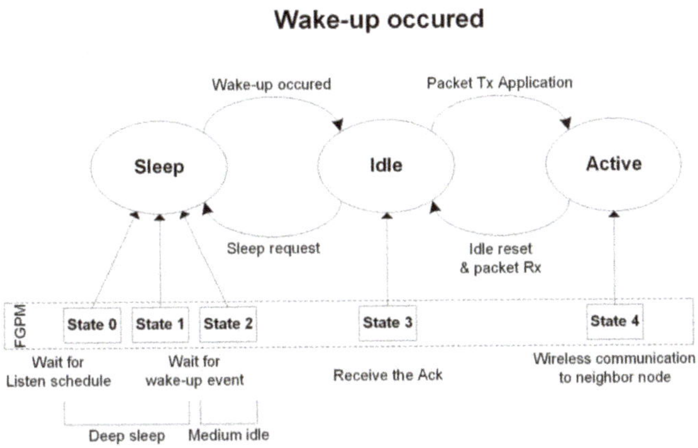

Figure 4. Tri-state power mode control is modified to five states power mode to reduce power consumption on WSN.

Table 2 shows modules and power consumption for each state [22,23]. As the state increases, the power consumption, as well as the time (T_{TR}) and energy (E_{TR}) for changing between modes also increases. Each state is classified based on the processing state of the sensor node. This power mode distinction is similar to Advanced Configuration and Power Interface (ACPI) [5,13,24]. The paper [9] also lists power consumption parameters for the preselected microcontrollers stated in the datasheet. However, it is estimated at the MCU level.

Table 2. Power consumption of the function blocks of the sensor module.

	MCU	Memory	Sensing Unit	Radio	Wake Ctrl	Power
State 0	Sleep	sleep	off	off	off	95 uW
State 1	Sleep	sleep	on	off	on	203 uW
State 2	Sleep	on	on	wRx	on	28.2 mW
State 3	Idle	on	off	wTx, pRx, Tx	off	88 mW
State 4	Run	on	off	wTx, pRx, Tx	off	385.5 mW

3.2. FGPM Operation

Figure 5 shows the transition between power modes using the proposed FGPM management technique, which controls the five states of various power modes [25]. It is represented as a finite state machine initializing at state 0. The communication schedule in Figure 3 shows how to switch between states. The sensors of the sensor module periodically switch between ON/OFF to check for events, which is the same as a state transition between State 0 and State 1. State 2 uses the wRx channel, and also periodically switches ON/OFF. This is identical to the medium idle state described above [12,20]. A node detects an event by periodically transitioning between State 0 $\rightarrow$ State 1, and State 2 $\rightarrow$ State 0 within a short period of time. When an event is detected in State 1 or State 2, it moves to State 4 through State 3. If an event is detected in State 1, then State 4 broadcasts a wake-up message and transmits the data packet. Finally, it switches to State 3 using only a packet Rx (pRx) at the receiving Ack. When an event is detected in State 2, the system switches to State 4 and returns the Ack signal. The sensor module goes back to State 0 when the scheduling of each node is completed. The inter-node communication performs the same procedure in a subsequent cycle.

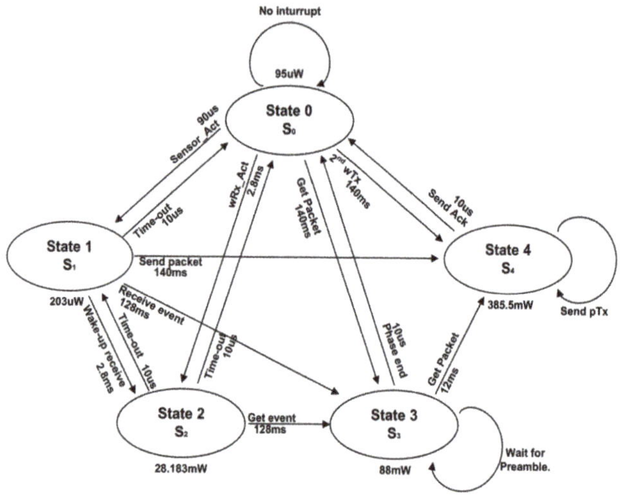

Figure 5. Power mode state transition for fine-grain (with five states) power control on WSN. State 0 is the initial state which is sleep state.

3.3. Power Mode Control on the Sensor Node Platform

Figure 6 shows the power control of functional blocks for states of a node by using the power management procedure of the microprocessor on the module. The sensing unit and wRx communicate their signals to the MCU through the wake-up controller. By using a separate wake-up controller, the MCU can sleep while receiving those signals thus preserving the battery life for longer than using an extended standby state. The MCU controls the power management mode that it supplies, or blocks the power of each function block. Table 2 shows power consumption of each functional block for each state [26,27].

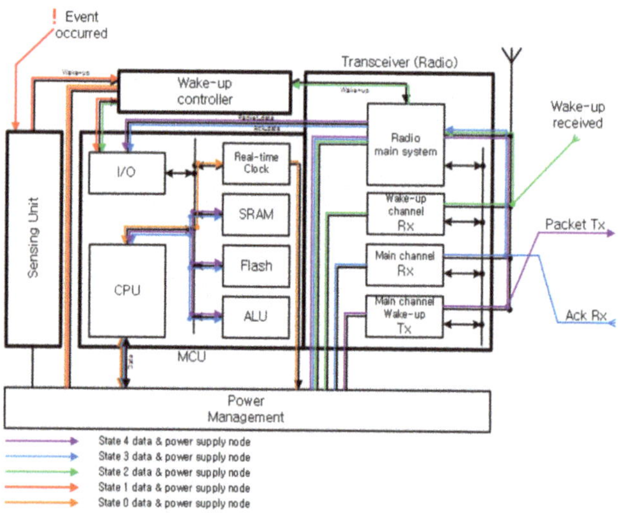

Figure 6. Self-waking paths on the sensor node module.

Figure 7 shows the wake-up signal process. When the sensor node switches from State 0 to State 1 and State 2 and detects an event, it transmits a wake-up message to the

MCU. This enables the sensing unit and wRx channel. After the sensor node processes the signal from the sensor and antenna, they are transmitted to the wake-up controller which is the intermediate manager [28]. A controller that uses a separated power supply sends an interrupt signal to the MCU. Having completed its task, the activated MCU resets the blocks.

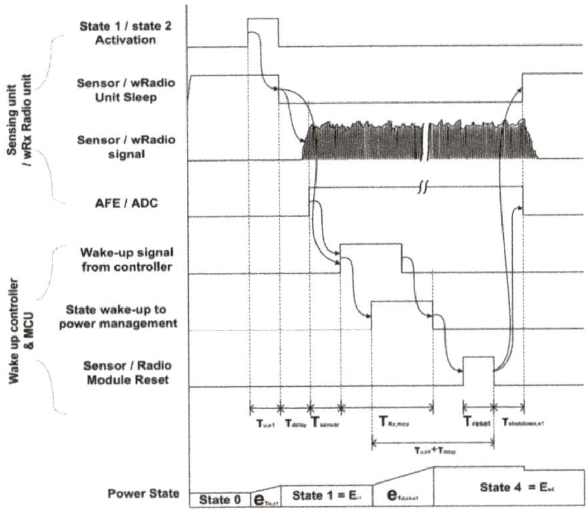

Figure 7. Modeling of node wake-up signal processing.

4. Energy Analysis of FGPM

Five-State Energy Consumption

The power consumption at each state of FGPM is shown in Figure 8. When the state increases, it is transmitted through the intermediate state. As described in Section 2, node communication is processed according to a defined protocol. Even if there is a time delay, it does not affect communication when adhering to a dedicated schedule [29]. FGPM can be applied with significant savings for WSNs that have long latency and short-length data communications, such as gas sensors [13,27].

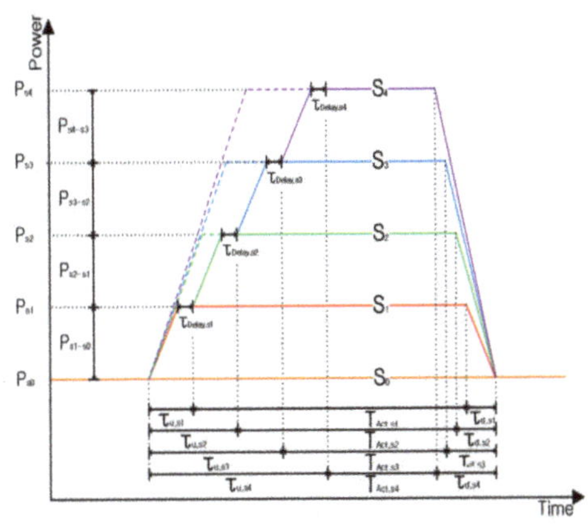

Figure 8. Modeling of node wake-up signal processing.

The energy consumed at a state k is shown in Equation (1). It is the sum of the energy consumed by the transition from State 0 to State k, E_{TR}, and the energy at State k, $E_{Act,k}$.

The energy consumption, E_{TR}, while activating State k is equal to the product of the power consumption, P_k, and the duration of State k, T_k.

$$E_k = P_k \times T_k = E_{TR} + E_{Act,k} \tag{1}$$

E_{TR} is an important factor in the calculation of energy consumption. By increasing power mode states, the energy consumption due to state transitions also contributes to a greater power consumption. Thus, too many power modes will result in diminishing returns. The equation for E_{TR} is given in Equations (2) to (7).

$$E_{TR,k} = P_{TR,k} \times T_{TR,k} \tag{2}$$

$$T_{TR,k} = \left(T_{s0 \leftarrow k} + T_{k \leftarrow s0} \right) = \frac{(\tau_{u,k} + \tau_{d,k})}{2} \tag{3}$$

where $\tau_{u,k}$ and $\tau_{d,k}$ are the time taken to activate the node in the previous State $k-1$ and to return to the previous power reduction mode, are given as follows:

$$\tau_{d,k} = \left(\tau_{d,k-1} + \tau_{d,(k-1) \leftarrow k} \right), \quad k \geq 2 \tag{4}$$

$$\tau_{u,k} = \left(\tau_{u,k-1} + \tau_{delay,k} + \tau_{u,k \leftarrow (k-1)} \right), \; k \geq 2 \tag{5}$$

Table 3 shows the time taken for WiseMAC protocol communication and the time required for state transitions when the node's power consumption state is changing.

Table 3. Power states and Time used in the sleep state.

Parameter	Value	Parameter	Value
T_w	1000 ms	$\tau_{u,s1}$	0.09 ms
T_D	16 ms	$\tau_{u,s2 \leftarrow s1}$	2.8 ms
T_c	3.2 ms	$\tau_{u,s3 \leftarrow s1}$	128 ms
T_T	0.4 ms	$\tau_{u,s4 \leftarrow s1}$	140 ms
T_P	120 ms	$\tau_{u,s4 \leftarrow s3}$	12 ms
L	1000 s	$\tau_{u,s3 \leftarrow s2}$	128 ms
θ	30 ppm	$\tau_{d,s1}$	0.01 ms

For an example, we calculate $\tau_{u,s2}$ using Equation (4), $\tau_{u,s2} = \tau_{u,s1} + \tau_{delay} + \tau_{u,s2 \leftarrow s1}$. The parameters are summarized in Table 3 as $\tau_{u,s1}$ = 90 us, $\tau_{u,s2 \leftarrow s1}$ = 2.8 ms, where $\tau_{delay,k}$ is the time for the state k to stabilize (set to 0 in this experiment). Therefore, $\tau_{u,s2}$ = 2.89 ms. The power $P_{TR,k}$ required to change State 0 to State k can be obtained using $T_{TR,k}$. We can derive $P_{TR,k}$ as shown in Equation (6) by substituting Equation (3) into Equation (2). This is used to obtain $E_{th,k}$ in Equation (7).

$$P_{TR,k} = \frac{T_{s0 \leftarrow k} P_{s0 \leftarrow k} + T_{k \leftarrow s0} P_{k \leftarrow s0}}{T_{TR}} = \frac{\frac{\tau_{u,k}}{2} |P_k - P_{s0}| + \frac{\tau_{d,k}}{2} |P_k - P_{s0}|}{T_{TR}} = \frac{P_k - P_{s0}}{\tau_{d,k}} + \frac{P_k - P_{s0}}{\tau_{u,k}} \tag{6}$$

$$E_{TR,k} = \frac{P_k - P_{s0}}{2} \tau_{u,k} + \frac{P_k - P_{s0}}{2} \tau_{d,k} \tag{7}$$

$T_{Act,k}$ contains the time parameters T_W, T_p and T_D shown in Table 1, which depend on the node state. The consumed energy $E_{Act,k}$ is given in Equation (8).

$$E_{Act,k} = T_{Act,k} \left(P_k - P_{s0} \right) \tag{8}$$

Thus, energy consumption E_k of State k during T_k is given by Equation (9).

$$E_k(T_k) = P_k T_{Act,k} + (P_k - P_{s0})T_{TR} = P_k T_{Act,k} + \frac{(P_k - P_{s0})}{2}(\tau_{d,k} + \tau_{u,k}) = P_k T_{Act,k} + \frac{(P_k - P_{s0})}{2}\tau_{u,k} + \frac{(P_k - P_{s0})}{2}\tau_{d,k} \quad (9)$$

Since State 4 and State 0 use the maximum and minimum power, respectively, it is clear that a lower number state will use less power. Equation (10) shows the energy saving for the case where $P_k(T_k) = P_{s4} - P_k$ in Equation (9) and subtracting the additional power consumption E_{TR}. Equation (11) shows the amount of power saving.

$$E_{saved,k}(T_k) = (P_{s4} - P_k)T_{Act,k} - E_{TR} \quad (10)$$

$$E_{saved,k} = \frac{(P_{s4} - P_k)T_{Act,k} - E_{TR}}{T_k} \quad (11)$$

Please note that State ($k - n$) (where $n \leq k - 1$) uses less energy than State k, but absolute energy reduction is not guaranteed. To save energy at State ($k - n$), the energy $E_{TR,n}$ used in the transition from the existing State ($k - n$) to State k must be greater than the energy $E_{th,k}$ given in Equation (12).

$$E_{th,n} = P_n \times T_{th,n} \quad (12)$$

We can find the minimum time $T_{th,k}$ at State k keeping $E_{save,k} > 0$ using $P_{TR,k}$ described above is given as:

$$T_{th,k-n} = \frac{1}{2}\left[\tau_{d,n} + \frac{(P_k + P_{k-n})}{(P_k - P_{k-n})}\right]\tau_{u,n} \quad (13)$$

5. Energy Consumption in Wireless Sensor Node

5.1. Node Energy Consumption in Sleep Mode

The proposed FGPM has three highly granular power modes in sleep mode. In the standby state, wRadio and sensor modules consume power to detect input signals. The proposed structure repeats the use of State 0 and State 1 in order to switch the sensing unit ON/OFF. In the medium idle state, the FGPM makes a state transition from State 3 to State 2 to reduce energy consumption. This section describes the total energy consumption considering all scenarios for the state of dormancy. Table 4 shows power consumption and timing parameters for the sleep mode shown in Figure 9. P_k is power consumption in the FGPM complying with a communication protocol. The duration T_k for State k has many variables as shown in Figure 9.

Table 4. Power states and Time used in the sleep states.

	Scenario	Power	Time
Conv	wRadio term	P_{s3}, P_{s1}	$T_{wRx} + n\left(T_{slp} + T_{sensing}\right)$
	Medium idle	P_{s2}	$T_{wRx,conv} = T_{TR,s3 \leftarrow s1} + T_P$
	Sensor term	P_{s1}	$T_{slp} + T_{sensing}$
Prop	wRadio term	P_{s2}, P_{s0}	$T_{wRx} + n\left(T_{slp} + T_{sensing}\right)$
	Medium idle	P_{s2}	$T_{wRx,prop} = T_{TR,s2} + T_P$
	Sensor term	P_{s1}, P_{s0}	$T_{slp} + T_{TR,s1} + T_{sensing}$

Each sensor node is in a sleep state before detecting an event which is as long as i $\times$ T_W. During this time, there is one wRx term for sensing a wake-up Rx signal and the sensing time is $n = (T_w - T_{wRx})/T_{st}$. Equations (14) to (16) show energy consumption while the node awaits the onset of an event.

$$E_{sleeps} = \sum_{T_W=1}^{j} (E_{wakeRx}(T_P) + n \cdot E_{sensor})_{T_W} + E_{sensed} \quad (14)$$

where j represents the total sleep time of a node. It is equivalent to the sum of the energy consumed from sensing, $j \times T_W$, and j times of energy of the medium idle state. E_{wakeRx} is the energy consumption of the medium idle state of a node while receiving a wake-up signal with wireless communication. It is the largest cause for energy consumption for a node in the sleep state is given as Equation (15).

$$E_{wakeRx,save}(T_P) = P_{s0}T_{slp} + (\ P_{s3} - P_{s2}\ )T_P + (E_{TR,s3 \leftarrow s1} - E_{TR,s2}) \tag{15}$$

E_{sensor} has two types of energy consumption. Based on the proposed method, $E_{sensor} \ni e_{\tau_{d,s1}} = (P_{s1} - P_{s0})\tau_{d,s1}$ because the node sleeps again if it fails to detect an event. If it detects an event, it sends an event signal to the MCU through the wake-up controller, so the energy is the same as $E_{sensor} \ni e_{\tau_{u,s4 \leftarrow s1}} = (P_{s4} - P_{s1})\tau_{u,s4 \leftarrow s1}$. Please note that $T_{Ack,s1} \geq T_{th,s1}$ can be obtained from Equation (13). Thus, the sensor energy saving is given in Equation (16).

$$E_{sensor,saved} = (P_{s1} - P_{s0})T_{slp} - (E_{TR,s1} + (\ P_{s1} - P_{s0}\ )\tau_{u,s1}) \tag{16}$$

Thus, the saved energy $E_{sleeps,saved}$ by the proposed method can be written as

$$E_{sleeps,saved} = \sum_{i=1}^{\frac{T_{wait}}{T_W}} \left(T_{slp}(n+1)(P_{s1} - P_{s0}) + T_P(P_{s3} - P_{s2}) + E_{TR,s3 \leftarrow s1} - (E_{TR,s2} + n \cdot E_{TR,s1})\right)_i \tag{17}$$

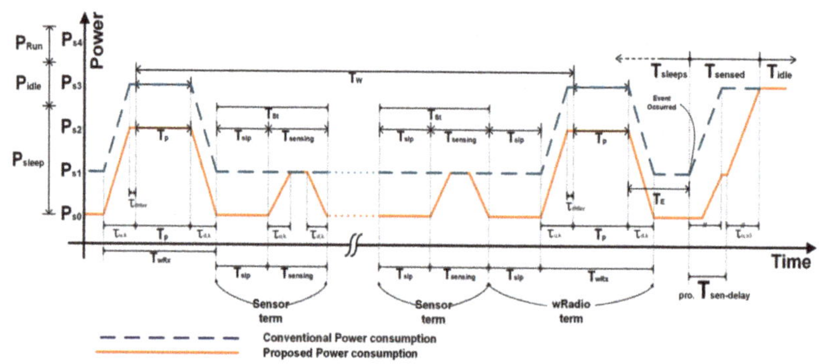

Figure 9. Transforming of energy states in sleep mode of a node.

5.2. Energy Consumption in Sleep Mode

Figure 10 shows energy consumption in sleep mode for both the conventional and proposed method. Based on the Mica2 node, the energy consumption when using the conventional continuous sensing unit and when the ON/OFF is repeatedly used is 178.64 uJ and 95. 96 uJ [27]. The energy saving is approximately 46.3%. Employing State 2 reduces energy consumption from 11.38 mJ to 3.5 mJ when the node is in the medium idle state, which is a 69.2% energy reduction. By adding these two factors, the Mica2 node consumes 7.96 mJ for 1 s, which results in a total energy saving of 74.2% [20,26,27].

Figure 11 shows energy consumption in sleep states for 4 s to compare the existing and proposed methods. The sensing interval shows a very slow rise of energy, and the medium idle interval shows a faster rate of energy consumption increase. This is because the energy consumed by the node's radio subsystem is large. Therefore, it can be seen that energy reduction in the radio system is emphasized. Compared with the proposed method, energy consumption varies greatly over time. Figure 11 presents a staircase energy consumption difference, but it increases linearly with respect to time.

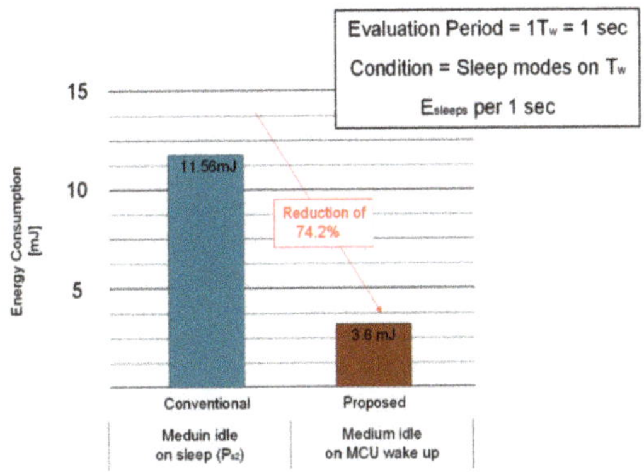

Figure 10. Comparison of energy consumption in the sleep mode.

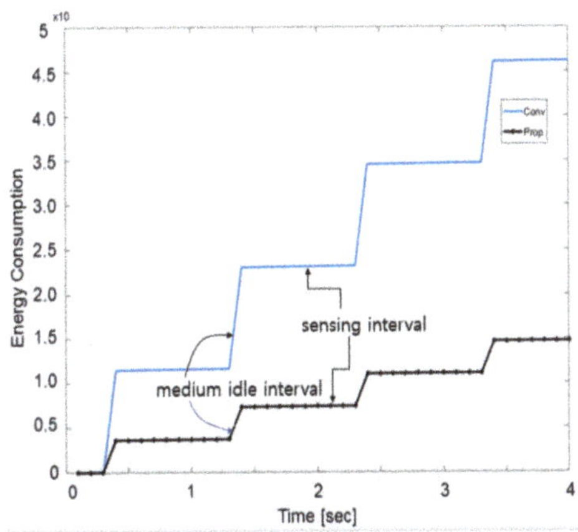

Figure 11. Node energy consumption in the sleep mode.

5.3. Sensor Node Energy Consumption in Communication

In most cases, the sensor node is in the sleep state for event sensing. When a node senses an event at the sensing unit, it becomes a sender node as the first wake up node. Then, it transmits a wake-up message and data packet to the neighboring nodes and receives an acknowledge signal. In these scenarios, power consumption in a node varies with each time schedule. The WiseMAC protocol has several power consumption levels in the node's communication scenario; however, the state of power consumption was defined only into three types: sleep, doze (idle), and communication state (RUN). The proposed approach introduces FPGM with five states, where partitioning the states enables fine-grained power management and reduced energy consumption. Figure 12 shows the change of energy consumption in communication between nodes for WiseMAC in each communication scenario. Node 1 wakes up from the sleep state by sensing an event of the sensing units and starts to communicate with Node 2. As illustrated in Figure 2, a node can have feedback after communicating at least twice for a single event. We show the advantages of the proposed fine-gained partitioned power mode that are analyzed using energy consumption benchmarks in conventional communication.

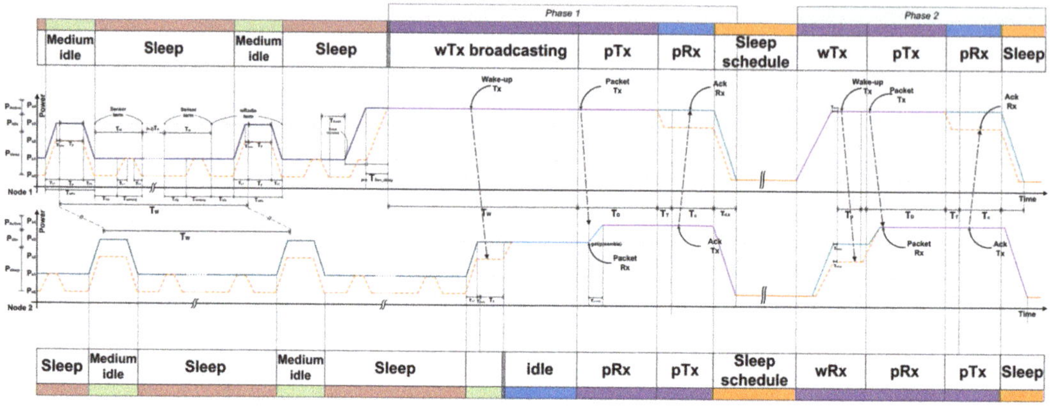

Figure 12. Energy state transition in WiseMAC protocol-based node communications.

Each node sends a wake-up message and data packet to neighboring nodes. The node that first wakes up through a sensor event from the sleep mode receives an Ack. This is phase 1 communication. A time delay of $\tau_{sendelay}$ may occur due to repeated use of the sensor switching ON/OFF that uses energy, though power consumption can be reduced by using State 3 in the Rx state receiving Ack. Energy consumption is calculated in Equations (18) and (19).

$$E_{conv,node1,ph1} = P_{s4}T_W + P_{s4}T_D + P_{s4}T_T + P_{s4}T_c + E_{TR,s4} \quad (18)$$

$$E_{prop,node1,ph1} = \tau_{sendelay}P_{s1} + P_{s4}T_W + P_{s4}T_D + P_{s3}T_T + P_{s3}T_c + E_{TR,s4} \quad (19)$$

Nodes can communicate repeatedly even after one data packet transmission. For example, in the case of a gas sensor node, the sensor detects events periodically for a given duration, and the number of times is set by the user for checking the current physical environment and the state of the node. It is periodically fed back to the upper node, which is phase 2 communication. In phase 2, the parameter value T_W of Node 1 is changed to T_P. Additionally, $\tau_{sendelay}$ is deleted because the medium idle state does not use the sensor. The energy consumption is given in Equations (20) and (21).

$$E_{conv,node1,ph2} = P_{s4}T_P + P_{s4}T_D + P_{s4}T_T + P_{s3}T_c + E_{TR,s4} \quad (20)$$

$$E_{prop,node1,ph2} = P_{s4}T_P + P_{s4}T_D + P_{s3}T_T + P_{s3}T_c + E_{TR,s4} \quad (21)$$

Node 2 wakes up from a long sleep state upon an incident wRx event. It therefore only plays the receiver role until communication with the wRx sender is completed. Node 2 becomes a receiver node that wakes up through a wireless signal without detecting an event from the sensor. In the initial communication, Node 2 can send a piggy-back Ack to inform the sending node that it is in a medium idle mode. Therefore, the energy used for the first communication of the transmitting node and the energy used in the next communication are different for $m \geq 2$ in $T_{idle,m}$, where m represents the number of communication events for a range of $0 \leq \mathrm{T_{idle,m}} \leq T_W - T_P$ [8,10,13]. Equations (18) and (19) are modified to (22) and (23).

$$E_{conv,node2,ph1} = P_{s3}T_P + P_{s3}T_{idle,1} + P_{s4}(T_D - \tau_{u,s4\leftarrow s3}) + P_{s4}T_T + P_{s3}T_c + E_{TR,s4} \quad (22)$$

$$E_{prop,node2,ph1} = P_{s2}T_P + P_{s3}T_{idle,1} + P_{s4}(T_D - \tau_{u,s4\leftarrow s2}) + P_{s4}T_T + P_{s3}T_c + E_{TR,s4} \quad (23)$$

From Phase 2, Node 1 can check the detailed medium idle schedule of Node 2. Thus, the time error T_{idle} of T_P approaches zero. The energy is modified to Equations (24) and (25).

$$E_{conv,node2,ph2} = P_{s3}T_P + P_{s3}T_{idle,2} + P_{s4}(T_D - \tau_{u,s4\leftarrow s3}) + P_{s4}T_T + P_{s3}T_c + E_{TR,s4} \quad (24)$$

$$E_{prop,node2,ph2} = P_{s2}T_P + P_{s3}T_{idle,2} + P_{s4}(T_D - \tau_{u,s4\leftarrow s2}) + P_{s4}T_T + P_{s3}T_c + E_{TR,s4} \quad (25)$$

6. Experimental Results

Based on the Mica2, we compared the energy consumption of the conventional and proposed approach using the Equations (18) to (25). The ad hoc network is shown in Figure 1, which accounts for inter-node communication from Node 1 to Node 6 in the network. The dotted circles are communication pairs. Node 1 wakes up Node 2 and transmits its signal. Subsequently, Node 2 wakes-up Node 3 by broadcasting a wake-up signal for other nodes. Node 2 wakes up Node 4 separately from communication between Node 1 and Node 3. In the same order, Node 4 communicates with Node 5 and Node 6. There is no communication between Nodes 3 and 4, because Node 2 has a direct line of communication with Node 4. Figure 13 shows the amount of power consumption for the single communication of a dotted circle. The time required at the send Node is $T_W + T_D + T_T + T_c + \tau_{u,s4\leftarrow s1} + \tau_{d,s4\leftarrow s1}$, and at the receive node is $T_W + T_D + T_T + T_c + \tau_{u,s4\leftarrow s2} + \tau_{d,s4\leftarrow s1}$. However, the power consumption according to the time period is taken from Table 2. Phase 1 is for sleep mode and phase 2 is for run mode. The proposed method is not significantly different from the previous method in the sender node, but a relatively large difference can be seen in the receiver node.

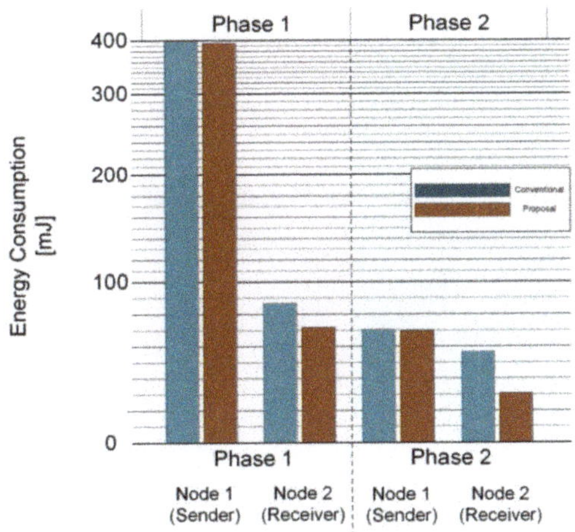

Figure 13. Comparison of node energy consumption for each communication scenario.

Figure 14 shows the energy saving as a function of the parameters time and the number of events in the node. The energy consumption of the waiting time in the proposed method is about 74.2% lower than that of conventional methods. However, the energy reduction effect caused by wake-up is less than 2%, so as the number of events increases, the energy reduction of the proposed method does not necessarily scale. We assumed that the number of events occurred from a minimum 0 to a maximum 16 in a space of two minutes. There is an energy saving of about 57 J for 0 events, but about 9.7 J of energy saving is less than 15% of that for 16 events.

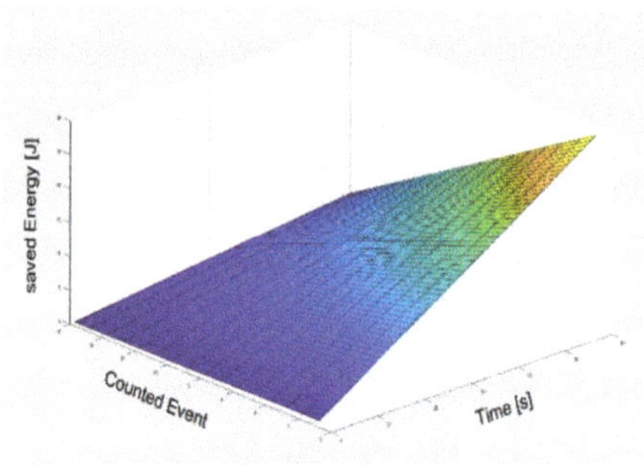

Figure 14. Reduced energy consumption with the parameters time and number of events.

Figure 15 shows the reduced energy consumption parameter space over one minute for the number of nodes and event probability in the WSN. This shows the energy saving effect when it is assumed that each node wakes up once due to a single event. The energy saving rises as a 2nd order function for the event probability, but it is linearly increasing for the number of nodes. Based on 100 nodes, the sensor node consumes significantly less energy at 20 mJ for a 0% event probability, but energy consumption exceeds 1100 mJ at the 100% event probability. As in Figure 14, this shows that our approach can save more energy for systems of low event probability [27].

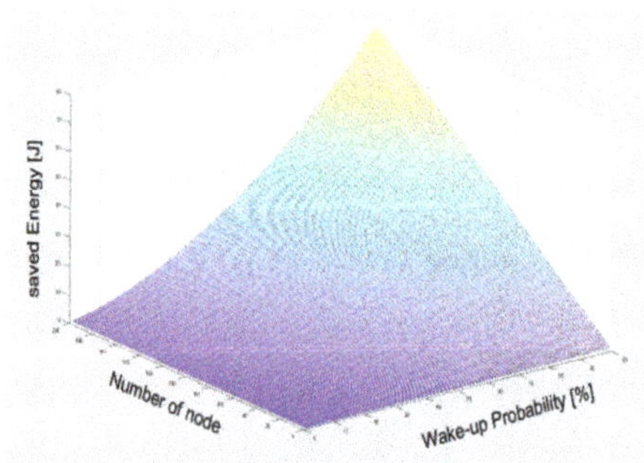

Figure 15. Comparison of node energy consumption for each communication scenario.

7. Conclusions

In this paper, we proposed a method to reduce power consumption according to the protocol usage by introducing fine-grain power modes to the WSN node. The conventional sensor system has three kinds of power modes: sleep, idle, and run. We divide the sleep mode into three states: a deep sleep mode (State 0), a medium-idle state of the sensor (State 1), and a medium-idle state of the wake-up Rx channel (State 2). Thus, the proposed WSN platform has five states of power mode, including idle and run. Even in sleep mode, power consumption is very different depending on WSN status. For example, the medium-idle mode for the sensor consumes eight times more power than the medium-idle state for wRx. We can manage the scheduling of power modes at the protocol level using a frame pending bit in the header of data packets. This power mode can be applied to the scenario of node communication, event detection, and standby according to the environment. The sleep state, which is a standby state, is an event that has the most impact on the battery life of a sensor board and has the longest time occupancy in power mode. Event recognition in sleep mode is not continuous, but periodic sensing. Thus, the minimum time to operate the sensor was calculated and we obtained the energy consumption. Besides, periodic radio signal receiving consumes the most energy in sleep mode. However, the proposed system saves energy during radio communication because the node is not used until the idle mode. The experimental results are simulations based on the architecture in Figure 1, using parameters in Tables 2 and 4, and Figure 5. We assumed the use of gas sensors on the Mica2 platform in WSNs. The node module senses gas at state 1 of a node. The sensor node is in sleep mode and the event sensing period is set to 1 s. The proposed FGPM controls the node status of the sleep mode finely that offers energy savings of 74.2% compared to the conventional approach. This can be seen as a significant contribution to battery saving since the sensor nodes are idle in sleep mode for majority of time. However, when events occur consecutively without a sleep time, the power reduction is less than 2%. As a result, the proposed method can be expected to save power more effectively in a wireless sensor network with a low event probability or a small number of events in which most networks operate.

Author Contributions: Conceptualization, S.Y. and K.C.; Validation, S.Y. and J.K.E.; formal analysis, S.Y. and K.C.; writing—original draft preparation, S.Y. and K.C.; writing—review and editing, J.K.E. and H.C.I.; supervision, H.C.I. and K.C.; All authors have read and agreed to the published version of the manuscript.

Funding: This work was supported by the National Research Foundation of Korea grant funded by the Korean government (MSIT) (No. 2020R1F1A1069381) and Basic Science Research Program through the National Research Foundation of Korea (NRF) funded by the Ministry of Education (No. 2020R1A6A1A12047945)

Institutional Review Board Statement: Not applicable.

Informed Consent Statement: Not applicable.

Data Availability Statement: Data sharing not applicable.

Conflicts of Interest: The authors declare no conflict of interest.

References

1. Jelicic:, V.; Magno, M.; Brunelli, D.; Paci, G.; Benini, L. Contextadaptive multimodal wireless sensor network for energy-efficient gas monitoring. *IEEE Sens. J.* **2013**, *13*, 328–338. [CrossRef]
2. Folea, G.; Mois, D. Lessons learned from the development of wireless environmental sensors. *IEEE Trans. Instrum. Meas.* **2020**, *69*, 3470–3480. [CrossRef]
3. Baniata, M.; Reda, H.; Chilamkurti, N.; Abuadbba, A. Energy-Efficient Hybrid Routing Protocol for IoT Communication Systems in 5G and Beyond. *Sensors* **2021**, *21*, 537. [CrossRef] [PubMed]
4. Piyare, R.; Oikonomou, G.; Elsts, A. TSCH for Long Range Low Data Rate Applications. *IEEEE Access* **2020**, *8*, 228753–228766. [CrossRef]
5. Sinha, A.; Chandrakasan, A. Dynamic power management in wireless sensor networks. *IEEE Des. Test Comput.* **2001**, *18*, 62–74. [CrossRef]
6. Del-Valle-Soto, C.; Mex-Perera, C.; Nolazco-Flores, A.; Rodríguez, A.; Rosas-Caro, J.; Martínez-Herrera, A.F. A Low-Cost Jamming Detection Approach Using Performance Metrics in Cluster-Based Wireless Sensor Networks. *Sensors* **2021**, *21*, 1179. [CrossRef] [PubMed]
7. Karakus, C.; Gurbuz, A.C.; Tavli, B. Analysis of Energy Efficiency of Compressive Sensing in Wireless Sensor Networks. *IEEE Sens. J.* **2013**, *13*, 1999–2008. [CrossRef]
8. Bdiri, S.; Derbel, F.; Kanoun, O. An 868 MHz 7.5 μW wake-up receiver with −60 dBm sensitivity. *J. Sens. Sens. Syst.* **2016**, *5*, 433–446. [CrossRef]
9. Bachir, A.; Dohler, M.; Watteyne, T.; Leung, K.K. MAC essentials for wireless sensor networks. *IEEE Commun. Surv. Tutor.* **2010**, *12*, 222–248. [CrossRef]
10. David, G.; Dailos, R.; Sunil, L.; Javier, P. Low-Power RFED Wake-Up Receiver Design for Low-Cost Wireless Sensor Network Applications. *Sensors* **2020**, *20*, 6406. [CrossRef]
11. Bdiri, S.; Derbel, F.; Kanoun, O. A Tuned-RF Duty-Cycled Wake-Up Receiver with −90 dBm Sensitivity. *Sensors* **2017**, *18*, 86. [CrossRef] [PubMed]
12. Internet of Things: Wireless Sensor Networks. Available online: https://www.iec.ch/resource-centre/internet-things-wireless-sensor-networks (accessed on 10 January 2021).
13. Anastasi, G.; Conti, M.; Francesco, M.; Passarella, A. Energy conservation in wireless sensor networks: A survey. *Ad Hoc Netw.* **2009**, *7*, 537–568. [CrossRef]

14. Stojev, M.; Kosanovi, M.; Golubovi, L. Power management and energy harvesting techniques for wireless sensor nodes. In Proceedings of the 2009 9th International Conference on Telecommunication in Modern Satellite, Cable, and Broadcasting Services, Nis, Serbia, 7–9 October 2009; pp. 65–72. [CrossRef]
15. Maqbool, S.; Sabeel, S. Arising Issues In Wireless Sensor Networks: Current Proposals and Future Developments. *IOSR J. Comput. Eng.* **2013**, *8*, 56–73. [CrossRef]
16. Available online: https://www.espressif.com/sites/default/files/documentation/esp32-solo-1_datasheet_en.pdf (accessed on 30 April 2021).
17. Nikoofard, A.; Mandal, S. An 11.5 nW Broadband Wake-up RF Receiver with −60 dBm Sensitivity at 50 MHz. In Proceedings of the 2016 IEEE ISCAS, Montréal, QC, Canada, 22–25 May 2016; pp. 2787–2790.
18. Zhao, J.; Govindan, R.; Estrin, D. Sensor Network Tomography: Monitoring wireless sensor networks. *SIGCOMM Comput. Commun. Rev.* **2002**, *32*, 64. [CrossRef]
19. El-Hoiydi, A.; Decotignie, J.-D.; Enz, C.; Le Roux, E. WiseMAC, an ultra low power MAC protocol for the wiseNET wireless sensor network. In Proceedings of the 1st International Conference on Embedded Networked Sensor Systems, SenSys 2003, Los Angeles, CA, USA, 5–7 November 2003; pp. 302–303. [CrossRef]
20. El-Hoiydi, A.; Decotignie, J.-D. WiseMAC: An ultra low power MAC protocol for the downlink of infrastructure wireless sensor networks. In Proceedings of the ISCC 2004 Ninth International Symposium on Computers and Communications (IEEE Cat. No.04TH8769), Alexandria, Egypt, 28 June–1 July 2004; Volume 1, pp. 244–251. [CrossRef]
21. Milankovich, A.; Lendvai, K.; Imre, G.; Szabo, S. Wakeup signal length optimization combined with payload aggregation and FEC in WSNs. In Proceedings of the European Wireless 2015, 21th European Wireless, Budapest, Hungary, 20–22 May 2015.
22. Advanced Configuration and Power Interface Specification. Version 6.0. 2015. Available online: http://gauss.ececs.uc.edu/Courses/c4029/pdf/ACPI_6.0.pdf (accessed on 1 March 2019).
23. Shan, L.; Tang, Y.; Zhu, Q. Passive Wake-up Scheme for Wireless Sensor Networks. In Proceedings of the Second International Conference on Innovative Computing, Informatio and Control, ICICIC 2007, Kumamoto, Japan, 5–7 September 2007; p. 507. [CrossRef]
24. Ersoy, D.; Onur, E. Wake-up receivers for wireless sensor networks: Benefits and challenges. *IEEE Wirel. Commun.* **2009**, *16*, 88–96. [CrossRef]
25. Ehiagwina, F.; Kehinde, O.; Iromini, N.; Sidiq, N.; Punetha, D. Ultra-Low Power Wireless Sensor Networks: Overview of Applications, Design Requirements and Challenges. *ABUAD J. Eng. Res. Dev.* **2018**, *1*, 331–345.
26. Wikipedia List of Wireless Sensor Nodes. Available online: https://en.wikipedia.org/wiki/List_of_wireless_sensor_nodes (accessed on 1 March 2021).
27. Hurni, P.; Braun, T.; Anwander, M. Evaluation of WiseMAC and extensions on wireless sensor nodes. *Telecommun. Syst.* **2010**, *43*, 49. [CrossRef]
28. Shin, H.; Ahn, S.; Song, T.; Baeg, S. Analysis of Low Power Sensor Node Using Data Compression. *IFAC Proc.* **2009**, *42*, 34–39. [CrossRef]
29. Piyare, R.; Murphy, A.L.; Magno, M.; Benini, L. On-Demand LoRa: Asynchronous TDMA for Energy Efficient and Low Latency Communication in IoT. *Sensors* **2018**, *18*, 3718. [CrossRef] [PubMed]

sensors

MDPI

Article

A Low-Power Spiking Neural Network Chip Based on a Compact LIF Neuron and Binary Exponential Charge Injector Synapse Circuits

Malik Summair Asghar [1,2], Saad Arslan [3] and Hyungwon Kim [1,*]

1 Department of Electronics Engineering, Chungbuk National University, Chungdae-ro 1, Seowon-gu, Cheongju 28644, Korea; summair@chungbuk.ac.kr
2 Department of Electrical and Computer Engineering, Abbottabad Campus, COMSATS University Islamabad, University Road, Tobe Camp, Abbottabad 22044, Pakistan
3 Department of Electrical and Computer Engineering, COMSATS University Islamabad, Park Road, Tarlai Kalan, Islamabad 45550, Pakistan; saad.arslan@comsats.edu.pk
* Correspondence: hwkim@chungbuk.ac.kr

Abstract: To realize a large-scale Spiking Neural Network (SNN) on hardware for mobile applications, area and power optimized electronic circuit design is critical. In this work, an area and power optimized hardware implementation of a large-scale SNN for real time IoT applications is presented. The analog Complementary Metal Oxide Semiconductor (CMOS) implementation incorporates neuron and synaptic circuits optimized for area and power consumption. The asynchronous neuronal circuits implemented benefit from higher energy efficiency and higher sensitivity. The proposed synapse circuit based on Binary Exponential Charge Injector (BECI) saves area and power consumption, and provides design scalability for higher resolutions. The SNN model implemented is optimized for 9×9 pixel input image and minimum bit-width weights that can satisfy target accuracy, occupies less area and power consumption. Moreover, the spiking neural network is replicated in full digital implementation for area and power comparisons. The SNN chip integrated from neuron and synapse circuits is capable of pattern recognition. The proposed SNN chip is fabricated using 180 nm CMOS process, which occupies a 3.6 mm^2 chip core area, and achieves a classification accuracy of 94.66% for the MNIST dataset. The proposed SNN chip consumes an average power of 1.06 mW—20 times lower than the digital implementation.

Keywords: spiking neural network; leaky integrate and fire; neuromorphic; artificial neural networks; artificial intelligence; image classification; CMOS

Citation: Asghar, M.S.; Arslan, S.; Kim, H. A Low-Power Spiking Neural Network Chip Based on a Compact LIF Neuron and Binary Exponential Charge Injector Synapse Circuits. *Sensors* **2021**, *21*, 4462. https://doi.org/10.3390/s21134462

Academic Editor: Marcin Woźniak

Received: 12 June 2021
Accepted: 24 June 2021
Published: 29 June 2021

1. Introduction

In recent years, neuromorphic systems, which are comparable to a biological neural network, have been widely investigated for prospective computing systems [1]. These neuromorphic systems consume very little energy and provide parallel signal processing [2]. Deep Neural Networks (DNN) are receiving much attention as the chosen classifier for various machine learning and computer vision applications due to their high classification accuracy [3]. However, for applications requiring a real environment using conventional von Neuman computing systems, the DNNs involve an immense number of computations and memory requirements. The bottleneck for DNN efficiency is the excessive and repeated update of data [4], high power consumption and memory bandwidth, making them not suitable for mobile applications where area and power are big constraints [5].

The driving force in field of computing has been to outperform the human brain using the von Neumann architecture. However, this architecture has significant differences with the human brain in terms of the organizational structure, computing methodology, area, and power constraints [6]. Today's digital microprocessors have a basic architectural difference with the central nervous system. The von Neuman-based microprocessors, constituted of logic gates, have distinct computational and storage devices [7]. However, the latter is composed of a massive parallel arrangement of neurons, densely interconnected to one another in a vast network through links called synapses [8]. Moreover, in the central nervous system of the human brain, the processing elements (neurons) are placed very close to the memory units (synapses). The human brain emerges as a vast energy-efficient network by increasing the neuron–synapse interconnection, while consuming just 20 W and performing computations unmatched by modern computers [9]. Due to power and memory bottlenecks of conventional computers along with inspiration from biological achievements a new paradigm of neuromorphic architectures have evolved. Fundamentally, these are complementary to von Neumann architectures and have shown promising results in specific applications [10–12]. Neuromorphic architectures are not only energy efficient, but perform parallel signal processing, fault tolerance, and can be configurable. Moreover, they can be realized by numerous silicon-based technologies, large-scale architectures, and computational models of neural elements [13,14].

Although Artificial Neural Networks (ANN) can offer higher efficiency in training and inference tasks, they generally require higher complexity in hardware and power consumption than SNNs. Among various ANNs inspired from the human brain, Spiking Neural Networks aim to bridge the gap between neuroscience and machine learning. SNNs offer massive parallel computations on hardware, mimicking the human brain. SNNs process information via numerous neurons and communicate that information through a web of synapses. The transfer of information in the shape of small electric currents is carried out by a sequence of action potentials called spike trains. In light of the Shannon Theory, some studies suggest that the information contained in spike trains in the form of temporal codes are more energetically efficient than rate codes [15–17]. Neuromorphic architectures based on SNN benefit from computation organization—achieving high energy efficiency by co-locating computing (neurons) and memory (synapse) elements, and information representation—less power consumption by event-driven spikes encoding information [18]. The power constraints of mobile applications suffer due to the exponentially increasing processing complexity of large-scale ANN-based neuromorphic hardware and thus necessitate energy-efficient SNN-based neuromorphic hardware [19]. SNNs process the information by spike propagation, which enables it to accelerate the computational speed and improve energy efficiency [20]. It incorporates biologically plausible neuron models in acquiring the temporal dynamics of the neural membrane [21]. To realize the advantages of SNNs on a chip, it is crucial to optimize the neuron and synapse circuits for low power and a compact area. Therefore, to realize large-scale SNN hardware requires power and area optimized computational (neuron) and memory (synapse) elements.

When Carver Mead coined the idea of implementing neural networks in hardware, a new paradigm in neuromorphic engineering was established. The neural characteristics can be captured on hardware by the integration of robust and less power-hungry analog components [1]. Confronting challenges in the field of neuromorphic engineering requires a multidisciplinary approach, as the next generation VLSI technology can realize better hardware [22]. Implementing neuromorphic hardware requires numerous transistors to be integrated on the chip. This can be achieved to some extent by the scalability feat of the CMOS technology, although still incomparable with the integration density of the brain. Therefore, hybrid techniques that have merged conventional robust CMOS with newly developed technologies like memristors are attracting interest in research [23,24]. However, memristor crossbar architectures still require additional process requirements and are currently under study for simpler problems with discrete devices, whereas conventional CMOS, by virtue of its robustness and scalability has clear advantages for realizing large-scale neuromorphic hardware [25].

Since the emergence of neuromorphic architecture, several studies have been performed considering different large-scale architecture, technologies, and neuronal models. Neurogrid [26], a mixed signal and system, employs subthreshold circuits to model neuronal elements. This allows for a compact neuron area, but with all neurons sharing the same parameters and therefore becoming less configurable and power hungry. In BrainScaleS [21], a mixed signal approach was adopted where all the neurons have dedicated parameters. This allows high tunability, reconfigurability, and reliable parameter mapping at the cost of higher power consumption. The analog–digital neural network classifier [5] consists of an analog core with a current multiplier and environmental noise compensation circuits; it achieves area efficiency due to its recurrent operations. However, its excessive read and write operations via the digital controller makes it less energy efficient.

This paper presents the implementation of a large-scale SNN Artificial Intelligence (AI) hardware based on analog CMOS for real time IoT applications. The SNN hardware, optimized for area, power, and accuracy, is realized on a chip based on the preliminary prototype design [27]. The choice of SNN model is dictated by the analog design constraints for hardware. A four-layer fully connected SNN model is optimized for a reduced number of neurons, input image size, and weight resolution which can achieve targeted classification accuracy along with a minimal chip area. The proposed SNN comprises synapse and neuron circuits optimized for power and area using a 180 nm process design kit. Furthermore, a fully digital implementation of SNN is replicated to analyze and compare the power consumption and estimated chip area with the analog counterpart. This work elucidates our SNN model and design methodology, elemental circuit designs, simulations and calibrations, and measurement results from the test chip on board. Section 2 describes the overall SNN model architecture and design pre-considerations. In Section 3, the underlying analog CMOS circuits design and implementation is explained and validated by simulation results along with a description of its digital implementation. In Section 4, the measurements and performance result of analog and digital SNN are analyzed. Finally, a discussion and comparison of results achieved with other state-of-the-art large-scale neural networks is performed in Section 5, before concluding the paper in Section 6.

2. Architecture and Design

2.1. Spiking Neural Network Model

The choice of neural network on chip for this work is SNN, regarded as the third-generation neural network, due to its biological feats and efficiency in the spatial–temporal signal coding [28]. In SNN, a web is weaved by the interconnections of neurons and synapses to perform inference and training tasks. Contrary to the continuous behavior of ANN, SNN works by making use of discrete events called spikes, which appear at specific time intervals. The occurrence of a spike event (0/1) mirrors the biological-chemical process of information delivery between different neurons. The neuron (post-synaptic) in a layer receives an input spike train from another neuron (pre-synaptic) in the previous layer, which are interconnected through synapses. The Figure 1a demonstrates a simple model of an SNN with a neuron, connected to a multitude of input synapses that receive input spike trains from pre-synaptic neurons. These input spike trains are modulated according to the respective synaptic strength (weight) and converted into current. The proportionate charge from all the synapses is then accumulated on the neuronal membrane as membrane potential. When membrane potential accumulates up to a predefined threshold value then the neuron emits or fires an output spike [29]. Thus, neurons act as an accumulation and comparison processing unit, while synapses are formed as memory with a communication interface.

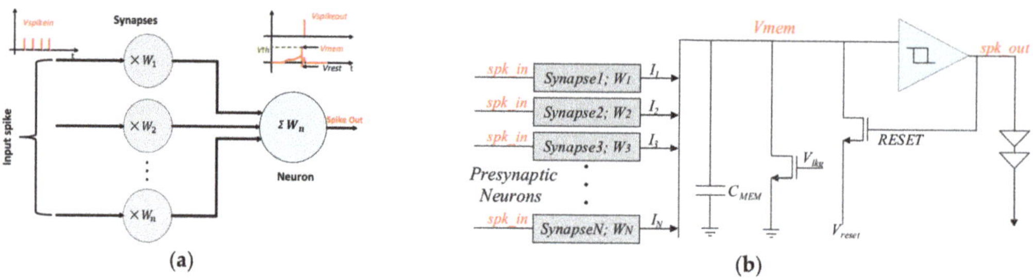

Figure 1. (**a**) Spiking Neural Network Model with a single neuron connected to multiple input synapses; (**b**) Leaky Integrate and Fire Model of the CMOS neuron cell.

2.2. Leaky Integrate and Fire Model

The choice of neuronal model for our large scale SNN is influenced by the pre-considerations of optimizing the design for area and power consumption, along with capturing the biological and temporal dynamics of the neuronal membrane. This requires that the neuronal model to be imitated replicates most of the computational studies, along with its simple design, which can unify a multitude of neurons on a chip. Therefore, for the implementation of large-scale SNN, our choice of model is the Leaky Integrate and Fire (LIF). This model possesses refractoriness and adaptation features to effectively mimic the biological computational features of neurons with good accuracy and adopts simpler circuit designs [30,31]. Contrary to other models [32,33], the LIF model for large scale neuromorphic architectures still draw the interest of researchers by virtue of the robustness of the CMOS design and its compact silicon implementation. Figure 1b models the LIF, where the neuron is represented by simple CMOS devices. The evolution of the neuronal membrane potential V_{mem} can be modeled by a Resistor–Capacitor circuit, which is composed of a membrane capacitor C_{mem} and a membrane resistance R_{mem} (leakage path). A spiking neuron receives input spikes from the several pre-synaptic neurons interconnected via synapses. Each synapse acts a current source and injects a current equivalent to its strength called weights (W). The state (potential) of C_{mem} is updated by injecting current through multiple current sources. Upon reaching a predefined threshold value, the comparator decides to evoke an output spike and resets the V_{mem}.

The parallel combination of C_{mem} and R_{mem} can be defined by Kirchhoff's current law [21] and is modeled by Equation (1). When an input current $I(t)$ flows into a neuron, it will charge the C_{mem} with $I_C(t)$ and discharge through the resistance with the current $I_R(t)$.

$$I(t) = I_R(t) + I_C(t), \tag{1}$$

$$I(t) = \frac{V_{mem} - V_{reset}}{R_{mem}} + C_{mem}\frac{\partial V_{mem}}{\partial t}, \tag{2}$$

$$C_{mem}\frac{\partial V_{mem}}{\partial t} = I(t) - \frac{V_{mem} - V_{reset}}{R_{mem}}. \tag{3}$$

When $V_{mem} \geq V_{th}$, then $V_{mem} = V_{reset}$. In Equation (3) above, V_{mem} is the membrane potential, V_{reset} is the resting potential, and V_{th} is the predefined threshold potential. When there is no input current, then the capacitive charge will decay by leaking through the resistance until it reaches the resting potential. In the resting state without any input current, the V_{mem} stays at the resting potential V_{reset}. In Equation (3), the $I(t)$ is the summation of the excitatory and inhibitory input synaptic currents and is expressed as:

$$I(t) = \sum_i \sum_f W_i \times I_{ref}, \tag{4}$$

where W_i denotes the weight strength of the ith synapse and 'f' is the number of the spike. I_{ref} is the external reference current to be injected upon each input spike activity. To exhibit the inhibitory behavior of the synapse, the weights may have negative values, thus discharging the I_{ref} upon each input spike activity.

3. SNN Implementation and Circuit Design

Learning in the neuromorphic systems can be categorized into on-chip [34,35] and off-chip learning [36,37]. The former mimics the biological neural systems, while the latter can benefit from the use of pretrained weights for achieving the same performance results as that of a software-based ANN [38]. Some SNN implementations employ their SNN architecture equivalent to an ANN, to determine trained weights using the ANN model in software and map the weights into the SNN hardware implementation [39]. Other works such as the STBP method in [40] have proposed direct training techniques that propagate spiking signals through the synapse and neuron models of the target SNN iteratively while converging the weights of each synapse for higher accuracy. Our proposed SNN architecture is based on the latter technique.

3.1. BSRC-Based SNN Architecture

For the SNN implementation proposed in this paper, the SNN model for the MNIST dataset is first optimized using a low-cost spike signal representation technique called Binary Streamed Rate Coding (BSRC), which was presented in our previous work [41]. While the proposed SNN employs the off-chip training technique, it determines the SNN's floating point weights by propagating sequences of spikes through an accurate model of the target SNN chip, instead of an ANN counterpart as in most other SNNs. It then quantizes the floating-point weights into integer values of minimum bit-width while satisfying the target accuracy goal. In this paper, the target accuracy of 94% or higher is chosen, which leads to 4-bit quantized weights for our SNN chip implementation. The equivalence between our BSRC-based SNN model and its hardware implementation enables us to train the SNN with very high accuracy while allowing for a very small chip size.

Figure 2 illustrates the overall block diagram for the proposed SNN hardware whose architecture is optimized using the BSRC-based SNN model. The SNN under study comprises four fully connected layers constituting synapses, neurons, and flipflops for weight and image memories. In order to minimize the chip size, our BSRC SNN optimization method reduced the input image size to 9×9 pixels from 28×28 pixels of the MNIST dataset. This proved to be the smallest size that could satisfy our target goal for accuracy of 94% or higher. The Input layer has 81 neurons for utilizing the 9×9 grayscale images as an input to the SNN. Our BSRC SNN optimization also determined the structure of each layer as follows: The output layer consists of 10 neurons for classifying the image, while two hidden layers consist of 30 and 20 neurons. All the neurons in each layer are fully connected with one another via 3311 synapses. After many iterations of the BSRC SNN optimization process, an SNN of four layers, namely (81-30-20-10), is determined, which consists of an input layer of 81 nodes, 1st hidden layer of 30 nodes, 2nd hidden layer of 20 nodes, and an output layer of 10 nodes.

The minimum weights are determined to be 4-bit wide, which are trained by a supervised training technique based on the BSRC-based backpropagation using the MNIST dataset. The 4-bits are stored in flip flop memories located close to the associated synapses in each layer, which leads to a short wire design, and consequently to faster operation and lower power consumption. The BSRC SNN optimization also chose 4-bits to represent the sequence of spike signals, which consists of a maximum of 15 spike pulses—no spike indicates a pixel value of zero, while 15 spikes denote a pixel value of 15. These spike pulses are provided to the inputs of synapse circuits in each layer, and also produced as outputs by neuron circuits of each layer. Each MNIST image is converted to an input spike sequence consisting of 15 possible spike pulses, proportional to the 4-bit pixel value of the image. In the output layer, a digital controller counts the number of spikes in each spike sequence and classifies the type of image on the measured spiking activity. The Analog CMOS-based implementation of the SNN's constituent circuits and a fully digital implementation of SNNs are described in upcoming subsections.

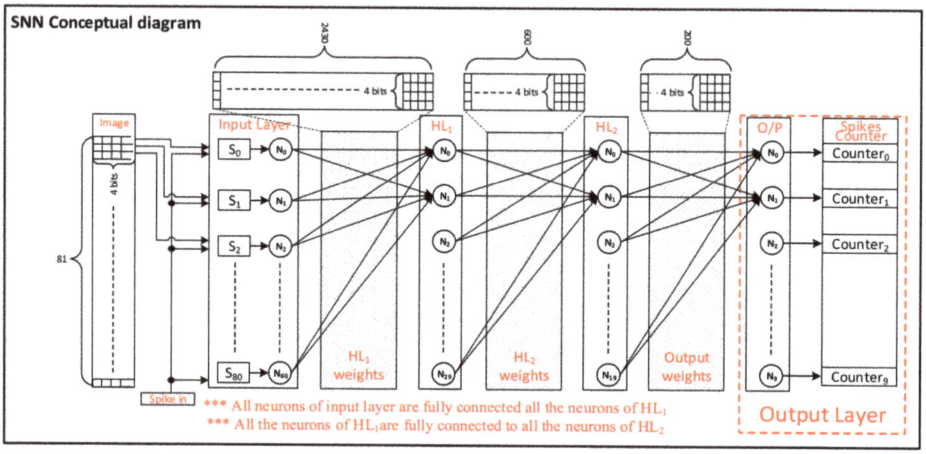

Figure 2. Block diagram of the SNN implementation consisting of four fully connected layers.

3.2. *Analog Spiking Neural Network*

To realize a large-scale SNN on hardware requires not only the minimal SNN architectures, but also highly optimized CMOS circuits for minimal area and power consumption. The SNN model (81-30-20-10) in Figure 2 consist of 4 layers, where each layer has been built up from many unit cells that are constituted from Synapse and neuron circuits. These unit cells are designed from CMOS devices based upon the principles of the LIF model explained earlier. Keeping in mind the aforementioned design goals and constraints, a prototype synapse and neuron circuits were designed earlier [42], which can be integrated in the development of a large-scale SNN hardware while replicating most of the biological neuromorphic characteristics.

3.2.1. Neuron Circuit

The LIF-based neuron circuit design is explained in Figure 3a. The neuronal membrane is simply modeled by a capacitor C_{mem}, which integrates the synaptic currents in order to generate V_{mem} across it. C_{mem} is realized by the Metal–Insulator–Metal Capacitor (MIMCAP) to achieve better linearity along with less power consumption. The leakage path of the neuronal membrane is modeled by a resistor that is implemented via partially-on NMOS transistor to save the area. The decision of a neuron to fire an output spike is performed by a comparator for which we have implemented a Schmitt Trigger circuit as shown in Figure 3b. The Schmitt Trigger compares the V_{mem} with the predefined threshold voltages. When the V_{mem} exceeds the threshold voltage, the Schmitt Trigger fires an output spike, and the feedback path resets the membrane potential to the initial resting potential (V_{reset}) via a separate NMOS transistor. As an important part of a neuron circuit, it needs careful designing in order to achieve optimum threshold voltages to ensure high speed and low power operations. Traditional neuron circuits often employ digital comparators requiring a global clock, which are power hungry as compared to asynchronous circuits. Analog asynchronous circuit implementations can achieve higher energy efficiency [43]. Therefore, the choice of asynchronous Schmitt Trigger for our neuron circuit provides a significant advantage by consuming less power and providing higher speed. Moreover, its simpler circuit implementation leads to small chip size, while its high sensitivity results in a highly accurate performance. The output buffers are implemented to provide the reshaping of the output spike pulses and isolation from the synapse circuits of the next layer.

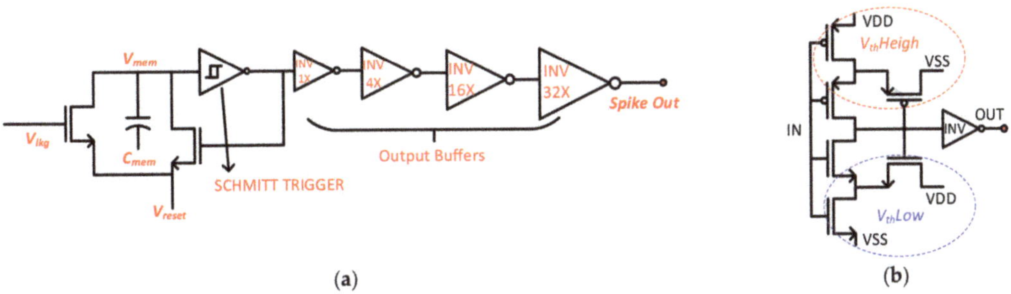

Figure 3. (**a**) The circuit schematic of the LIF-based neuron; (**b**) The circuit schematic of the Schmitt Trigger used inside the neuron circuit.

3.2.2. Synapse Circuit

The synapse circuit for our SNN implementation is shown in Figure 4, which is based on a Binary Exponential Charge Injector structure and uses a 4-bit value of weight for each synapse. Each synapse circuit comprises an excitatory synapse and an inhibitory synapse. The excitatory synapse is composed of a pull-up PMOS network, while the inhibitory synapse consists of a pull-down NMOS network. Each network has 3 branches to realize the 3-bits of weight and are binary exponential sized for equivalent charge injection. The MSB of the 4-bit weight is used to switch between excitatory and inhibitory behavior through digital gates. If the MSB is 0, then the weight is positive and thus activates the excitatory synapse by injecting charge onto C_{mem}. On the other hand, if the MSB is 1, then the weight is negative and thus activates the inhibitory synapse by discharging C_{mem}. Upon receiving a spike event in each synapse branch, the weight of each synapse determines the amount of injected current from the excitatory synapse or the discharged current of the inhibitory synapse. The injected current of all branches is accumulated onto the membrane potential capacitor C_{mem}. The BECI synapse circuit is inspired by the efficient structure of current mirroring DACs, which provides advantages including minimal number of transistors, minimum size for saving area, reduced power consumption, and design scalability for higher resolutions. For storing pretrained weight values on synapses, standard cell flip-flops collocated with each synapse circuit are used. Using this design method, not only can area be saved but energy required for read and write operations can also be reduced effectively.

3.2.3. Circuit Simulation

The proposed SNN chip employs two types of synapse-neuron (one neuron cell) circuits: input layer circuit and hidden layer circuit. First, the input layer circuit is introduced, which consists of one synapse and one neuron to convert input pixels to spike sequences. Then, the hidden layer circuit comprising of multiple synapses and one neuron is described, which fires output spike signals based on weights and input spikes.

Figure 5 shows a circuit simulation result for one neuron cell of input layer, composed of a synapse and neuron circuits explained in the former sections. It verifies our BSRC spike generation mechanism upon receiving an input spiking event, which is designed based on the LIF model using 4-bit weight values. Upon receiving an input spiking signal, the BECI synapse circuit injects an amount of charge onto V_{mem}, which is equivalent to its weight value. For this simulation, an input spike train acts as an initial enable signal, while the synapse is configured with both positive and negative weight values. Figure 5a shows an example of an excitatory behavior, where the synapse is configured with a positive weight value of +1, which slowly charges V_{mem} with every input spike activity. Subsequently, the neuron fires an output spike when V_{mem} reaches V_{th}. Similarly, Figure 5b illustrates an example of an inhibitory behavior, where the synapse is configured with a weight value of -1, which discharges the V_{mem} from its resting potential and the neuron refrains from firing an output spike. When the weight value is zero or when there is no input spike, then V_{mem} discharges (leaks) towards V_{reset} and the neuron does not fire an output spike.

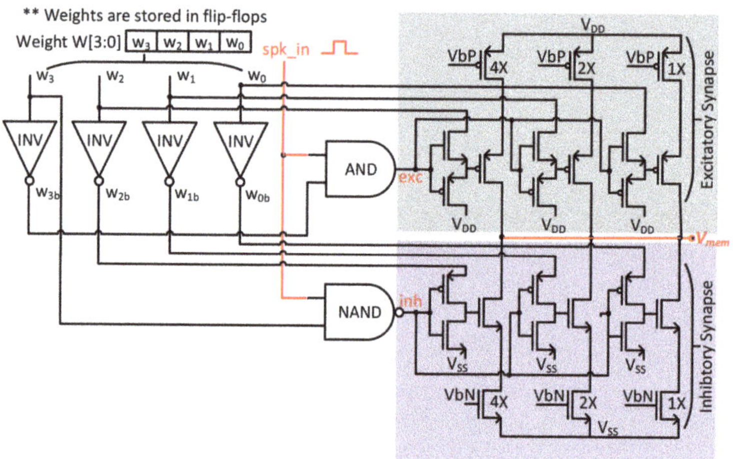

Figure 4. The circuit schematic for the BECI-based synapse with three branches and digital gates.

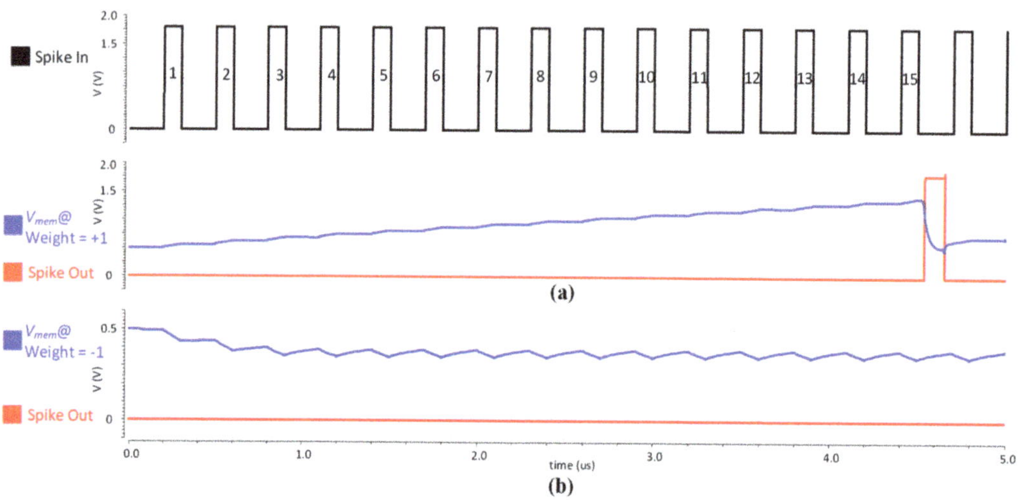

Figure 5. The circuit simulation results for one neuron cell of an input layer with an input spike train of 15 pulses: (**a**) shows the Vmem and spike-out when the weight value is +1; (**b**) shows the Vmem and spike-out when weight value is -1.

Figure 6a illustrates the general structure of a hidden layer, while Figure 6b depicts a circuit diagram of an example of a hidden layer with two synapses and one neuron. Each synapse has its own distinguished weight value, which can be either excitatory or inhibitory. Due to large number of input synaptic activity, the neuron and synapses parameters need to be configured accordingly. To realize such a fully connected SNN, the circuit shown in Figure 6b was simulated. Here, two synapses with weight values W1 and W2 are connected to a single neuron constituting one neuron cell of hidden layers.

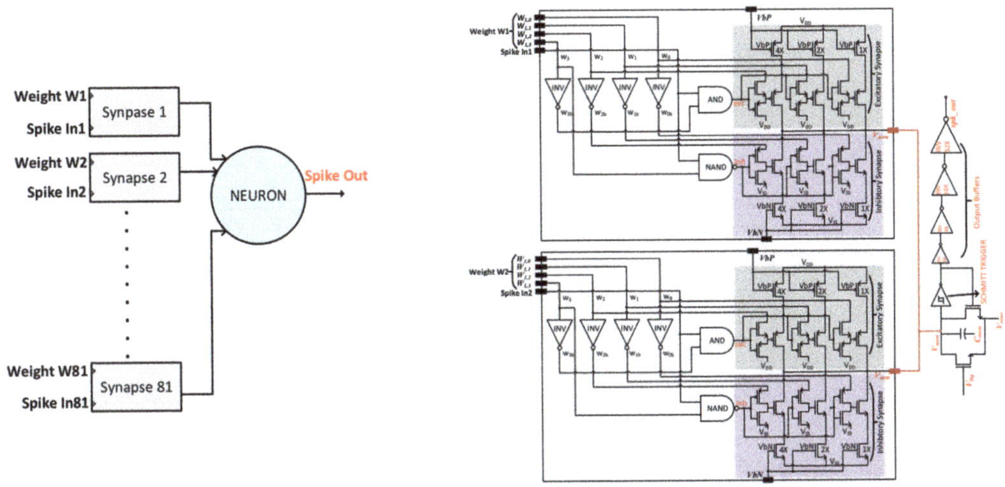

(a) (b)

Figure 6. (**a**) Fully connected hidden layer with a large number of synapses connected to a single neuron; (**b**) One neuron cell of the hidden layers with two synapses connected to a single neuron.

Figure 7 shows the circuit simulation result for the hidden layer circuit given by Figure 6b. In the top graph of Figure 7, the spike-in signal of the current layer is the output spike signal from the previous layer. Although individual synapses receive their spike-in signals from different neurons' outputs of the previous layer, in the simulation of Figure 7, for simplicity, the same spike-in signal is applied to the two synapses of Figure 6b. The synapses connected to a neuron in the hidden layer can have either positive or a negative weight value. The simulation verifies in three cases with two synapses connected to a neuron having a different combination of weight values. The first case in Figure 7a is when one synapse has a weight value W1 = +1, and the other has a weight value W2 = 0. This case demonstrates an excitatory behavior, where the neuron fires an output spike after receiving a sequence of input spikes. The second case illustrated in Figure 7b is when one synapse has W1 = −1 and the other has W2 = 0. It demonstrates an inhibitory behavior, where the neuron does not fire any output spike. The third case is depicted in Figure 7c, which shows the case when one synapse has W1 = +1 and the other has W2 = −1 weight value. Here the excitatory synapse injects the charge on V_{mem}, while the inhibitory synapse discharges the V_{mem} at the same amount. Therefore, V_{mem} remains at resting potential V_{reset} resulting in no output spiking activity. The simulation elucidates the biological neural behavior by the proposed circuit implementation of LIF model. The important simulation parameters along with neuron and synapse area and power estimation are summarized in Table 1.

Table 1. Neuron Cell Simulation Specifications.

Parameters	Values
Neuron cell Area	2022.72 μm^2
Neuron cell Power Consumption	25 μW
Resting potential	500 mV
Threshold Voltage	1.4 V
Resolution	4-bit
VDD/VSS	1.8 V/0 V
Membrane Capacitance	35 fF
Off-chip Bias voltages	8

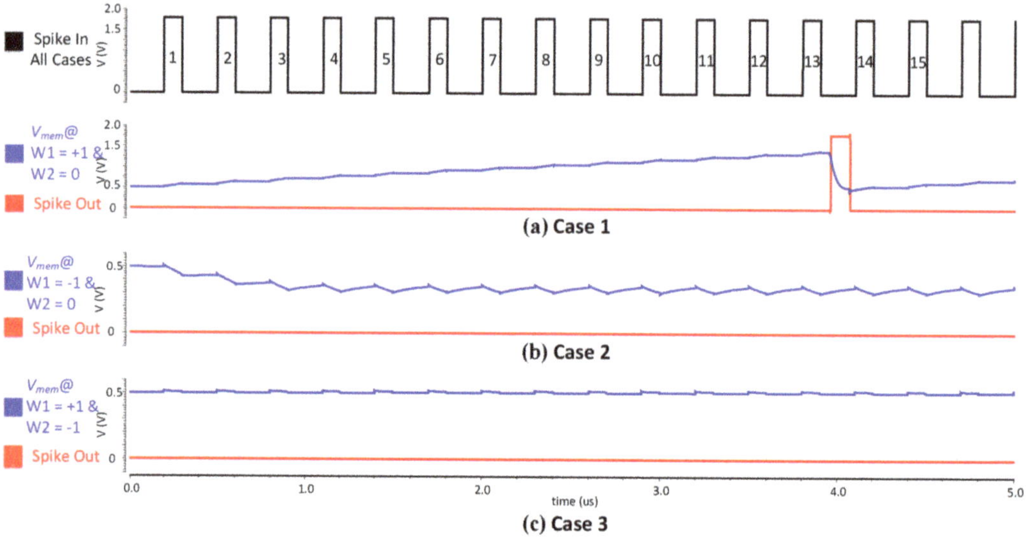

Figure 7. Circuit simulations for a neuron cell (one neuron connected to two synapses) where 15 input spikes are same for all the three cases: (**a**) shows Vmem and an output spike when weight values are W1 = +1 and W2 = 0; (**b**) shows Vmem and spike-out when weight values are W1 = −1 and W2 = 0; (**c**) shows Vmem and spike-out when weight values are W1 = +1 and W2 = −1.

3.3. Fully Digital Implementation of Spiking Neural Network

To evaluate the advantage of our analog implementation in terms of area and power, the BSRC-based SNN model is also implemented in a full digital design. The full digital design of SNN is verified on FPGA, while its area and power estimates are obtained using Synopsys Design Compiler using Standard Cell Library of the same process technology as the analog counterpart—TSMC 180 nm. The digital SNN realizes the same network shown in Figure 2 using Verilog HDL.

The input layer converts each input pixel value to a sequence of spikes using a design implementation of the integrate and fire (without leakage) neuron logic. The input layer converts 81 pixel values using 81 synapses in a way similar to the analog SNN implementation presented in Section 3.1. Figure 8 shows the detailed block diagram of a hidden layer implementation. Each neuron logic in the SNN has a membrane potential storage register and has a number of synapses attached as inputs. The neurons and synapses operate at discrete events represented by the rising edge of the spike-event clock. The spike-event clock is generated by dividing the system clock by an adjustable factor, which is set to 3 for the measurements results provided in Section 4.2. Each synapse, upon receiving a spike, adds its weight value to the membrane register. Whenever, the membrane potential exceeds the threshold value, the neuron generates an output spike and reset its membrane register value. The spikes are generated and received at the spike-event clock, while the width of each spike is equal to the system clock period.

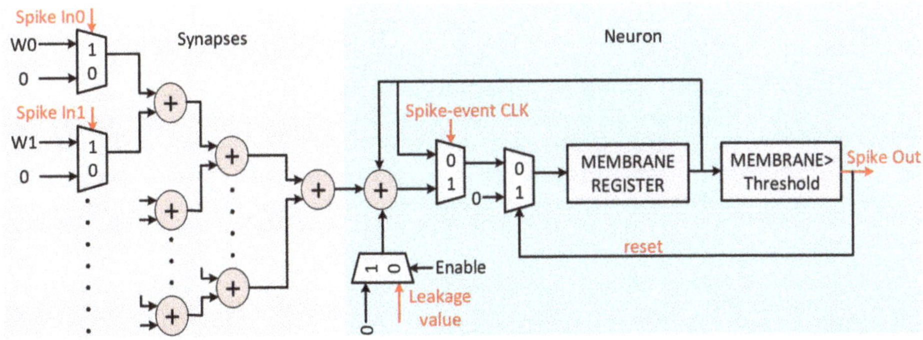

Figure 8. The block diagram of the full digital implementation for the output layer. A number of synapses accumulate in membrane register and the comparator fires the digital output spike pulses.

4. Measurement Results and Analysis

This section analyzes the performance and measurements results of analog SNN and digital SNN implementations for determining an optimized hardware for the area and power constraints.

4.1. Implementation of Analog SNN

The full-chip layout (including pads) of the analog SNN is described in Figure 9a. The test chip of the analog SNN is fabricated using the 180 nm CMOS process as part of a multiple project wafer. The fabricated chip's micrograph with bonding wires is given in Figure 9b. The test chip occupies an active area of 3.6 mm^2 within the allocated die area of 2700 µm × 1550 µm. The layout of the chip distinguishes the four fully connected layers of the SNN as input layer (IPL), hidden layer1 (HL1), hidden layer2 (HL2), output layer (OPL), and a Digital Controller (DC). The compactness is achieved via the tight placement of individual cells and smaller routing paths. The layers are placed side by side so that interconnection between two layers is of minimal length. Moreover, decoupling capacitors are added inside every neuron cell to allow for scalable and reliable design.

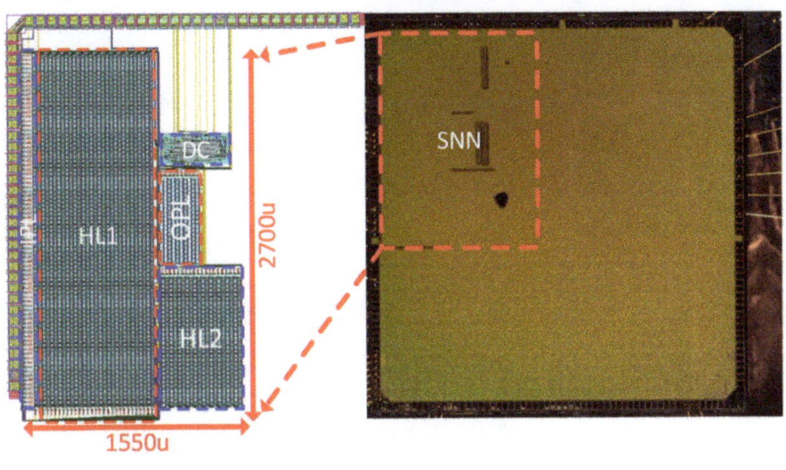

Figure 9. (**a**) The complete layout of the SNN implementation constituting of 4 fully connected layers with area estimation; (**b**) The bonded die micrograph highlighting the fabricated SNN chip.

Figure 10a shows a test printed circuit board (PCB) for testing the analog SNN chip and the measurement setup. A host CPU board (Raspberry Pi in our test setup) was used to download weight data and configuration parameters, and then provide input image to the SNN chip via a Serial Peripheral Interface (SPI). The on-chip digital controller activates the input layer to take pixel values from the image memory and convert them to input spike signals for SNN. Once all the spike signals are propagated through each layer, the output layer counts the final output spikes, while the digital controller sends back the data to the host CPU board for an evaluation of the classification results. Figure 10b shows the measurement setup for the fully digital implementation, which is described in Section 4.2.

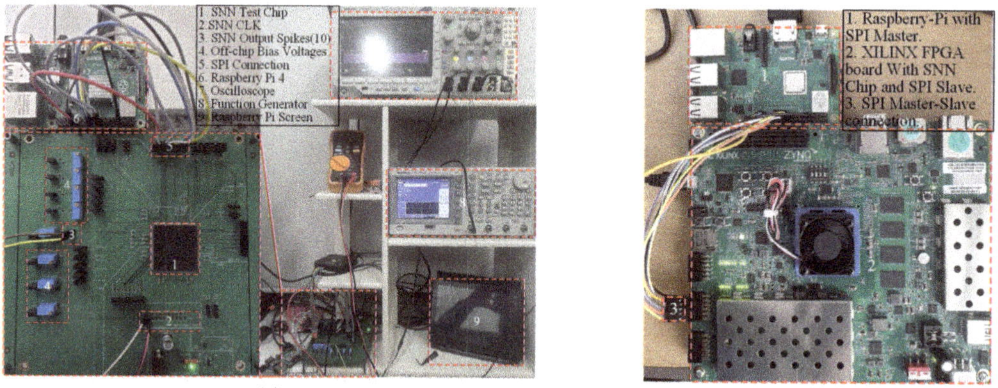

(**a**) (**b**)

Figure 10. Measurement Setups: (**a**) analog SNN test chip mounted on test PCB is measured using oscilloscope, function generator, and a host CPU board (Raspberry Pi 4); (**b**) digital SNN implementation is measured via FPGA board interfaced with a host CPU board (Raspberry Pi 4).

The measured results are shown in Figure 11. The spike propagation phenomenon is achieved successfully, as observed in circuit simulations. Figure 11a is the case when one input spike applied at the first neuron cell propagates through the first neuron cell of all the layers. This propagation is achieved by having maximum weight (7 in our implementation) for the first synapse of the first neuron in every layer, while all other weight values are set to zero. With one input spike applied to the first layer and the maximum weight value in every layer, the output layer generates one output spike. Similarly, in Figure 11b, 7 input spikes are applied to the first layer and propagated by using the same weight configuration as of Figure 11a. The measured results of Figure 11 verify the biological propagation of spikes from one layer to another layer for the SNN chip under test. The MNIST data set was applied during testing and the average on-chip current was measured for different applied images. The measured average current was 592 µW, which gives the power consumption of the test chip as 1.06 mW, while operating at a 10 MHz clock signal.

4.2. Implementation of Digital SNN

The aforementioned implementation of a fully digital SNN was tested using the measurement set-up described in Figure 10b. The fully digital SNN chip was implemented on an FPGA board (Xilinx ZYNQ Ultrascale), interfaced with Raspberry Pi 4 for the digital SNN measurements. Similar to analog SNN, the input image, weights data, and the configuration parameters for the digital controller are provided via SPI to the digital SNN. The controller generates stimulus input spike signals for SNN, counts the output spikes from the output layer, and sends back the data for further classification. The measured results obtained via Xilinx Integrated Logic Analyzer from FPGA are illustrated in Figure 12. The results show different input test images with respective to the classifier's output spiking activity for correct and failed classification. For example, when applied with an input image of 7, the 7th node of the output layer exhibits the maximum spiking activity. The digital SNN has a measured area of 3.5 mm^2 including routing overheads and a power consumption of 21.2 mW. This power consumption of the fully digital SNN chip was estimated by Synopsys Design compiler, while the power consumption of the analog SNN chip was measured from the fabricated chip. The estimated power consumption is accurate enough for our purpose of comparison, since the EDA tool was configured with an accurate 180 nm CMOS process PDK database provided by the same foundry.

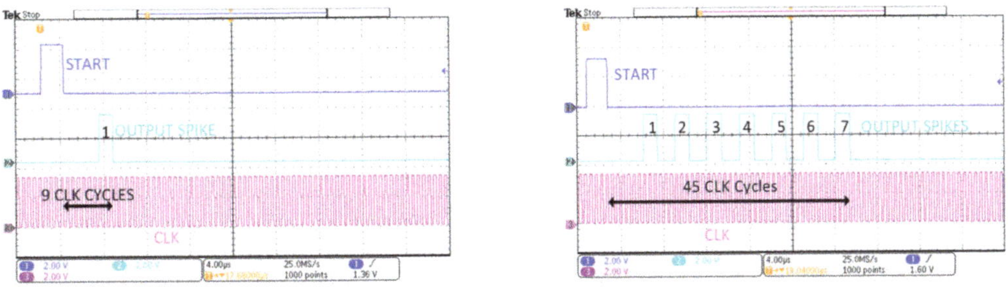

(a) (b)

Figure 11. Measurement results for the spike propagation: (**a**) One input spike propagates through all layers; (**b**) seven input spikes propagates through all layers.

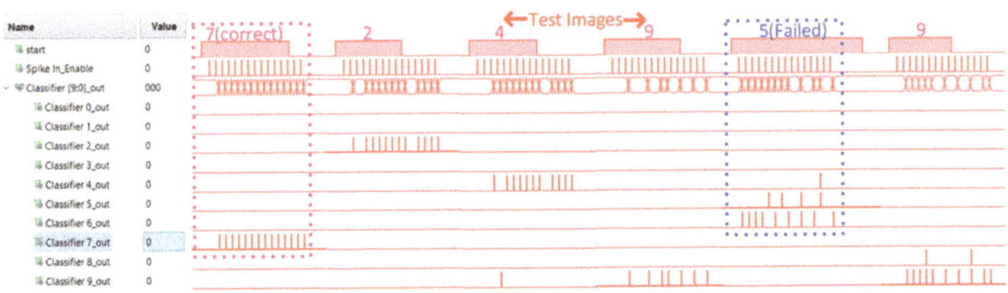

Figure 12. The measurement results for Digital SNN showing the correct classification of different input images along with the failed classification of input image '5'.

Table 2 compares the area and power consumption between our analog and digital SNN implementations for the same optimized SNN model. The digital SNN implementation benefits from less design complexity and is easier to implement as compared to the CMOS-based analog SNN implementation. However, the analog SNN finds massive advantage over its digital counterpart in terms of power consumption while occupying nearly the equivalent area. This makes the analog SNN design a superior choice for a low-power implementation of SNNs for mobile applications.

Table 2. Comparison between the Proposed Analog SNN and Fully Digital SNN.

SNN Chip	Area	Power
Proposed Analog SNN Chip	3.6 mm^2	1.06 mW
Fully Digital SNN Chip	3.5 mm^2	21.2 mW

5. Performance Analysis

The classification analysis for implemented SNN is performed using MNIST—a handwritten digit dataset comprising of 10,000 test images. Firstly, an optimized SNN model is implemented in the Python framework considering its analog circuit architecture and chip size. Secondly, the SNN model is trained in the Python framework that follows the detailed architecture of the analog SNN structure. Finally, the trained weights are downloaded to the analog SNN chip to conduct the inference operation using the MNIST validation dataset. Figure 13 shows the case when the applied input image is '7', the SNN predicts the correct classification result of '7' by producing the maximum spiking activity at the 7th output node. The average classification accuracy achieved for the SNN is 94.66%, which is very close to the accuracy of 94.69% for the optimized SNN model in Python [41].

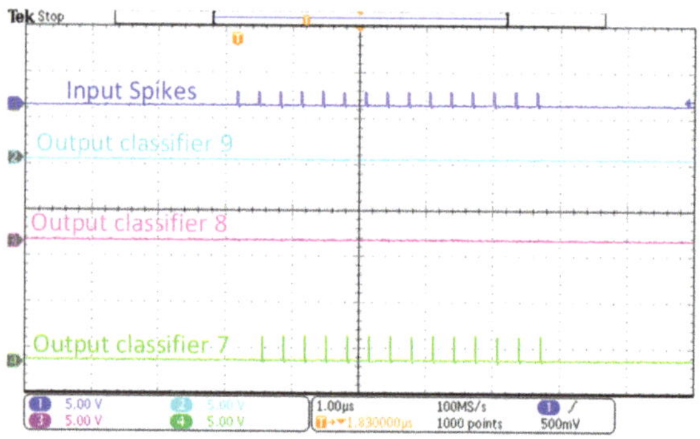

Figure 13. The measured results on the oscilloscope for the analog SNN. The applied input image to the SNN is '7' and the 7th output classifier shows maximum spiking activity.

The performance of the SNN test chip has been compared with state-of-the-art works previously reported, as shown in Table 3. The comparison has been made among various neuromorphic chips to identify the optimum implementation in terms of area, power consumption, and classification accuracy. As the physical dynamics of various neuromorphic chips is different for diverse applications, the complexity of the system is therefore defined by the total number of weights. The complexity is then divided by total occupied area, and by the power consumed for the area, and the power efficiency (η) analysis, respectively. The large-scale biological plausible analog Neurogrid [26] implementation benefits from the compact neuron size (1800 μm^2), which is based on the quadratic I & F model, achieves good area efficiency. Whereas our SNN based on the LIF model neuron cell occupies an almost similar area of 2022.72 μm^2, it consumes much less power and achieves 40$\times$ higher power efficiency. The second generation analog BrainScaleS neuromorphic systems [21], comprised of neuronal array prototype chip with tunable parameters, has a compact area but is power hungry. The mixed-mode neural network classifier [5], based upon the Radial Biased Function Network (RBFN) and the Multilayer Perception (MLP), occupies less area due to its analog core implementation but is less power efficient. In contrast, the proposed SNN provides optimum area and power efficiencies and is thus suitable for AI mobile applications.

Table 3. Comparison of Performance of the SNN Chip with Other State of the Art Neuromorphic Chips.

Parameters	[26]	[21]	[5]	[44]	This Work
CMOS tech. [nm]	180	65	130	800	180
Architecture	Analog	Analog	Mixed-Mode	Mixed-Mode	Analog
Classifier type	SNN	SNN	MLP/RBFN	SNN	SNN
Neuron Model	Quad. I & F	LIF	Current mode	LIF	LIF
Chip Area [mm^2]	168	3.6	0.140	1.6	3.6
Power [mW]	3100	48.62	2.20	40 μ [1]	1.06
Energy/Spike [pJ]	941 [2]	790	-	900	900
Accuracy (%)	-	-	92	-	94.60
Complexity [Total # of weights]	256 K	1024	750	256	3311
Area η [Complexity/Area]	1523	284.5	5360	160	920
Power η [Complexity/power]	82.5	21.06	341	-	3123

[1] For one neuron. [2] Energy per synaptic operation.

The energy consumption per spike was estimated for the current SNN chip by computing the total energy consumed for processing input spiking event divided by the number of the processed input spiking events [45]. As shown earlier in Figure 11a, for a single input spike it takes 9 clock cycles to evoke an output spike. It therefore consumes 900 pJ of energy per spike while operating at 10 MHz, which is different from energy per synaptic activation calculated as 941 pJ for [26]. The energy per spike for the proposed SNN is quite comparable with the 790 pJ energy per spike calculated for [21] and 900 pJ for [44]. Data for the accuracy was not available for the [26] and [21] analog implementations. Compared with the previous implementations [5], the proposed SNN implementation achieves an optimum classification accuracy at lower cost and power consumption, which makes it a strong candidate for mobile AI applications.

6. Conclusions

This work presents a hardware implementation of a large-scale SNN optimized for area and power, which is aimed at real-time AI/IoT applications. The SNNs allow for compact hardware implementation that is better suited for mobile or edge AI applications, if compact synapse and neuron circuits are used. Area and power efficient synapse and neuron circuits are proposed and an example SNN chip of 4 layers is constructed by integrating the synapse and neuron circuits. The SNN chip was implemented with a 180 nm CMOS process, which occupies a die area of 3.6 mm^2 and consumes a power of around 1 mW. The analog SNN chip has an advantage over its digital counterpart in terms of power consumption while occupying almost same area. The SNN chip achieves a classification accuracy of 94.60%, which is comparable with its software model, while consuming 900 pJ of energy per spike, which is 20 times lower than the digital SSN chip. Moreover, the prototype SNN implementation can be easily expanded for higher resolutions and number of classes. As future work, we plan to develop a large-scale SNN chip that can be an alternative solution to ANNs for increased image size and number of classes.

Author Contributions: Conceptualization, M.S.A. and H.K.; methodology, M.S.A., S.A., and H.K.; software, M.S.A. and S.A.; validation, M.S.A. and S.A.; formal analysis, M.S.A.; investigation, M.S.A.; writing—original draft preparation, M.S.A.; writing—review and editing, S.A. and H.K.; supervision, H.K.; project administration, H.K.; funding acquisition, H.K. All authors have read and agreed to the published version of the manuscript.

Funding: This work was supported by IITP grant (No. 2020-0-01304), Development of Self-learnable Mobile Recursive Neural Network Processor Technology Project, and also supported by the Grand Information Technology Research Center support program (IITP-2020-0-01462) supervised by the

IITP and funded by the MSIT (Ministry of Science and ICT), Korean government. It was also supported by Industry coupled IoT Semiconductor System Convergence Nurturing Center under System Semiconductor Convergence Specialist Nurturing Project funded by the National Research Foundation (NRF) of Korea (2020M3H2A107678611) and sponsored partly by Institute of Information & communications Technology Planning & Evaluation (IITP) grant funded by the Korea government (MSIT) (2020-0-01077, Development of Intelligent SoC having Multimodal IOT Interface for Data Sensing, Edge computing analysis and Data sharing).

Institutional Review Board Statement: Not applicable.

Informed Consent Statement: Not applicable.

Data Availability Statement: Not applicable.

Acknowledgments: We appreciate the collaboration and help provided by Ali A. Al-Hamid for the optimization of the SNN model and the quantization of the weights.

Conflicts of Interest: The authors declare no conflict of interest.

References

1. Mead, C. Neuromorphic electronic systems. *Proc. IEEE* **1990**, *78*, 1629–1636. [CrossRef]
2. Prezioso, M.; Merrikh-Bayat, F.; Hoskins, B.D.; Adam, G.C.; Likharev, K.K.; Strukov, D.B. Training and operation of an integrated neuromorphic network based on metal-oxide memristors. *Nature* **2015**, *521*, 61–64. [CrossRef] [PubMed]
3. Alex, K.; Ilya, S.; Geoffrey, E.H. ImageNet classification with deep convolutional neural networks. In Proceedings of the 25th International Conference on Neural Information Processing Systems, Lake Tahoe, NV, USA, 3–6 December 2012; pp. 1097–1105.
4. Lee, J.H.; Delbruck, T.; Pfeiffer, M. Training Deep Spiking Neural Networks Using Backpropagation. *Front. Neurosci.* **2016**, *10*, 508. [CrossRef] [PubMed]
5. Kyuho, L.; Junyoung, P.; Hoi-Jun, Y. A Low-power, Mixed-mode Neural Network Classifier for Robust Scene Classification. *J. Semicond. Technol. Sci.* **2019**, *19*, 129–136. [CrossRef]
6. Schuman, C.D.; Potok, T.E.; Patton, R.M.; Birdwell, J.D.; Dean, M.E.; Rose, G.S.; Plank, J.S. A Survey of Neuromorphic Computing and Neural Networks in Hardware. Available online: http://arxiv.org/abs/1705.06963 (accessed on 25 January 2021).
7. Von Neumann, J. First draft of a report on the EDVAC. *IEEE Ann. Hist. Comput.* **1993**, *15*, 27–75. [CrossRef]
8. Kandel, E.R.; Schwartz, J.H.; Jessell, T.M. *Principles of Neural Science*, 5th ed.; McGraw-Hill, Health Professions Division: New York, NY, USA, 2012.
9. Meier, K. Special report: Can we copy the brain?—The brain as computer. *IEEE Spectr.* **2017**, *54*, 28–33. [CrossRef]
10. Mahowald, M.A.; Mead, C. The silicon retina. In *Scientific American*; Springer Nature: New York, NY, USA, 1991; Volume 264, pp. 76–82. [CrossRef]
11. Kim, S.; Kim, H.; Hwang, S.; Kim, M.H.; Chang, Y.F.; Park, B.G. Analog synaptic behavior of a silicon nitride memristor. *ACS Appl. Mater. Interfaces* **2017**, *9*, 40420–40427. [CrossRef]
12. Rashvand, P.; Ahmadzadeh, M.R.; Shayegh, F. Design and Implementation of a Spiking Neural Network with Integrate-and-Fire Neuron Model for Pattern Recognition. *Int. J. Neural Syst.* **2021**, *31*, 2050073. [CrossRef]
13. Mead, C. *Anlaog VLSI and Neural Systems*, 1st ed.; Addison-Wesley: Reading, PA, USA, 1989.
14. Merolla, P.; Arthur, J.; Akopyan, F.; Imam, N.; Manohar, R.; Modha, D.S. A Digital Neurosynaptic Core Using Embedded Crossbar Memory with 45pJ per Spike in 45 nm. In Proceedings of the IEEE Custom Integrated Circuits Conference, San Jose, CA, USA, 19–21 September 2011; pp. 1–4. [CrossRef]
15. Islas, C.; Padilla, P.; Prado, M.A. Information Processing in the Brain as Optimal Entropy Transport: A Theoretical Approach. *Entropy* **2020**, *22*, 1231. [CrossRef]
16. Pregowska, A.; Kaplan, E.; Szczepanski, J. How Far can Neural Correlations Reduce Uncertainty? Comparison of Information Transmission Rates for Markov and Bernoulli Processes. *Int. J. Neural Syst.* **2019**, *29*, 1950003. [CrossRef]
17. Crumiller, M.; Knight, B.; Kaplan, E. The Measurement of Information Transmitted by a Neural Population: Promises and Challenges. *Entropy* **2013**, *15*, 3507–3527. [CrossRef]
18. Frenkel, C.; Lefebvre, M.; Legat, J.D.; Bol, D. A 0.086-mm^2 12.7-pJ/SOP 64k-Synapse 256-Neuron Online-Learning Digital Spiking Neuromorphic Processor in 28-nm CMOS. *IEEE Trans. Biomed. Circuits Syst.* **2019**, *13*, 145–158. [CrossRef]
19. Miyashita, D.; Kousai, S.; Suzuki, T.; Deguchi, J. Time-Domain Neural Network: A 48.5 TSOp/s/W neuromorphic chip optimized for deep learning and CMOS technology. In Proceedings of the IEEE Asian SSC Conference, Toyama, Japan, 7–9 November 2016; pp. 25–28. [CrossRef]
20. Harris, J.J.; Jolivet, R.; Engl, E.; Attwell, D. Energy-Efficient Information Transfer by Visual Pathway Synapses. *Curr. Biol.* **2015**, *25*, 3151–3160. [CrossRef]
21. Aamir, S.A.; Stradmann, Y.; Müller, P.; Pehle, C.; Hartel, A.; Grübl, A.; Schemmel, J.; Meier, K. An Accelerated LIF Neuronal Network Array for a Large-Scale Mixed-Signal Neuromorphic Architecture. *IEEE Trans. Circuits Syst. I Regul. Pap.* **2018**, *65*, 4299–4312. [CrossRef]

22. Indiveri, G.; Horiuchi, T. Frontiers in Neuromorphic Engineering. *Front. Neurosci.* **2011**, *5*, 118. [CrossRef]
23. Ebong, I.E.; Mazumder, P. CMOS and memristor-based neural network design for position detection. *Proc. IEEE* **2012**, *100*, 2050–2060. [CrossRef]
24. Chua, L. Memristor-the missing circuit element. *IEEE Trans. Circuit Theory* **1971**, *18*, 507–519. [CrossRef]
25. Li, C.; Belkin, D.; Li, Y.; Yan, P.; Hu, M.; Ge, N.; Jiang, H.; Montgomery, E.; Lin, P.; Wang, Z.; et al. Efficient and self-adaptive in-situ learning in multilayer memristor neural networks. *Nat. Commun.* **2018**, *9*, 2385. [CrossRef]
26. Benjamin, B.V.; Gao, P.; McQuinn, E.; Choudhary, S.; Chandrasekaran, A.R.; Bussat, J.M.; Alvarez-Icaza, R.; Arthur, J.V.; Merolla, P.A.; Boahen, K. Neurogrid: A Mixed-Analog-Digital Multichip System for Large-Scale Neural Simulations. *Proc. IEEE* **2014**, *102*, 699–716. [CrossRef]
27. Park, D.; Lee, J.; Asghar, M.S.; Hong, J.; Arslan, S.; Kim, H. Compact spiking neural network chip design for image classification. *J. RICIC* **2020**, *28*, 21–27.
28. Camuñas-Mesa, L.A.; Linares-Barranco, B.; Serrano-Gotarredona, T. Neuromorphic Spiking Neural Networks and Their Memristor-CMOS Hardware Implementations. *Materials* **2019**, *12*, 2745. [CrossRef] [PubMed]
29. Ankit, A.; Sengupta, A.; Panda, P.; Roy, K. RESPARC: A Reconfigurable and Energy-Efficient Architecture with Memristive Crossbars for Deep SNN. In Proceedings of the 54th ACM/EDAC/IEEE Design Automation Conference, Austin, TX, USA, 18–22 June 2017; pp. 1–6. [CrossRef]
30. Jolivet, R.; Rauch, A.; Lüscher, H.R.; Gerstner, W. Integrate-and-fire models with adaptation are good enough: Predicting spike times under random current injection. In Proceedings of the NIPS 18, Vancouver, BC, Canada, 5–8 December 2005; pp. 595–602.
31. Gerstner, W.; Kistler, W.M.; Naud, R.; Paninski, L. *Neuronal Dynamics*; Cambridge University Press: Cambridge, UK, 2014. [CrossRef]
32. Izhikevich, E.M. Simple model of spiking neurons. *IEEE Trans. Neural Netw.* **2003**, *14*, 1569–1572. [CrossRef] [PubMed]
33. Hodgkin, A.L.; Huxley, A.F. A quantitative description of membrane current and its application to conduction and excitation in nerve. *Bull. Math. Biol.* **1990**, *52*, 25–71. [CrossRef]
34. Diehl, P.U.; Cook, M. Unsupervised learning of digit recognition using spike-timing-dependent plasticity. *Front. Comput. Neurosci.* **2015**, *9*, 99. [CrossRef] [PubMed]
35. Kim, S.; Choi, B.; Lim, M.; Yoon, J.; Lee, J.; Kim, H.D.; Choi, S.J. Pattern recognition using carbon nanotube synaptic transistors with an adjustable weight update protocol. *ACS Nano* **2017**, *11*, 2814–2822. [CrossRef]
36. Merrikh-Bayat, F.; Guo, X.; Klachko, M.; Prezioso, M.; Likharev, K.K.; Strukov, D.B. High-performance mixed-signal neurocomputing with nanoscale floating-gate memory cell arrays. *IEEE Trans. Neural Netw. Learn. Syst.* **2018**, *29*, 4782–4790. [CrossRef]
37. Woo, J.; Padovani, A.; Moon, K.; Kwak, M.; Larcher, L.; Hwang, H. Linking conductive filament properties and evolution to synaptic behavior of RRAM devices for neuromorphic applications. *IEEE Electron. Device Lett.* **2017**, *38*, 1220–1223. [CrossRef]
38. Kim, H.; Hwang, S.; Park, J.; Yun, S.; Lee, J.H.; Park, B.G. Spiking Neural Network Using Synaptic Transistors and Neuron Circuits for Pattern Recognition with Noisy Images. *IEEE Electron. Device Lett.* **2018**, *39*, 630–633. [CrossRef]
39. O'Connor, P.; Welling, M. Deep Spiking Networks. Available online: https://arxiv.org/abs/1602.08323 (accessed on 20 January 2021).
40. Wu, Y.; Deng, L.; Li, G.; Zhu, J.; Shi, L. Spatio-Temporal Backpropagation for Training High-Performance Spiking Neural Networks. *Front. Neurosci.* **2018**, *12*, 331. [CrossRef]
41. Al-Hamid, A.A.; Kim, H. Optimization of Spiking Neural Networks Based on Binary Streamed Rate Coding. *Electronics* **2020**, *9*, 1599. [CrossRef]
42. Asghar, M.S.; Arslan, S.; Kim, H. Low Power Spiking Neural Network Circuit with Compact Synapse and Neuron Cells. In Proceedings of the 2020 International SoC Design Conference, Yeosu, Korea, 21–24 October 2020; pp. 157–158. [CrossRef]
43. Dutta, S.; Kumar, V.; Shukla, A.; Mohapatra, N.R.; Ganguly, U. Leaky Integrate and Fire Neuron by Charge-Discharge Dynamics in Floating-Body MOSFET. *Sci. Rep.* **2017**, *7*, 8257. [CrossRef]
44. Indiveri, G.; Chicca, E.; Douglas, R. A VLSI Array of Low-Power Spiking Neurons and Bistable Synapses with Spike-Timing Dependent Plasticity. *IEEE Trans. Neural Netw.* **2006**, *17*, 211–221. [CrossRef]
45. Tang, H.; Kim, H.; Cho, D.; Park, J. Spike Counts Based Low Complexity Learning with Binary Synapse. In Proceedings of the 2018 International Joint Conference on Neural Networks, Rio de Janeiro, Brazil, 8–13 July 2018; pp. 1–8. [CrossRef]

sensors

MDPI

Article

Design of a Capacitance-to-Digital Converter Based on Iterative Delay-Chain Discharge in 180 nm CMOS Technology

Mattia Cicalini, Massimo Piotto, Paolo Bruschi and Michele Dei *

Department of Information Engineering, University of Pisa, 56122 Pisa, Italy; mattia.cicalini@phd.unipi.it (M.C.); massimo.piotto@unipi.it (M.P.); paolo.bruschi@unipi.it (P.B.)
* Correspondence: michele.dei@dii.unipi.it

Abstract: The design of advanced miniaturized ultra-low power interfaces for sensors is extremely important for energy-constrained monitoring applications, such as wearable, ingestible and implantable devices used in the health and medical field. Capacitive sensors, together with their correspondent digital-output readout interfaces, make no exception. Here, we analyse and design a capacitance-to-digital converter, based on the recently introduced iterative delay-chain discharge architecture, showing the circuit inner operating principles and the correspondent design trade-offs. A complete design case, implemented in a commercial 180 nm CMOS process, operating at 0.9 V supply for a 0–250 pF input capacitance range, is presented. The circuit, tested by means of detailed electrical simulations, shows ultra-low energy consumption ($\leq$1.884 nJ/conversion), excellent linearity (linearity error 15.26 ppm), good robustness against process and temperature corners (conversion gain sensitivity to process corners variation of 114.0 ppm and maximum temperature sensitivity of 81.9 ppm/°C in the -40 °C, $+125$ °C interval) and medium-low resolution of 10.3 effective number of bits, while using only 0.0192 mm^2 of silicon area and employing 2.93 ms for a single conversion.

Keywords: capacitance-to-digital converter; iterative-delay-chain discharge; CMOS capacitive sensor interface

Citation: Cicalini, M.; Piotto, M.; Bruschi, P.; Dei, M. Design of a Capacitance-to-Digital Converter Based on Iterative Delay-Chain Discharge in 180 nm CMOS Technology. *Sensors* **2022**, 22, 121. https://doi.org/10.3390/s22010121

Academic Editor: Youfan Hu

Received: 26 November 2021
Accepted: 23 December 2021
Published: 24 December 2021

1. Introduction

Capacitive sensing technologies underpin many sensory applications, including industrial, automotive, consumer [1] and life-science electronics [2]. At the same time, dedicated and power-optimized readout interfaces have been proposed to take full advantage of this technology. In this sense, capacitance-to-digital converters (CDCs) represent a class of integrated interfaces capable of delivering a digital output readout of the capacitive sensor. Many architectures of CDCs are demonstrated in the literature, exploiting the principles of phase/pulse modulation (PM) [3–7], $\Delta\Sigma$ modulation ($\Delta\Sigma$M) [8–10] and capacitive successive approximation register (CSAR) [11–15]. A detailed review of these techniques can be found in [16].

Recently, a simple and compact solution, which presents a significant number of innovations over other kinds of CDCs, was proposed in [17]. The most relevant innovations regard that (i) the CDC implementation is based on basic digital gates (inverters, Nands and Xors); (ii) an external clock signal is not required; and, (iii) as it will be clear in the remainder of this paper, the scaling of the capacitance full scale, i.e., the maximum capacitance value that can be converted, does not affect the internal state variables range in terms of voltage headroom and/or current intensity, as it usually occurs in many other CDC architectures. This fact allows for the extension of the CDC dynamic range (DR) relying only on the length extension of the digital output register. However, the inner working principles of the iterative delay-chain discharge (IDCD) architecture are poorly explained, leaving the designer with numerous unknowns hindering the adoption of this architecture despite its excellent performance in terms of power.

In this work, we address this issue by providing a deeper insight into this new architecture by giving a formal (rather than heuristic) explanation of the CDC operating principle.

This discloses the CDC's intrinsic limits, thus providing awareness of the fundamental trade-offs. Moreover, the analysis paves the way for different implementations of the same architecture that better adapts to specific cases within the voltage-headroom/signal-bandwidth design space.

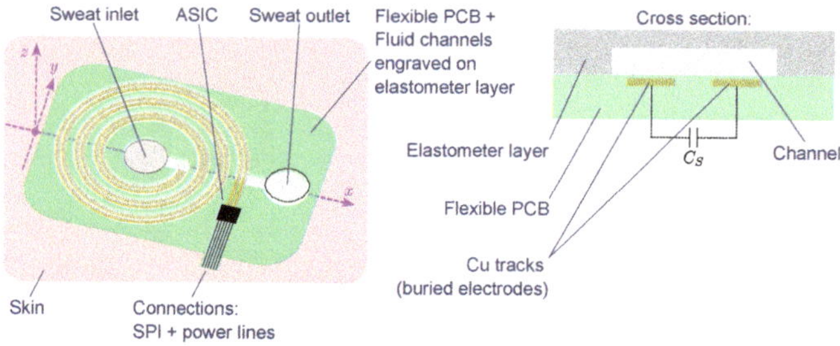

Figure 1. Concept of a wearable platform for volumetric sweat-rate sensing.

The target capacitive sensor considered in this work derives from the wearable platform for sweat-rate sensing sketched in Figure 1. This device is intended to be used for activity tracking in sport applications, and it consists of (i) a flexible printed-circuit board (FPCB) layer, typically a polyimide film; (ii) a decorated elastometer layer, typically polydimethylsiloxane (PDMS), and (iii) an application-specific integrated circuit (ASIC) [18–21]. The fluidic pathway is then formed by sealing the two layers together and providing an inlet and an outlet, facing, respectively, the skin and the air. In correspondence to the fluidic pathway, two buried electrodes, implemented by the FPCB Cu tracks, work as electrostatically coupling electrodes, providing the capacitive transduction mechanism for the volume occupied by the sweat within the channel. By taking successive capacitance measurements, the volumetric sweat flow can be reconstructed. The measurement readout control is provided by the ASIC, which is placed in close proximity to the sensor in order to avoid interference and excessive parasitic coupling. The ASIC may also provide a standard digital interface, e.g., a serial peripheral interface (SPI), for communication with an external wireless communication module. Preliminary estimation of the capacitance range of structures, such as those in Figure 1, suggests values between 10 and 250 pF, depending on the specific channel geometries and constitutive materials. Similar capacitance range can also be found in other capacitive sensors [22,23].

A 0–250 pF capacitive sensor interface, applying the design rules resulting from the theoretical analysis, is implemented in the UMC 180 nm complementary metal–oxide–semiconductor (CMOS) technology. The chosen capacitive conversion range is compatible with a number of micro-electro-mechanical systems (MEMS) capacitive sensors. Detailed electrical simulations show the following converter performance: systematic input offset of 255.6 fF, linearity error of 15.26 ppm, worst-case process-corner sensitivity on the conversion gain of 114 ppm, temperature sensitivity of 81.9 ppm/°C, maximum signal-to-noise ratio (SNR) of 63.9 dB and maximum conversion energy of 1.884 nJ when operated at 0.9 V supply. In the discussion section of this work, these figures are compared to those of [17] in order to provide insight into the porting of this architecture across different CMOS technological nodes.

2. Materials and Methods

Electrical simulations were performed on a 3.3 GHz 14 core CPU x86-64 workstation, operated through CentOS 7, and Cadence IC6.1.7 (ADEXL, Spectre simulator and AMS simulator). The CMOS design kit from UMC 180 nm mixed mode/RF was made available from the Europractice IC Service to European academic and research institutions. Graphical data preparation and presentation were performed by means of Python 3.5.2 importing the following modules: Numpy 1.17.0 and Matplotlib 3.0.3.

3. Results

The CDC operation principle is analysed for the first time in Section 3.1, while its implementation in the commercial 180 nm CMOS technology is presented in Section 3.2, followed by detailed electrical simulation in Section 3.3.

3.1. Principle of Operation

The CDC operation consists of the discharge of the capacitance C_S between two voltage levels, V_H and V_L, with V_H being the precharge value and V_L the value assumed at the end of the conversion (see Figure 2). For the sake of a clearer explanation, let us assume that C_S has one of its terminals connected to the ground. The conversion operation starts by the falling edge of the `precharge` signal. The discharging of C_S supplies the attached ring oscillator (RO), simply implemented by inverter gates, which starts oscillating at a frequency determined by its supply voltage (V_C). The output of the RO is the frequency modulated two-level signal $p(t)$, whose instantaneous oscillation frequency encodes the amplitude V_C. The integral of this quantity is the phase φ, which is updated at every cycle as shown in Figure 2b. The oscillation frequency decreases by decreasing V_C since the overdrive voltages of the logical gates are decreasing, thus slowing the charge of the next gate in the ring.

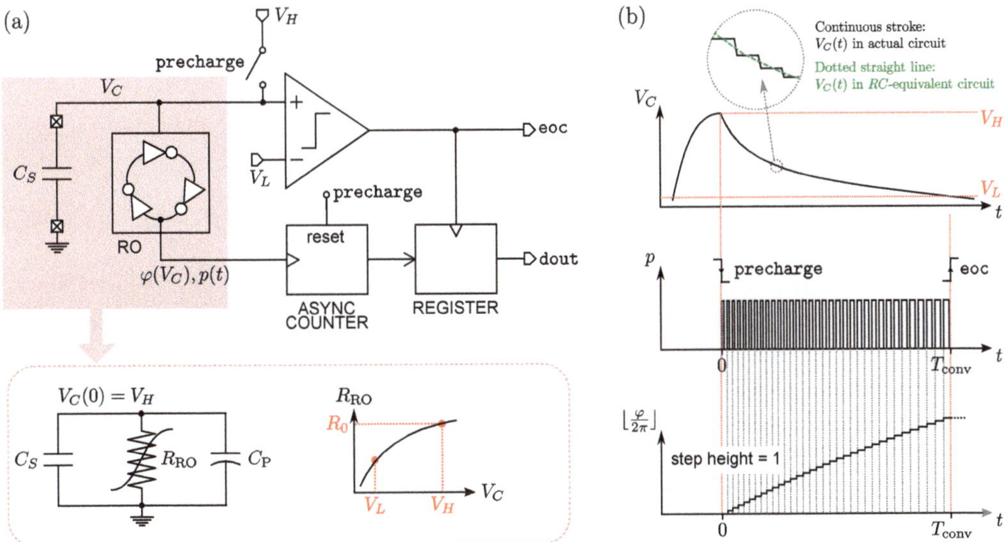

Figure 2. Simplified CDC operation based on a voltage level comparator: (**a**) block-level schematic diagram comprising an RC-circuit equivalent of the RO; (**b**) chronograms of the most important signals.

While the oscillation edge completes a loop, i.e., φ completes a full cycle, the RO absorbs a certain amount of charge from C_S, which causes V_C to decrease in time. An asynchronous counter keeps track of the number of loops. Finally, V_C reaches the V_L level, eventually detected by a voltage comparator set to the V_L threshold, which, in turn, produces the end-of-conversion signal (`eoc`) used also to strobe the counter value (`dout`) into an output register.

Since each loop consumes a certain quantity of charge $q[i]$ (at i-th loop), the following relationship must hold:

$$\sum_{i=1}^{N} q[i] + q_\epsilon = C_S(V_H - V_L), \tag{1}$$

where N is the number of loops during the discharge, and q_ϵ is the residual error due to the last incomplete loop.

The heuristic conclusion drawn in [17] is that N is proportional to C_S apart from the quantization error $q_\epsilon/(V_H - V_L)$. Nevertheless, Equation (1) does not give any support to this conclusion since the relation between N and C_S is not explicit. Moreover, since the RO supply voltage-to-frequency characteristic is generally non-linear, the capacitance-to-digital conversion law is not evident. An explanation of the principle of conversion is given in [24]; however, some unverified assumptions were made to simplify the analysis, which, on the other hand, may lead to wrong interpretations about the linearity of the conversion characteristic.

In order to show the linear relationship between C_S and N, let us consider the RC circuit represented in Figure 2a, where the parameters R_{RO} and C_{P} were introduced. The parameter C_{P} represents any parasitic capacitance due to the RO and the precharge switch added to the discharge node, while R_{RO} models the charge absorption rate at each voltage value V_{C}. It is important to note that during the full RO cycle, charge is impulsively absorbed due to the sequential switching of the digital gates, causing $V_{\text{C}}(t)$ to resemble a staircase shape. Hence, an effective current I_{C} *per* cycle can be defined, accounting for the amount of charge q in the interval of time defined by the p period. In our approach, $V_{\text{C}}(t)$ interpolates the actual staircase, allowing for a continuous-time description of the circuit behaviour as in Figure 2b. Hence, R_{RO} is simply the ratio between the interpolated V_{C} and I_{C}. It is convenient to express R_{RO} and C_{P} as

$$R_{\text{RO}}(V_{\text{C}}) = R_0\, u_{\text{RO}}(V_{\text{C}}) \qquad \text{and} \qquad C_{\text{P}}(V_{\text{C}}) = C_0\, u_{\text{P}}(V_{\text{C}}), \tag{2}$$

being that $R_0 = R_{\text{RO}}(V_H)$, $C_0 = C_{\text{P}}(V_H)$ and the functions $u_{\text{RO}}(V_{\text{C}})$ and $u_{\text{P}}(V_{\text{C}})$ are positive and continuous in the (V_L, V_H) interval such that $u_{\text{RO}}(V_H) = 1$ and $u_{\text{P}}(V_H) = 1$. The charge absorption rate modelled by R_{RO} is determined basically by two mechanisms: (i) charge is absorbed due to inter-stage charging within the RO, and (ii) charge is absorbed due to short-circuit currents in the digital gates of the RO at transition times.

The Kirchhoff's law of currents applied to the simple RC circuit of Figure 2 gives

$$\frac{d((C_S + C_{\text{P}})V_{\text{C}})}{dt} = -\frac{V_{\text{C}}}{R_{\text{RO}}}, \tag{3}$$

where the total charge $Q = (C_S + C_{\text{P}})V_{\text{C}}$ is subjected to variations in time due to both $V_{\text{C}}(t)$ and $C_S(t)$, being that the latter is the dynamic component of the capacitive sensor (i.e., the capacitively transduced signal to be converted). This can be neglected when

$$\frac{1}{C_S + C_{\text{P}}}\left|\frac{dC_S}{dt} + \frac{dC_{\text{P}}}{dt}\right| \ll \left|\frac{1}{V_{\text{C}}}\frac{dV_{\text{C}}}{dt}\right|, \tag{4}$$

meaning that at any time point during the conversion, the variations of C_S and C_{P} relative to the total capacitance $C_S + C_{\text{P}}$ are much smaller than the relative variation of V_{C}. Such a condition is typically found in a large class of capacitive sensors, where the capacitively transduced signal varies slowly compared to the conversion time T_{conv}. Under this hypothesis, (3) can be simplified in order to obtain

$$\left(1 + \frac{C_0}{C_S} u_{\text{P}}(V_{\text{C}})\right) u_{\text{RO}}(V_{\text{C}})\, \frac{dV_{\text{C}}}{V_{\text{C}}} = -\frac{dt}{\tau} \quad \text{and} \quad \tau = R_0 C_S. \tag{5}$$

Note that in a linear RC circuit, i.e., where both R_{RO} and C_{P} are independent from V_{C}, Equation (5) describes the known exponential relaxation of $V_{\text{C}}(t)$, determined by the time-constant τ. The analytical and/or numerical solution of Equation (5) is, in principle, viable once $u_{\text{RO}}(V_{\text{C}})$ and $u_{\text{P}}(V_{\text{C}})$ are known, either from an analytical insight on a particular RO topology, or directly from fitting simulation data.

The number of counts N, stored in dout, is determined by the accumulation of cycles during T_{conv}, which is related to the accumulated phase φ as follows:

$$\varphi(T_{\text{conv}}) = 2\pi \int_0^{T_{\text{conv}}} f_{osc}(t)\,dt \qquad \text{and} \qquad N = \lfloor \frac{\varphi(T_{\text{conv}})}{2\pi} \rfloor, \tag{6}$$

being that f_{osc} is the instantaneous oscillation frequency of $p(t)$. The operator $\lfloor x \rfloor$ indicates the floor operation on the variable x. Since f_{osc} is dependent on V_C, we can elaborate Equation (6) as

$$N = \lfloor \int_0^{T_{\text{conv}}} f_{osc}(t)\,dt \rfloor = \lfloor R_0 C_S \int_{V_L}^{V_H} \left(1 + \frac{C_0}{C_S} u_P(V_C)\right) u_{\text{RO}}(V_C) f_{osc}(V_C) \frac{dV_C}{V_C} \rfloor, \tag{7}$$

where the differential dt and the time constant τ are substituted with their respective expressions given in Equation (5). For better readability, Equation (7) can be rewritten as

$$N = \lfloor k_G C_S + k_{G0} C_0 \rfloor, \tag{8}$$

where

$$k_G = R_0 \int_{V_L}^{V_H} \frac{u_{\text{RO}}(V_C) f_{osc}(V_C)}{V_C}\,dV_C; \quad k_{G0} = R_0 \int_{V_L}^{V_H} \frac{u_P(V_C) u_{\text{RO}}(V_C) f_{osc}(V_C)}{V_C}\,dV_C. \tag{9}$$

The expressions in Equations (8) and (9) remarkably show that N is linearly dependent to the input C_S through the conversion gain k_G regardless of the oscillator implementation, as long as $f_{osc} > 0$. An offset term, $k_{G0}C_0$, is also present due to any parasitic capacitance added to the precharge node.

The quantization error ϵ_Q is

$$\epsilon_Q = \frac{\varphi(T_{\text{conv}})}{2\pi} - N = k_G C_S + k_{G0} C_0 - \lfloor k_G C_S + k_{G0} C_0 \rfloor. \tag{10}$$

Clear design guidelines can be obtained from the expression of k_G of Equation (9) under the following simplifying assumptions. First, let us assume the following relationship between f_{osc} and V_C, describing the linearised behaviour of the RO:

$$f_{osc} = f_0 + k_{osc} V_C, \tag{11}$$

where f_0 is the frequency bias and k_{osc}, given in [$s^{-1}V^{-1}$], is the voltage sensitivity coefficient. A second simplification regards the $u_{\text{RO}}(V_C)$ function introduced in Equation (2), which is approximated to an effective constant value $u^\star_{\text{RO}} \geq 1$ across the whole interval (V_L, V_H):

$$u_{\text{RO}}(V_C) = u^\star_{\text{RO}}, \quad \text{for} \quad V_L \leq V_C \leq V_H. \tag{12}$$

Under the assumptions (11)–(12), the integral of Equation (9) is simplified to

$$k_G = R_0 u^\star_{\text{RO}} \cdot \left(f_0 \log \frac{V_H}{V_L} + k_{osc} \cdot (V_H - V_L)\right). \tag{13}$$

The quantization error referred to as C_S, i.e., ε_Q/k_G, is reduced by increasing k_G (Equation (10)). Therefore, the simplified expression of k_G suggests the following design guidelines:

1. k_G is increased by increasing the $R_0 u^\star_{\text{RO}}$ term, which is related to both the W/L aspect ratio and the area WL of the digital ports and the number of delay stages of the inverter-based RO. The short-circuit current, which contributes to I_C, is reduced by increasing L; however, the short-circuit time interval is minimized by reducing the total area. So for a given gate area WL, it is convenient to reduce the W/L ratio.

Clearly, incrementing the number of delay stages increases the discharge rate in each cycle, thus reducing $R_0 u^{\star}_{\mathrm{RO}}$.

2. k_G is increased by increasing f_0, which can be attained for minimum-sized transistors, i.e., $W = W_{min}$ and $L = L_{min}$. The parameter k_{osc} depends on the chosen linearisation point, being strongly dependent on the V_L-V_H range. However, as it will be clear in the following discussion, the fully-digital implementation of the CDC rules out this parameter from the design space. As in point 1, reducing the number of delay stages is beneficial to increase k_G.
3. k_G is increased by maximizing V_H and minimizing V_L as can be seen in the logarithm argument and in the difference term. V_H is limited by the available supply voltage value, while V_L is limited by the minimum viable supply voltage for the correct operation of the digital gates.

Points 1 and 2, in principle, may lead to divergent design indications as far as the L of the digital gates is concerned. For this reason, the optimal solution can be obtained by performing electrical simulations, where L is swept across a reasonable interval that includes L_{min}.

Regarding the contribution of the comparator physical noise affecting the architecture shown in Figure 2, we can consider the comparator root-mean-square noise $V_{n,\mathrm{cmp}}$. At the end of the conversion, V_C will pass the V_L threshold with a certain slope, so

$$N_{n,\mathrm{comparator}} \simeq f_{osc}(V_L)\frac{V_{n,\mathrm{cmp}}}{dV_C/dt|_{t=T_{\mathrm{conv}}}} \simeq \tau\, f_{osc}(V_L) u_{\mathrm{RO}}(V_L)\frac{V_{n,\mathrm{cmp}}}{V_L}, \quad (14)$$

The last part of the approximation is found elaborating Equation (5)—which also gives the definition of τ—and neglecting, for the sake of simplicity, the contribution of C_P. Equation (14) describes the relationship between the comparator noise and the fluctuation on the conversion code, but most importantly, it establishes also a linear relationship between this fluctuation and the capacitance value through $\tau = R_0 C_S$. This is a very remarkable property of this converter type since the effects of one of the most important sources of physical noise scale proportionally with the quantity to be converted. This also suggests that no particular effort is to be put in the comparator design.

The architecture represented in Figure 2 is based on a continuous-time voltage-domain comparator whose noise effects are analysed in Equation (14). The next step in our analysis is the introduction of the time-domain comparator used in [17], which allows for a fully-digital implementation of the CDC—clearly advantageous since it nulls any static current consumption (except leakage current components).

In order to understand this step, let us consider the synchronized delay-chain RO shown in Figure 3b, derived from the simple RO of Figure 3a. Here, the time-encoded signals, $A1$ and $A2$, are originated by two separate delay chains. The following Xnor gate asserts the Boolean "$A1 == A2$" condition, i.e., both signals present the same logic level, so allowing the propagation of the oscillator travelling edge. In a scenario where the travelling edge of $A2$ lags the one of $A1$, this assertion permits their synchronization at the Nand gate before closing the feedback loop. Figure 3c shows the chronogram details of the oscillator signals A, $A1$, $A2$ and B.

In the actual CDC operation, $A2$ is generated by the reference delay chain fed at V_L, while $A1$ is generated by the sensing delay chain, fed at V_C. So, while $V_C > V_L$, the reference delay chain always lags behind the sensing delay chain. Ideally, both chains are synchronized for $V_C = V_L$, while the lagging condition is inverted as soon as $V_C < V_L$, marking the end-of-conversion condition.

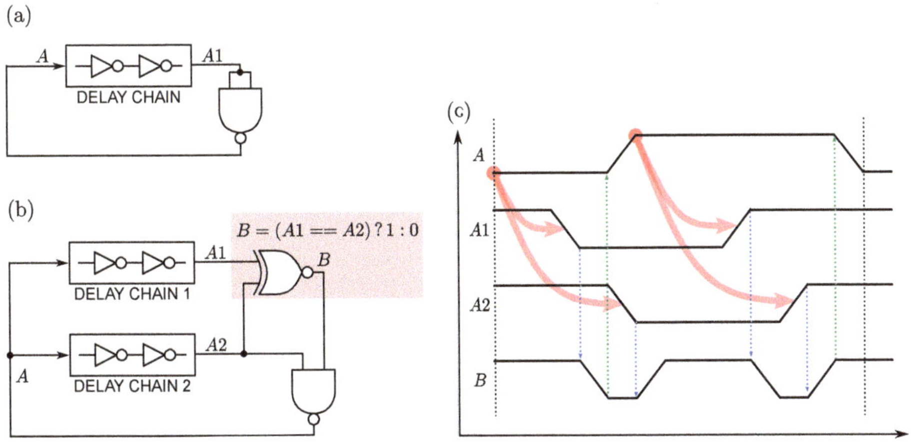

Figure 3. Derivation of a RO with synchronized delay chains: (**a**) starting point representation of a generic RO; (**b**) synchronization principle by a Xnor gate; (**c**) synchronized delay-chain oscillator chronogram.

The time-delay comparator, proposed in [17] and depicted in Figure 4a, provides the same synchronizing function of the Xnor/Nand gates of Figure 3b, while also signalling the end of conversion. It is based on a Nand-type set–reset latch and simple combinational logic to produce the two output signals, B and `finish`. The operation of such circuit is described in Figure 4c considering the following conditions: (i) $A1$ leads $A2$, and (ii) $A2$ leads $A1$. In both conditions, B acts as a synchronization gate, while `finish` is an active-low signal that pulses only after the first occurrence of the $A2$-leads-$A1$ condition. It is important to observe at this point that, while the voltage-domain comparator of Figure 2a is placed outside the RO, the time-domain comparator will be part of the RO, thus contributing to the oscillator parameters, such as f_0 and the conversion gain k_G (see Equation (13)).

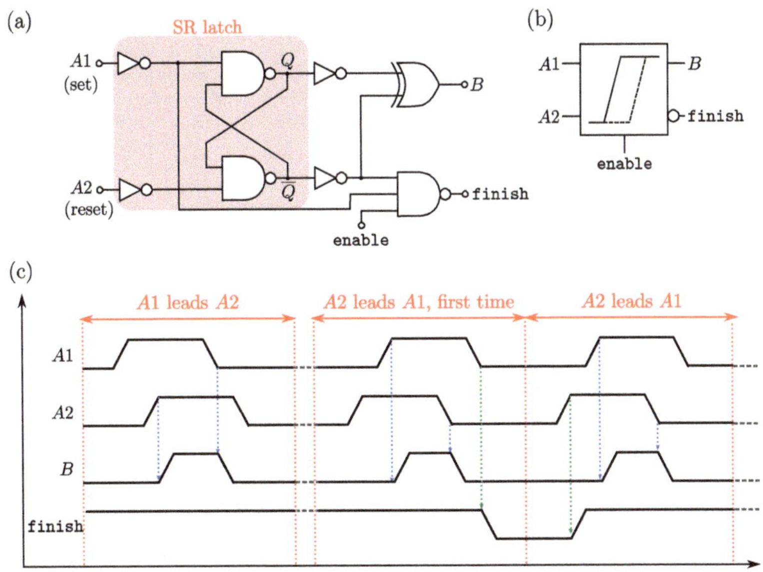

Figure 4. Time-delay comparator: (**a**) schematic diagram, (**b**) symbol view and (**c**) chronogram when operated inside the synchronized delay chains loop.

Figure 5 shows the effect of noise on the decision process of the comparator, both voltage-level based and time-delay based, when V_C crosses the decision threshold V_L. The figure shows how a lower value of C_S makes the decision process less prone to error since for a constant amount of charge dissipated by the RO in a single cycle, the voltage step (the delay between the travelling edges of $A1$ and $A2$) is higher for smaller C_S values. This observation is in accordance with Equation (14) and its related discussion on the contribution of comparator noise.

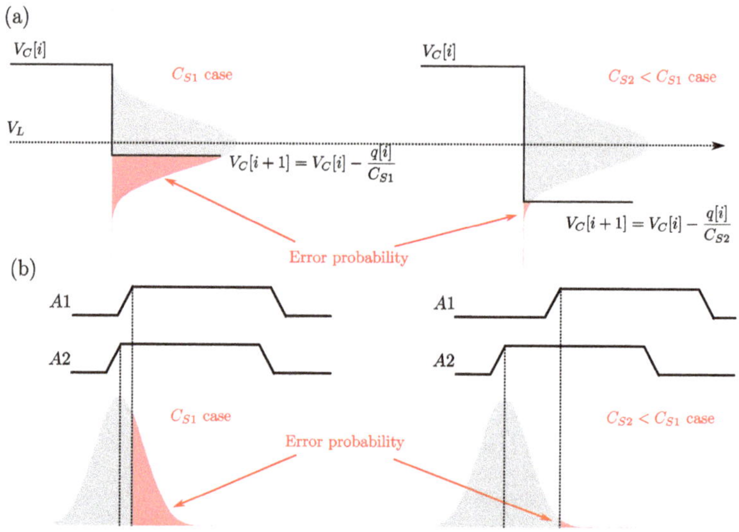

Figure 5. Comparison of comparator noise effects for two different values of C_S: (**a**) classic voltage-level based comparator as in Figure 2; (**b**) time-delay comparator of Figure 4.

Regarding the rest of noise sources in the circuit, it is well known that a standard voltage-fed RO presents a typical phase-noise spectrum characterized by the $1/f^3$ and $1/f^2$ behaviours, corresponding to the flicker and thermal noise sources, respectively [25–27]. In the synchronized-delay-chains case of Figure 3b, however, part of this noise is rejected due to the synchronization between the travelling edges of $A1$ and $A2$. Intuitively, every disturbance (i.e., phase lag or lead) produced after B and before A, affects both $A1$ and $A2$ in the same way, thus showing up as a common-mode noise, rejected by the differential-input nature of the time-delay comparator.

The residual differential-mode phase noise is generated once the RO path is split, corresponding to the separate delay-chain paths before the time-delay comparator. The effects of such noise on the final conversion count are influenced by the interval of time Δt between the $A1$ and $A2$ edges. We observe that at the end of conversion, this temporal difference tends to zero; however, the time-domain comparators are less affected by metastability (less prone to error) if the sensitivity of Δt with respect to V_C, i.e., the quantity $d(\Delta t)/dV_C$, is high.

The complete IDCD-CDC is shown in Figure 6, which features also a noise reduction technique, also proposed in [17], based on correlated averaging on a three-comparators system.

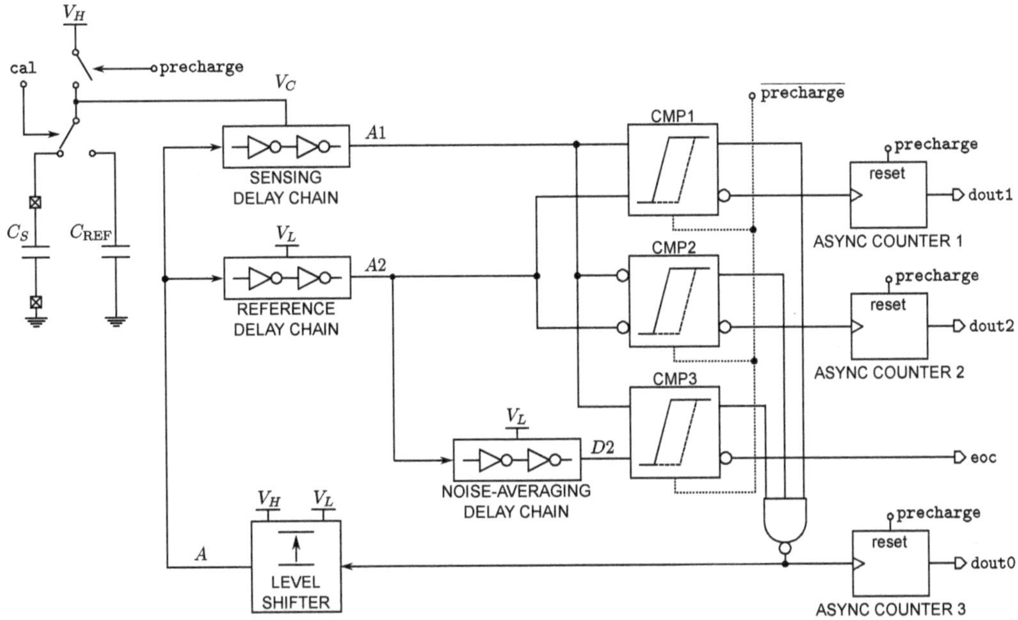

Figure 6. Complete CDC schematic including time-delay comparator-noise averaging and the one-point calibration network.

The comparator-noise averaging operates as follows: CMP1 and CMP2 are respectively fed with $A1$ and $A2$ and their inverted correspondents, while CMP3 is fed by $A1$ and a delayed version of $A2$ ($D2$). While CMP1's `finish` will detect the lagging condition on the rising edges, CMP2's `finish` will detect the same condition on the falling edges of $A1$ and $A2$. The travelling edges at comparator output are synchronized by a three-input Nand gate. The complete RO loop includes a V_L-to-V_H level shifter that guarantees the correct level transmission to both sensing and reference delay chains. The `eoc` signal pulses when the $A1$ travelling edge lags that of $D2$. Before this condition occurs, the `finish` outputs of CMP1 and CMP2 have pulsed a certain number of times depending on the amount of extra delay provided by the noise-averaging delay chain. These `finish` pulses of CMP1 and CMP2 are registered by dedicated counters, which provide `dout1` and `dout2`, respectively. The final conversion code is given by

$$N = 2 \times \texttt{dout0} - (\texttt{dout1} + \texttt{dout2}). \qquad (15)$$

The multiplicative factor of 2 before `dout0` accounts for both the rising and falling edges. To give a better understanding of the noise averaging mechanism, let us consider in the first instance that all the delay chains of Figure 6 are identical and their individual delay on the travelling edge dominates over the rest of the elements in the RO, i.e., the time-delay comparators, the Nand gate and the level shifter.

In such a scenario and in absence of noise, if we artificially set $V_C = V_L$, CMP1 and CMP2 have 50% probability to pulse their `finish` signals, while CMP3's `finish` will not pulse. In order to force CMP3's `finish` to pulse, we need to further lower V_C to a certain value $V_C = V_L^\star < V_L$. At this point, the conversion ends, meaning that the effective voltage step explored by the sensing chain is $V_H - V_L^\star$, and thus, some excess count was made. Nevertheless, the `finish` signals of both CMP1 and CMP2 start to pulse as soon as V_C is slightly below V_L, thus $\texttt{dout1} = \texttt{dout2}$, accounting for the excess of counts.

When the comparator noise is considered, the probability of CMP1 and CMP2 to make the wrong decision goes from 50% when $V_C = V_L$ to much lower values, as soon $V_C < V_L$.

By repeating the comparison process a certain number of times at different V_C values below V_L, the probability of error, and thus the noise effect, is reduced. In practice, starting from a certain value of $V_L^\star$ far from V_L, the probability of decision error can be neglected; thus, the decision redundancy only adds up to power consumption. So, in terms of power vs. resolution trade-off, an optimum value $V_L^\star$ exists, which can be tuned by the sizing of the noise-averaging delay chain of Figure 6. It must be observed that the crossing of the zone between V_L and $V_L^\star$ occurs at different slopes, depending of the value of C_S to be converted and also depending on $d(\Delta t)/dV_C$, as previously discussed. As a consequence, the number of excess counts increases for higher values of C_S, having a beneficial effect on the maximum attainable SNR.

The one-point calibration scheme is also shown in Figure 6, implemented through the C_{REF} capacitance and a switch controlled by the signal `cal`. The CDC calibration is obtained on demand by operating a conversion on a known value of C_{REF}, obtaining from Equations (6) and (10)

$$N_{\mathrm{REF}} = k_G C_{\mathrm{REF}} + k_{G0} C_0 + \epsilon_{Q,\mathrm{REF}}. \tag{16}$$

The parameters k_G, k_{G0} and C_0 may be strongly dependent on process corners and the operating temperature. While the former can be addressed by a one-time calibration at the beginning of the CDC operation, the latter can be addressed by occasionally performing a calibration conversion.

The calibrated value of the conversion, neglecting the physical noise, is obtained by the following formula:

$$C_S^{\mathrm{calibrated}} = C_{\mathrm{REF}} \frac{N}{N_{\mathrm{REF}}} = C_S \frac{1 + \dfrac{k_{G0} C_0}{k_G C_S} + \epsilon_Q}{1 + \dfrac{k_{G0} C_0}{k_G C_{\mathrm{REF}}} + \epsilon_{Q,\mathrm{REF}}}. \tag{17}$$

The rightmost side of Equation (17) reveals the residual error after calibration that can be minimized once $C_S \gg C_0$ and $C_{\mathrm{REF}} \gg C_0$ for acceptable quantization errors ϵ_Q and $\epsilon_{Q,\mathrm{REF}}$. Clearly, this calibration method relies profoundly on the stability of the absolute value of C_{REF}. Any process-related dispersion on the nominal value of C_{REF} affects the conversion value, despite the calibration. From the system-level point of view, two alternative solutions can be adopted. On one side, C_{REF} can be a very reliable external component, which, however, is affected by connection parasitics. On the other side, C_{REF} can be integrated all together with the converter circuitry using a metal–oxide–metal (MOM) or a metal–insulator–metal (MIM), when available from the process, capacitor. Nevertheless, the solution concerning the monolithic integration will be affected by the process corners spread. This former hindrance can be overcome by dedicated C_{REF} testing structures at the wafer level.

3.2. 180 nm-CMOS Implementation

Following the design indication explained in Section 3.1, a monolithic implementation of a IDCD-CDC is done in a standard 0.18 µm 1-poly 6-metal-level MIM CMOS technology. In this case study, we aim at optimizing the energy efficiency of the CDC while maintaining 10 effective number of bits (ENOB) of resolution and a total area $\leq$0.02 mm^2. Regarding the operating conditions, we aim for a button-cell operated system; thus, the specification $V_H = 0.9$ V applies for the rest of the discussion.

Referring to Figure 6, all inverters in the delay chain have $W = 240$ nm, $L = 180$ nm. All the delay chains (sensing, reference and noise-averaging) are implemented with 2 stages. With these values, k_G results to be $246.468 \times 10^{-12}\,\mathrm{F}^{-1}$, and the output code can be stored in a 16-bit output register. The digital gates of CMP1–CMP3, all identical, have all minimal $W = 240$ nm, and $L = 180$ nm.

The level shifter topology is adopted from [28]. Its schematic together with the sizes of transistor parameters are shown in Figure 7. Among other possible circuital solutions [29–32], that of Figure 7 provides the best energy efficiency when operating across subthreshold and super-threshold regions, defined by V_L and V_H. It is important to note that, in this design, the circuit propagation delay is of minor concern since it only affects the conversion time.

The C_{REF} capacitance is implemented by a MIM capacitor of 10 pF, which is the largest component of the CDC. However, since it is implemented between the two highest top-metal layers, the area underneath is used for the rest of the digital circuitry, using the rest of the metal layers for signal routing.

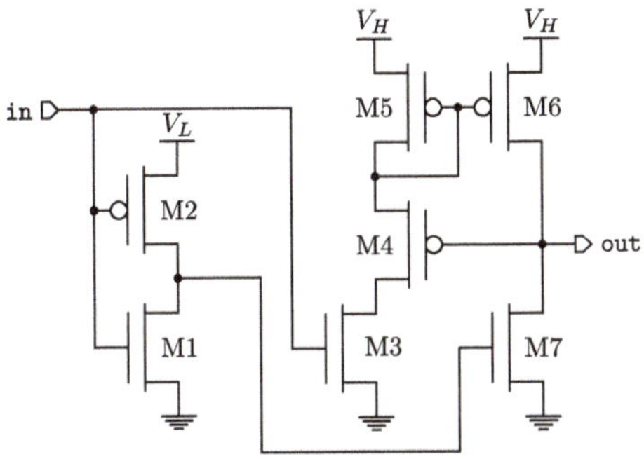

MOS	W [nm]	L [nm]
M1	440	250
M2	330	200
M3	250	440
M4	200	240
M5	240	3750
M6	400	200
M7	250	440

Figure 7. Wilson current-mirror based level shifter and transistor optimized geometrical parameters values for the design case in Section 3.2.

The total energy per conversion, E_{tot}, and the conversion time, T_{conv}, are evaluated as function of V_L in order to find an acceptable trade-off between quantization error and energy consumption. Figure 8a shows that a shallow optimum is found for $V_L = 0.5$ V. This is due to the fact that E_{tot} accounts for currents supplied by the V_H and V_L sources, respectively E_H and E_L, during the precharge and the conversion phases:

$$E_{tot} = E_H^{\text{precharge}} + E_H^{\text{conversion}} + E_L^{\text{conversion}}, \quad (18)$$

where the precharge energy supplied by V_H is

$$E_H^{\text{precharge}} = C_S V_H (V_H - V_L) \quad (19)$$

and $E_H^{\text{conversion}}$ is supplied to the level shifter.

Both Equations (18) and (19) neglect any leakage components, which add up to the total energy balance proportionally to T_{conv}. Equation (19) depicts a monotonically decreasing function of V_L. The terms $E_H^{\text{conversion}}$ and $E_L^{\text{conversion}}$, related to the conversion phase, depend on T_{conv}, which increase by lowering V_L, as shown in Figure 8b, where V_H is fixed to 0.9 V. Intuitively, we may expect that both $E_H^{\text{conversion}}$ and $E_L^{\text{conversion}}$ should follow the same trend as T_{conv}. This is true for $E_H^{\text{conversion}}$, but $E_L^{\text{conversion}}$ actually has the opposite behaviour as shown in Figure 8c. This is due to the dominant contribution of comparators activity happening at higher V_L values: the higher the V_L, the higher the $E_L^{\text{conversion}}$.

In our design, V_L is set to 0.5 V. For such value and for $C_S = 250$ pF, $E_{tot} = 1884$ fJ, accounting for the following contributions: $E_H^{\text{precharge}} = 90$ fJ, $E_H^{\text{conversion}} = 224$ fJ (due to the operation of the level shifter) and $E_L^{\text{conversion}} = 1570$ fF. The latter is the major contribution since V_L supplies also the time-domain comparators and the asynchronous counters.

The behaviour of R_{RO} as a function of V_C, introduced in Figure 2a, is shown in Figure 8d along with $\Delta t(V_C)$. The $R_{\text{RO}}(V_C)$ trend is to increase by increasing V_C. This is due to the dominant short-circuit currents contributions (transition time shorten as V_C increases) over the RO interstage-charging contribution. On the other hand, $\Delta t(V_C)$ shows a quite noticeable non-linear behaviour. The relatively high value of $d(\Delta t)/dV_C$ in the vicinity of V_L, resulting to be 71.8 ns/V, favours the CDC immunity against the noise introduced by the split path of the sensing and reference delay chains, as discussed previously.

Finally, the layout of the implemented CDC is shown in Figure 9 showing a silicon area occupancy of 0.0192 mm^2 (excluding pads).

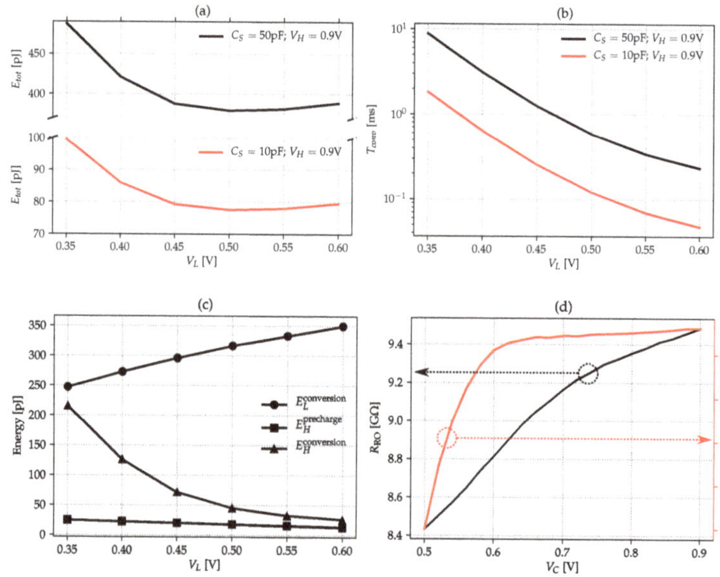

Figure 8. Key design parameters: (**a**) total energy per conversion E_{tot} as function of V_L for fixed $V_H = 0.9$ V; (**b**) energy balance as from Equation (18) as function of V_L for $C_S = 50$ pF and fixed $V_H = 0.9$ V; (**c**) conversion time T_{conv} as function of V_L for fixed $V_H = 0.9$ V; (**d**) R_{RO} and Δt as function of V_C.

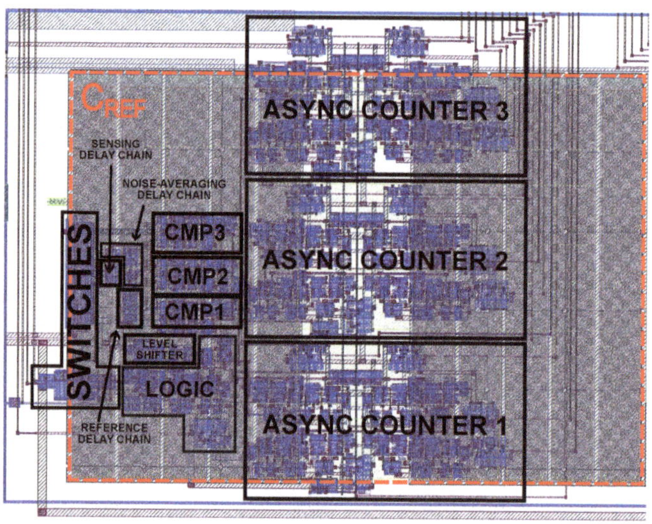

Figure 9. Layout of the CDC in a standard 0.18 μm 1-poly 6-metal-level-MIM CMOS technology. Bounding box size is 160 μm (width) × 120 μm (height).

3.3. Prototype Performance

The CDC DNL, calculated with respect to the end-points characteristic, when operated at $V_L = 0.5\,\text{V}$ and $V_H = 0.9\,\text{V}$, is shown in Figure 10, and tested against process corners. In all cases, the maximum observed code deviation falls within the ± 12 counts interval over an output register of 16 bits, corresponding to an equivalent capacitance LSB of 3.82 fF ($k_G = 246.468 \times 10^{12}\,\text{F}^{-1}$). The total energy per conversion scales linearly with C_S, resulting to be 1.884 nJ at full-scale $C_{S,FS} = 250\,\text{pF}$. As far as process corner sensitivity is concerned, E_{tot} presents small variations around its nominal value (worst case: +2.9% in the Fast-NMOS Slow-PMOS corner), while at the same time, T_{conv} shows quite large variations: 2.93 ms in the nominal case vs. 0.99 ms and 10.71 ms in the fast-NMOS fast-PMOS and slow-NMOS slow-PMOS, respectively.

The effectiveness of the one-point calibration against process corners is reported in Table 1, where the relative error ϵ_{k_G}, defined as

$$\epsilon_{k_G} = \frac{k_G^{\text{nominal}} - k_G}{k_G^{\text{nominal}}} \tag{20}$$

is evaluated, showing a $\times 30$ error reduction when calibrated. The systematic offset of the CDC, as in Equation (6), is <255.6 fF. Hence, the CDC shows an input capacitance range from 255.6 fF to 250 pF with a small linearity error of 15.26 ppm.

Table 1. Conversion-gain relative error ϵ_{k_G} (see Equation (20)) againts process corners: before and after calibration. Offset code, for $C_S = 0$, is also reported. Nominal offset code is 63.

Process Corner	Uncalibrated $\epsilon_{k_G}^{\text{corner}}$ [%]	Calibrated $\epsilon_{k_G}^{\text{corner}}$ [%]	Offset Code
Fast NMOS, Fast PMOS	3.126	−0.099	59
Slow NMOS, Slow PMOS	−3.342	0.112	67
Fast NMOS, Slow PMOS	0.437	−0.114	62
Slow NMOS, Fast PMOS	0.213	−0.063	62

For the sake of equal comparison, the figure of merit (FoM), as defined in [17], is evaluated:

$$\text{FoM} = \frac{E_{tot}(C_{S,FS})}{2^{(20\log_{10}(\text{Input range}/2\sqrt{2}/\text{Resolution})-1.76)/6.02}} = 99.61\,\text{fJ/conversion-step}. \tag{21}$$

where the resolution is calculated only on the basis of nonlinearity effects, while noise is not taken into account.

Transient noise simulations are performed to determine the SNR, which results to be 63.9 dB (10.3 ENOB) at $C_{S,FS}$. The noise-related FoM, FoM_N, of this converter is calculated as

$$\text{FoM}_\text{N} = \frac{E_{tot}(C_{S,FS})}{2^{(\text{SNR}_{\text{max}}-1.76)/6.02}} = 1.47\,\text{pJ/conversion-step}. \tag{22}$$

Temperature sensitivity is also evaluated as shown in Figure 11, showing a $\times 20$ improvement, from 1696.5 ppm/°C without calibration to 81.9 ppm/°C after calibration, across the −40 °C, +125 °C range.

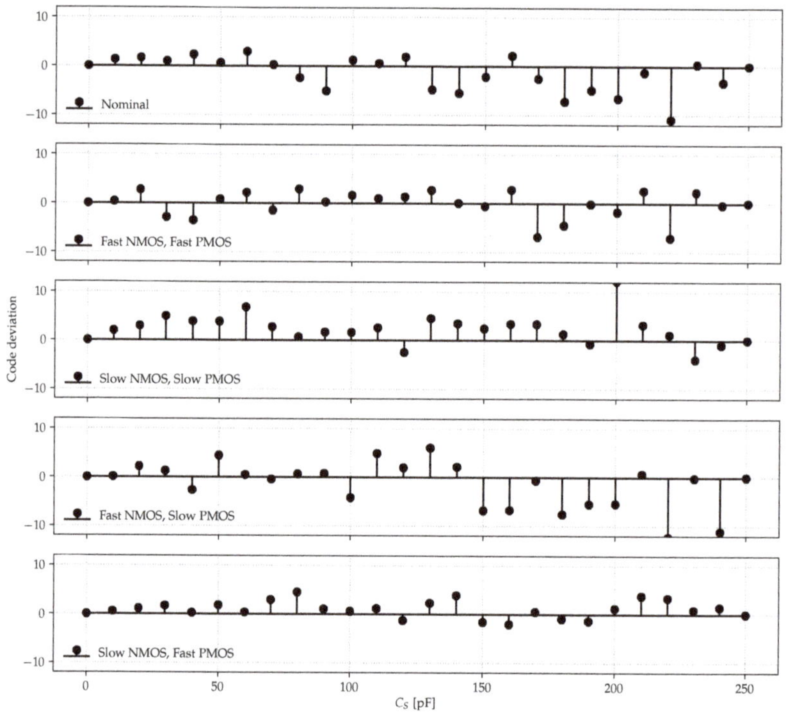

Figure 10. CDC differential non-linearity (DNL) against process corners. The output register width is 16 bits.

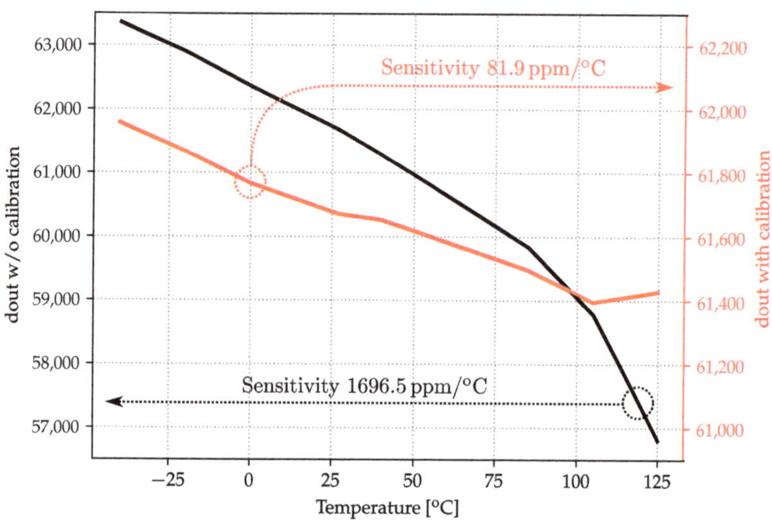

Figure 11. CDC output temperature sensitivity before and after one-point calibration.

4. Discussion

The CDC based on the IDCD architecture, introduced in [17], has the important characteristic to relay only on digital gates, thus being easily portable among different technological nodes once the fundamental design trade-offs, analysed for the first time in Section 3.1, are taken into account.

Here, we presented a design case, implemented on a low-cost commercial 180 nm-CMOS technology, capable of operating at button-cell supply voltages. Direct comparison with the original implementation of [17] is presented in Table 2. Energy figures are less favourable in the presented design case, as expected, due to the larger minimum feature size of the process used in this work with respect to the case of [17].

Table 2. Operative conditions and performance comparison table of IDCD CDCs.

	ISSCC'15 [17]	**This Work**
Technology	40 nm	180 nm
V_H, V_L	1.0 V, 0.45 V	0.9 V, 0.5 V
Input range	0.7 pF to 10 nF	255.6 fF to 250 pF
Linearity error	1090 ppm	15.26 ppm
Conversion time	19.06 µs at $C_S = 11.3$ pF	132.43 µs at $C_S = 11.3$ pF 2.93 ms at $C_S = 250.0$ pF
Conversion energy	35.1 pJ at $C_S = 11.3$ pF	85.2 pJ at $C_S = 11.3$ pF 1884.0 pJ at $C_S = 250.0$ pF
SNR	53.0 dB	63.9 dB
FoM (Equation (21))	141.0 fJ/conversion-step	99.6 fJ/conversion-step
FoM_N (Equation (22))	96.5 fJ/conversion-step	1.47 pJ/conversion-step
Temperature sensitivity	15.5 ppm/°C	81.9 ppm/°C
Core size	42 µm × 40 µm	160 µm × 120 µm including C_{REF}

The large difference between the FoM and FoM_N values clearly states that, in the current work, distortion effects are much less important than physical noise, while in [17], both distortion and noise contributed to the final resolution of the converter. These aspects confirm the analysis developed in Section 3.1 and give insights into energy efficiency vs. resolution trade-offs of the IDCD-CDC architecture when ported across different CMOS technological nodes.

In conclusion, the IDCD-CDC architecture proves to be a valid solution for capacitive sensor read-out interfaces in the medium/low resolution range. The IDCD-CDC fully exploits the benefits of miniaturization offered by more advanced CMOS technological nodes, while still providing competitive energy figures, even when implemented in low-cost 180 nm CMOS technology. In both cases, compatibility with low-voltage operation is maintained. When looking at evolutions of this architecture, capable of targeting more stringent resolution requirements, the inclusion of additional control circuitry needs to be investigated. Such circuitry should be devoted to the implementation of dynamic techniques for noise reduction and/or noise-shaping mechanisms.

Author Contributions: Conceptualization, M.D.; methodology, P.B. and M.P.; software, M.C.; validation, M.C.; formal analysis, M.D., P.B. and M.C.; resources, P.B. and M.P.; data curation, M.D., P.B., M.C. and M.P.; writing—original draft preparation, M.D.; writing—review and editing, M.D., M.C., P.B. and M.P.; supervision, P.B. and M.P.; project administration, M.D. and P.B.; funding acquisition, M.D. All authors have read and agreed to the published version of the manuscript.

Funding: This project received funding from the European Union's Horizon 2020 research and innovation programme under the Marie Skłodowska-Curie grant agreement No. 893544.

Institutional Review Board Statement: Not applicable.

Informed Consent Statement: Not applicable.

Data Availability Statement: Data contained in the text.

Conflicts of Interest: The authors declare no conflict of interest. The funders had no role in the design of the study; in the collection, analyses, or interpretation of data; in the writing of the manuscript, or in the decision to publish the results.

Abbreviations

The following abbreviations are used in this manuscript:

CDC	Capacitance-to-Digital Converter
PM	Phase/Pulse Modulation
ΔΣM	ΔΣ Modulation
CSAR	Capacitive Successive Approximation Register
DR	Dynamic Range
IDCD	Iterative Delay-Chain Discharge
FPCB	Flexible Printed-Circuit Board
PDMS	Polydimethylsiloxane
ASIC	Application-Specific Integrated Circuit
SPI	Serial Peripheral Interface
CMOS	Complementary Metal–Oxide–Semiconductor
MEMS	Micro-Electro-Mechanical Systems
RO	Ring Oscillator
VCO	Voltage-Controlled Oscillator
MOM	Metal–Oxide–Metal
MIM	Metal–Insulator–Metal
SNR	Signal-to-Noise Ratio
ENOB	Effective Number Of Bits
DNL	Differential Non-Linearity
LSB	Least-Significant Bit
FoM	Figure of Merit

References

1. Baxter Larry, K. *Capactive Sensors, Design and Applications;* John Wiley & Sons: Hoboken, NJ, USA, 1996.
2. Ebrahim Ghafar-Zadeh, S.F. *Emerging CMOS Capacitive Sensors for Biomedical Applications: A Multidisciplinary Approach;* The Institution of Engineering and Technology: London, UK, 2021.
3. Tan, Z.; Shalmany, S.H.; Meijer, G.C.M.; Pertijs, M.A.P. An Energy-Efficient 15-Bit Capacitive-Sensor Interface Based on Period Modulation. *IEEE J. Solid-State Circuits* **2012**, *47*, 1703–1711. [CrossRef]
4. Nizza, N.; Dei, M.; Butti, F.; Bruschi, P. A Low-Power Interface for Capacitive Sensors with PWM Output and Intrinsic Low Pass Characteristic. *IEEE Trans. Circuits Syst. Regul. Pap.* **2013**, *60*, 1419–1431. [CrossRef]
5. Mohan, R.; Zaliasl, S.; Gielen, G.G.E.; Hoof, C.V.; Yazicioglu, R.F.; Helleputte, N.V. A 0.6-V, 0.015-mm^2, Time-Based ECG Readout for Ambulatory Applications in 40-nm CMOS. *IEEE J. Solid-State Circuits* **2017**, *52*, 298–308. [CrossRef]
6. Yang, R.; Pertijs, M.A.P.; Nihtianov, S. A Precision Capacitance-to-Digital Converter with 16.7-bit ENOB and 7.5-ppm/°C Thermal Drift. *IEEE J. Solid-State Circuits* **2017**, *52*, 3018–3031. [CrossRef]
7. Woo, K.C.; Yang, B.D. 0.3-V *RC*-to-Digital Converter Using a Negative Charge-Pump Switch. *IEEE Trans. Circuits Syst. II Express Briefs* **2020**, *67*, 245–249. [CrossRef]
8. Zhang, T.T.; Law, M.K.; Mak, P.I.; Vai, M.I.; Martins, R.P. Nano-Watt Class Energy-Efficient Capacitive Sensor Interface with On-Chip Temperature Drift Compensation. *IEEE Sens. J.* **2018**, *18*, 2870–2882. [CrossRef]
9. Rogi, C.; Buffa, C.; Milleri, N.D.; Gaggl, R.; Prefasi, E. A Fully-Differential Switched-Capacitor Dual-Slope Capacitance-To-Digital Converter (CDC) for a Capacitive Pressure Sensor. *Sensors* **2019**, *19*, 3673. [CrossRef]
10. Tang, X.; Li, S.; Yang, X.; Shen, L.; Zhao, W.; Williams, R.P.; Liu, J.; Tan, Z.; Hall, N.A.; Pan, D.Z.; et al. An Energy-Efficient Time-Domain Incremental Zoom Capacitance-to-Digital Converter. *IEEE J. Solid-State Circuits* **2020**, *55*, 3064–3075. [CrossRef]
11. Ghanbari, M.M.; Tsai, J.M.; Nirmalathas, A.; Muller, R.; Gambini, S. An Energy-Efficient Miniaturized Intracranial Pressure Monitoring System. *IEEE J. Solid-State Circuits* **2017**, *52*, 720–734. [CrossRef]
12. Alhoshany, A.; Salama, K.N. A Precision, Energy-Efficient, Oversampling, Noise-Shaping Differential SAR Capacitance-to-Digital Converter. *IEEE Trans. Instrum. Meas.* **2019**, *68*, 392–401. [CrossRef]

13. Xin, H.; Andraud, M.; Baltus, P.; Cantatore, E.; Harpe, P. A 0.1-nW^{-1}-μW Energy-Efficient All-Dynamic Versatile Capacitance-to-Digital Converter. *IEEE J. Solid-State Circuits* **2019**, *54*, 1841–1851. [CrossRef]
14. Hussaini, S.; Jiang, H.; Walsh, P.; MacSweeney, D.; Makinwa, K.A.A. A 15-nW per Sensor Interference-Immune Readout IC for Capacitive Touch Sensors. *IEEE J. Solid-State Circuits* **2019**, *54*, 1874–1882. [CrossRef]
15. Narasimman, N.; Nag, D.; Chai, K.T.C.; Kim, T.T.H. A Capacitance to Digital Converter Using Continuous Time ΔΣ Modulator for Microphone-Based Auscultation. *IEEE Sens. J.* **2021**, *21*, 13373–13383. [CrossRef]
16. Tan, Z.; Jiang, H.; Zhang, H.; Tang, X.; Xin, H.; Nihtianov, S. Power-Efficiency Evolution of Capacitive Sensor Interfaces. *IEEE Sens. J.* **2021**, *21*, 12457–12468. [CrossRef]
17. Jung, W.; Jeong, S.; Oh, S.; Sylvester, D.; Blaauw, D. 27.6 A 0.7 pF-to-10 nF fully digital capacitance-to-digital converter using iterative delay-chain discharge. In Proceedings of the 2015 IEEE International Solid-State Circuits Conference-(ISSCC) Digest of Technical Papers, San Francisco, CA, USA, 22–26 February 2015. [CrossRef]
18. Seshadri, D.R.; Drummond, C.; Craker, J.; Rowbottom, J.R.; Voos, J.E. Wearable Devices for Sports: New Integrated Technologies Allow Coaches, Physicians, and Trainers to Better Understand the Physical Demands of Athletes in Real time. *IEEE Pulse* **2017**, *8*, 38–43. [CrossRef] [PubMed]
19. Dei, M.; Aymerich, J.; Piotto, M.; Bruschi, P.; del Campo, F.; Serra-Graells, F. CMOS Interfaces for Internet-of-Wearables Electrochemical Sensors: Trends and Challenges. *Electronics* **2019**, *8*, 150. [CrossRef]
20. Sharma, A.; Al-Dala'in, T.; Alsadoon, G.; Alwan, A. Use of Wearable Technologies for Analysis of Activity recognition for sports. In Proceedings of the 2020 5th International Conference on Innovative Technologies in Intelligent Systems and Industrial Applications (CITISIA), Sydney, Australia, 23–26 September 2020. [CrossRef]
21. Nithya, N.; Nallavan, G. Role of Wearables in Sports based on Activity recognition and biometric parameters: A Survey. In Proceedings of the 2021 International Conference on Artificial Intelligence and Smart Systems (ICAIS), Tamilnadu, India, 25–27 March 2021. [CrossRef]
22. Dobrzynska, J.A.; Gijs, M.A.M. Polymer-based flexible capacitive sensor for three-axial force measurements. *J. Micromech. Microeng.* **2012**, *23*, 015009. [CrossRef]
23. Anandan, N.; George, B. A Wide-Range Capacitive Sensor for Linear and Angular Displacement Measurement. *IEEE Trans. Ind. Electron.* **2017**, *64*, 5728–5737. [CrossRef]
24. Xu, Y.; Shang, D.; Xia, F.; Yakovlev, A. A smart all-digital charge to digital converter. In Proceedings of the 2016 IEEE International Conference on Electronics, Circuits and Systems (ICECS), Monte Carlo, Monaco, 11–14 December 2016. [CrossRef]
25. Razavi, B. A study of phase noise in CMOS oscillators. *IEEE J. Solid-State Circuits* **1996**, *31*, 331–343. [CrossRef]
26. Cardes, F.; Quintero, A.; Gutierrez, E.; Buffa, C.; Wiesbauer, A.; Hernandez, L. SNDR Limits of Oscillator-Based Sensor Readout Circuits. *Sensors* **2018**, *18*, 445. [CrossRef]
27. Pepe, F.; Andreani, P. An Accurate Analysis of Phase Noise in CMOS Ring Oscillators. *IEEE Trans. Circuits Syst. II Express Briefs* **2019**, *66*, 1292–1296. [CrossRef]
28. Lutkemeier, S.; Ruckert, U. A Subthreshold to Above-Threshold Level Shifter Comprising a Wilson Current Mirror. *IEEE Trans. Circuits Syst. II Express Briefs* **2010**, *57*, 721–724. [CrossRef]
29. Ishihara, F.; Sheikh, F.; Nikolic, B. Level conversion for dual-supply systems. *IEEE Trans. Very Large Scale Integr. (VLSI) Syst.* **2004**, *12*, 185–195. [CrossRef]
30. Chavan, A.; MacDonald, E. Ultra Low Voltage Level Shifters to Interface Sub and Super Threshold Reconfigurable Logic Cells. *IEEE Aerosp. Conf. Proc.* **2008**. [CrossRef]
31. Kumar, M.; Arya, S.K.; Pandey, S. Level Shifter Design for Low Power Applications. *Int. J. Comput. Sci. Inf. Technol.* **2010**, *2*, 124–132. [CrossRef]
32. Lanuzza, M.; Corsonello, P.; Perri, S. Low-Power Level Shifter for Multi-Supply Voltage Designs. *IEEE Trans. Circuits Syst. II Express Briefs* **2012**, *59*, 922–926. [CrossRef]

MDPI
St. Alban-Anlage 66
4052 Basel
Switzerland
Tel. +41 61 683 77 34
Fax +41 61 302 89 18
www.mdpi.com

Sensors Editorial Office
E-mail: sensors@mdpi.com
www.mdpi.com/journal/sensors

www.ingramcontent.com/pod-product-compliance
Lightning Source LLC
Chambersburg PA
CBHW061500210926
51850CB00008B/1773
9783036532646